T0338268

Intelligent Data Analysis

Intelligent Data Analysis

From Data Gathering to Data Comprehension

Edited by

Deepak Gupta
Maharaja Agrasen Institute of Technology
Delhi, India

Siddhartha Bhattacharyya
CHRIST (Deemed to be University)
Bengaluru, India

Ashish Khanna
Maharaja Agrasen Institute of Technology
Delhi, India

Kalpna Sagar
KIET Group of Institutions
Uttar Pradesh, India

Registered Offices
John Wiley & Sons, Inc., 111 River Street, Hoboken, NJ 07030, USA
John Wiley & Sons Ltd, The Atrium, Southern Gate, Chichester, West Sussex, PO19 8SQ, UK

Editorial Office
The Atrium, Southern Gate, Chichester, West Sussex, PO19 8SQ, UK

For details of our global editorial offices, customer services, and more information about Wiley products visit us at www.wiley.com.

Wiley also publishes its books in a variety of electronic formats and by print-on-demand. Some content that appears in standard print versions of this book may not be available in other formats.

Limit of Liability/Disclaimer of Warranty
MATLAB® is a trademark of The MathWorks, Inc. and is used with permission. The MathWorks does not warrant the accuracy of the text or exercises in this book. This work's use or discussion of MATLAB® software or related products does not constitute endorsement or sponsorship by The MathWorks of a particular pedagogical approach or particular use of the MATLAB® software.

While the publisher and authors have used their best efforts in preparing this work, they make no representations or warranties with respect to the accuracy or completeness of the contents of this work and specifically disclaim all warranties, including without limitation any implied warranties of merchantability or fitness for a particular purpose. No warranty may be created or extended by sales representatives, written sales materials or promotional statements for this work. The fact that an organization, website, or product is referred to in this work as a citation and/or potential source of further information does not mean that the publisher and authors endorse the information or services the organization, website, or product may provide or recommendations it may make. This work is sold with the understanding that the publisher is not engaged in rendering professional services. The advice and strategies contained herein may not be suitable for your situation. You should consult with a specialist where appropriate. Further, readers should be aware that websites listed in this work may have changed or disappeared between when this work was written and when it is read. Neither the publisher nor authors shall be liable for any loss of profit or any other commercial damages, including but not limited to special, incidental, consequential, or other damages.

Library of Congress Cataloging-in-Publication Data

Names: Gupta, Deepak, editor.
Title: Intelligent data analysis : from data gathering to data
    comprehension / edited by Dr. Deepak Gupta, Dr. Siddhartha
    Bhattacharyya, Dr. Ashish Khanna, Ms. Kalpna Sagar.
Description: Hoboken, NJ, USA : Wiley, 2020. | Series: The Wiley series in
    intelligent signal and data processing | Includes bibliographical
    references and index.
Identifiers: LCCN 2019056735 (print) | LCCN 2019056736 (ebook) | ISBN
    9781119544456 (hardback) | ISBN 9781119544449 (adobe pdf) | ISBN
    9781119544463 (epub)
Subjects: LCSH: Data mining. | Computational intelligence.
Classification: LCC QA76.9.D343 I57435 2020 (print) | LCC QA76.9.D343
    (ebook) | DDC 006.3/12–dc23
LC record available at https://lccn.loc.gov/2019056735
LC ebook record available at https://lccn.loc.gov/2019056736

Cover Design: Wiley
Cover Image: © gremlin/Getty Images

Set in 9.5/12.5pt STIXTwoText by SPi Global, Chennai, India
Printed and bound in Singapore by Markono Print Media Pte Ltd

10 9 8 7 6 5 4 3 2 1

Deepak Gupta would like to dedicate this book to his father, Sh. R.K. Gupta, his mother, Smt. Geeta Gupta, his mentors for their constant encouragement, and his family members, including his wife, brothers, sisters, kids and the students.

Siddhartha Bhattacharyya would like to dedicate this book to his parents, the late Ajit Kumar Bhattacharyya and the late Hashi Bhattacharyya, his beloved wife, Rashni, and his research scholars, Sourav, Sandip, Hrishikesh, Pankaj, Debanjan, Alokananda, Koyel, and Tulika.

Ashish Khanna would like to dedicate this book to his parents, the late R.C. Khanna and Smt. Surekha Khanna, for their constant encouragement and support, and to his wife, Sheenu, and children, Master Bhavya and Master Sanyukt.

Kalpna Sagar would like to dedicate this book to her father, Mr. Lekh Ram Sagar, and her mother, Smt. Gomti Sagar, the strongest persons of her life.

Contents

3 **Statistical Methods for Intelligent Data Analysis: Introduction and Various Concepts** *43*
Shubham Kumaram, Samarth Chugh, and Deepak Kumar Sharma

List of Contributors

Ambarish G. Mohapatra
Silicon Institute of Technology
Bhubaneswar
India

Anirban Mukherjee
RCC Institute of Information Technology
West Bengal
India

Aniruddha Sadhukhan
RCC Institute of Information Technology
West Bengal
India

Anisha Roy
RCC Institute of Information Technology
West Bengal
India

Arvinder Kaur
Guru Gobind Singh Indraprastha
University
India

Ayush Ahuja
Jaypee Institute of Information Technology
Noida
India

Biswajit Modak
Nabadwip State General Hospital
Nabadwip
India

R.S. Bhatia
National Institute of Technology
Kurukshetra
India

Bright Keswani
Suresh Gyan Vihar University
Jaipur
India

Dakun Lai
University of Electronic Science and
Technology of China
Chengdu
China

Deepak Kumar Sharma
Netaji Subhas University of Technology
New Delhi
India

Dhanushka Abeyratne
Yellowfin (HQ)
The University of Melbourne
Australia

Faijan Akhtar
Jamia Hamdard
New Delhi
India

Gihan S. Pathirana
Charles Sturt University
Melbourne
Australia

Huy V. Pham
Ton Duc Thang University
Vietnam

Malka N. Halgamuge
The University of Melbourne
Australia

Manashi De
Techno India
West Bengal
India

Manik Sharma
DAV University
Jalandhar
India

Manu Agarwal
Jaypee Institute of Information Technology
Noida
India

Manu Sood
University Shimla
India

Md Belal Bin Heyat
University of Electronic Science and
Technology of China
Chengdu
China

Mohd Ammar Bin Hayat
Medical University
India

Moolchand Sharma
Maharaja Agrasen Institute of Technology
(MAIT)
Delhi
India

Nabendu Chaki
University of Calcutta
Kolkata
India

Nisheeth Joshi
Banasthali Vidyapith
Rajasthan
India

Om Prakash Rishi
University of Kota
India

Poonam Keswani
Akashdeep PG College
Jaipur
India

Prableen Kaur
DAV University
Jalandhar
India

Pragya Katyayan
Banasthali Vidyapith
Rajasthan
India

Pratiyush Guleria
University Shimla
India

Prerna Sharma
Maharaja Agrasen Institute of Technology
(MAIT)
Delhi
India

Rachna Jain
Bharati Vidyapeeth's College of
Engineering
New Delhi
India

Rahul Johari
GGSIP University
New Delhi
India

Rajib Saha
RCC Institute of Information Technology
West Bengal
India

Rakesh Roshan
Institute of Management Studies
Ghaziabad
India

Ramneek Singhal
Bharati Vidyapeeth's College of
Engineering
New Delhi
India

Ravinder Ahuja
Jaypee Institute of Information Technology
Noida
India

Samarth Chugh
Netaji Subhas University of Technology
New Delhi
India

Samridhi Seth
GGSIP University
New Delhi
India

Sarthak Gupta
Netaji Subhas University of Technology
New Delhi
India

Shadab Azad
Chaudhary Charan Singh University
Meerut
India

Shafan Azad
Dr. A.P.J. Abdul Kalam Technical
University
Uttar Pradesh
India

Shajan Azad
Hayat Institute of Nursing
Lucknow
India

Shikhar Asthana
Jaypee Institute of Information Technology
Noida
India

Shivam Bachhety
Bharati Vidyapeeth's College of
Engineering
New Delhi
India

Shubham Kumaram
Netaji Subhas University of Technology
New Delhi
India

Shubhra Goyal
Guru Gobind Singh Indraprastha
University
India

Siddhant Bagga
Netaji Subhas University of Technology
New Delhi
India

Soma Datta
University of Calcutta
Kolkata
India

Tarini Ch. Mishra
Silicon Institute of Technology
Bhubaneswar
India

Than D. Le
University of Bordeaux
France

Vikas Chaudhary
KIET
Ghaziabad
India

Series Preface

Dr. Siddhartha Bhattacharyya, CHRIST (Deemed to be University), Bengaluru, India (Series Editor)

The Intelligent Signal and Data Processing (ISDP) book series is aimed at fostering the field of signal and data processing, which encompasses the theory and practice of algorithms and hardware that convert signals produced by artificial or natural means into a form useful for a specific purpose. The signals might be speech, audio, images, video, sensor data, telemetry, electrocardiograms, or seismic data, among others. The possible application areas include transmission, display, storage, interpretation, classification, segmentation, or diagnosis. The primary objective of the ISDP book series is to evolve future-generation scalable intelligent systems for faithful analysis of signals and data. ISDP is mainly intended to enrich the scholarly discourse on intelligent signal and image processing in different incarnations. ISDP will benefit a wide range of learners, including students, researchers, and practitioners. The student community can use the volumes in the series as reference texts to advance their knowledge base. In addition, the monographs will also come in handy to the aspiring researcher because of the valuable contributions both have made in this field. Moreover, both faculty members and data practitioners are likely to grasp depth of the relevant knowledge base from these volumes.

The series coverage will contain, not exclusively, the following:

1. Intelligent signal processing
   a) Adaptive filtering
   b) Learning algorithms for neural networks
   c) Hybrid soft-computing techniques
   d) Spectrum estimation and modeling
2. Image processing
   a) Image thresholding
   b) Image restoration
   c) Image compression
   d) Image segmentation
   e) Image quality evaluation
   f) Computer vision and medical imaging
   g) Image mining

    h) Pattern recognition

    i) Remote sensing imagery

    j) Underwater image analysis

    k) Gesture analysis

    l) Human mind analysis

    m) Multidimensional image analysis

3. Speech processing

    a) Modeling

    b) Compression

    c) Speech recognition and analysis

4. Video processing

    a) Video compression

    b) Analysis and processing

    c) 3D video compression

    d) Target tracking

    e) Video surveillance

    f) Automated and distributed crowd analytics

    g) Stereo-to-auto stereoscopic 3D video conversion

    h) Virtual and augmented reality

5. Data analysis

    a) Intelligent data acquisition

    b) Data mining

    c) Exploratory data analysis

    d) Modeling and algorithms

    e) Big data analytics

    f) Business intelligence

    g) Smart cities and smart buildings

    h) Multiway data analysis

    i) Predictive analytics

    j) Intelligent systems

Preface

Intelligent data analysis (IDA), knowledge discovery, and decision support have recently become more challenging research fields and have gained much attention among a large number of researchers and practitioners. In our view, the awareness of these challenging research fields and emerging technologies among the research community will increase the applications in biomedical science. This book aims to present the various approaches, techniques, and methods that are available for IDA, and to present case studies of their application.

This volume comprises 18 chapters focusing on the latest advances in IDA tools and techniques.

Machine learning models are broadly categorized into two types: white box and black box. Due to the difficulty in interpreting their inner workings, some machine learning models are considered black box models. Chapter 1 focuses on the different machine learning models, along with their advantages and limitations as far as the analysis of data is concerned.

With the advancement of technology, the amount of data generated is very large. The data generated has useful information that needs to be gathered by data analytics tools in order to make better decisions. In Chapter 2, the definition of data and its classifications based on different factors is given. The reader will learn about how and what data is and about the breakup of the data. After a description of what data is, the chapter will focus on defining and explaining big data and the various challenges faced by dealing with big data. The authors also describe various types of analytics that can be performed on large data and six data analytics tools (Microsoft Excel, Apache Spark, OpenRefine, R, Hadoop, and Tableau).

In recent years, the widespread use of computers and the internet has led to the generation of data on an unprecedented scale. To make an effective use of this data, it is necessary that data must be collected and analyzed so that inferences can be made to improve various products and services. Statistics deals with the collection, organization, and analysis of data. The organization and description of data is studied under these statistics in Chapter 3 while analysis of data and how to make predictions based on it is dealt with in inferential statistics.

After having an idea about various aspects of IDA in the previous chapters, Chapter 4 deals with an overview of data mining. It also discusses the process of knowledge discovery in data along with a detailed analysis of various mining methods including classification, clustering, and decision tree. In addition to that, the chapter concludes with a view of data visualization and probability concepts for IDA.

In Chapter 5, the authors demonstrate one of the most crucial and challenge areas in computer vision and the IDA field based on manipulating the convergence. This subject is divided into a deep learning paradigm for object segmentation in computer vision and visualization paradigm for efficiently incremental interpretation in manipulating the datasets for supervised and unsupervised learning, and online or offline training in reinforcement learning. This topic recently has had a large impact in robotics and autonomous systems, food detection, recommendation systems, and medical applications.

Dental caries is a painful bacterial disease of teeth caused mainly by Streptococcus mutants, acid, and carbohydrates, and it destroys the enamel, or the dentine, layer of the tooth. As per the World Health Organization report, worldwide, 60–90% of school children and almost 100% of adults have dental caries. Dental caries and periodontal disease without treatment for long periods causes tooth loss. There is not a single method to detect caries in its earliest stages. The size of carious lesions and early caries detection are very challenging tasks for dental practitioners. The methods related to dental caries detection are the radiograph, QLF or or quantitative light-induced fluorescence, ECM, FOTI, DIFOTI, etc. In a radiograph-based technique, dentists analyze the image data. In Chapter 6, the authors present a method to detect caries by analyzing the secondary emission data.

With the growth of data in the education field in recent years, there is a need for intelligent data analytics, in order that academic data should be used effectively to improve learning. Educational data mining and learning analytics are the fields of IDA that play important roles in intelligent analysis of educational data. One of the real challenges faced by students and institutions alike is the quality of education. An equally important factor related to the quality of education is the performance of students in the higher education system. The decisions that the students make while selecting their area of specialization is of grave concern here. In the absence of support systems, the students and the teachers/mentors fall short when making the right decisions for the furthering of their chosen career paths. Therefore, in Chapter 7, the authors attempt to address the issue by proposing a system that can guide the student to choose and to focus on the right course(s) based on their personal preferences. For this purpose, a system has been envisaged by blending data mining and classification with big data. A methodology using MapReduce Framework and association rule mining is proposed in order to derive the right blend of courses for students to pursue to enhance their career prospects.

Atmospheric air pollution is creating significant health problems that affect millions of people around the world. Chapter 8 analyzes the hypothesis about whether or not global green space variation is changing the global air quality. The authors perform a big data analysis with a data set that contains more than 1M (1 048 000) green space data and air quality data points by considering 190 countries during the years 1990 to 2015. Air quality is measured by considering particular matter (PM) value. The analysis is carried out using multivariate graphs and a k-mean clustering algorithm. The relative geographical changes of the tree areas, as well as the level of the air quality, were identified and the results indicated encouraging news.

Space technology and geotechnology, such as geographic information systems, plays a vital role in the day-to-day activities of a society. In the initial days, the data collection was very rudimentary and primitive. The quality of the data collected was a subject of verification and the accuracy of the data was also questionable. With the advent of newer technology, the problems have been overcome. Using modern sophisticated systems, space science has been changed drastically. Implementing cutting-edge spaceborne sensors has made it possible to capture real-time data from space. Chapter 9 focuses on these aspects in detail.

Transportation plays an important role in our overall economy, conveying products and people through progressively mind-boggling, interconnected, and multidimensional transportation frameworks. But, the complexities of present-day transportation can't be managed by previous systems. The utilization of IDA frameworks and strategies, with compelling information gathering and data dispersion frameworks, gives openings that are required to building the future intelligent transportation systems (ITSs). In Chapter 10, the authors exhibit the application of IDA in IoT-based ITS.

Chapter 11 aims to observe emerging patterns and trends by using big data analysis to enhance predictions of motor vehicle collisions using a data set consisting of 17 attributes and 998 193 collisions in New York City. The data is extracted from the New York City Police Department (NYPD). The data set has then been tested in three classification algorithms, which are k-nearest neighbor, random forest, and naive Bayes. The outputs are captured using k-fold cross-validation method. These outputs are used to identify and compare classifier accuracy, and random forest node accuracy and processing time. Further, an analysis of raw data is performed describing the four different vehicle groups in order to detect significance within the recorded period. Finally, extreme cases of collision severity are identified using outlier analysis. The analysis demonstrates that out of three classifiers, random forest gives the best results.

Neurological disorders are the diseases that are related to the brain, nervous system, and the spinal cord of the human body. These disorders may affect the walking, speaking, learning, and moving capacity of human beings. Some of the major human neurological disorders are stroke, brain tumors, epilepsy, meningitis, Alzheimer's, etc. Additionally, remarkable growth has been observed in the areas of disease diagnosis and health informatics. The critical human disorders related to lung, kidney, skin, and brain have been successfully diagnosed using different data mining and machine learning techniques. In Chapter 12, several neurological and psychological disorders are discussed. The role of different computing techniques in designing different biomedical applications are presented. In addition, the challenges and promising areas of innovation in designing a smart and intelligent neurological disorder diagnostic system using big data, internet of things, and emerging computing techniques are also highlighted.

Bug reports are one of the crucial software artifacts in open-source software. Issue tracking systems maintain enormous bug reports with several attributes, such as long description of bugs, threaded discussion comments, and bug meta-data, which includes BugID, priority, status, resolution, time, and others. In Chapter 13, bug reports of 20 open-source projects of the Apache Software Foundation are extracted using a tool named the Bug Report Collection System for trend analysis. As per the quantitative analysis of data, about 20% of open bugs are critical in nature, which directly impacts the functioning of the system. The presence of a large number of bugs of this kind can put systems into vulnerability positions and reduces the risk aversion capability. Thus, it is essential to resolve these issues on a high priority. The test lead can assign these issues to the most contributing developers of a project for quick closure of opened critical bugs. The comments are mined, which help us identify the developers resolving the majority of bugs, which is beneficial for test leads of distinct projects. As per the collated data, the areas more prone to system failures are determined such as input/output type error and logical code error.

Sentiments are the standard way by which people express their feelings. Sentiments are broadly classified as positive and negative. The problem occurs when the user expresses with words that are different than the actual feelings. This phenomenon is generally known to us as sarcasm, where people say something opposite the actual sentiments. Sarcasm detection is of great importance for the correct analysis of sentiments. Chapter 14 attempts to give an algorithm for successful detection of hyperbolic sarcasm and general sarcasm in a data set of sarcastic posts that are collected from pages dedicated for sarcasm on social media sites such as Facebook, Pinterest, and Instagram. This chapter also shows the initial results of the algorithm and its evaluation.

Predictive analytics refers to forecasting the future probabilities by extracting information from existing data sets and determining patterns from predicted outcomes. Predictive analytics also includes what-if scenarios and risk assessment. In Chapter 15, an effort has been made to use principles of predictive modeling to analyze the authentic social network data set, and results have been encouraging. The post-analysis of the results have been focused on exhibiting contact details, mobility pattern, and a number of degree of connections/minutes leading to identification of the linkage/bonding between the nodes in the social network.

Modern medicine has been confronted by a major challenge of achieving promise and capacity of tremendous expansion in medical data sets of all kinds. Medical databases develop huge bulk of knowledge and data, which mandates a specialized tool to store and perform analysis of data and as a result, effectively use saved knowledge and data. Information is extracted from data by using a domain's background knowledge in the process of IDA. Various matters dealt with regard use, definition, and impact of these processes and they are tested for their optimization in application domains of medicine. The primary focus of Chapter 16 is on the methods and tools of IDA, with an aim to minimize the growing differences between data comprehension and data gathering.

Snoozing, or sleeping, is a physical phenomenon of the human life. When human snooze is disturbed, it generates many problems, such as mental disease, heart disease, etc. Total snooze is characterized by two stages, viz., rapid eye movement and nonrapid eye movement. Bruxism is a type of snooze disorder. The traditional method of the prognosis takes time and the result is in analog form. Chapter 17 proposes a method for easy prognosis of snooze bruxism.

Neurodegenerative diseases like Alzheimer's and Parkinson's impair the cognitive and motor abilities of the patient, along with memory loss and confusion. As handwriting involves proper functioning of the brain and motor control, it is affected. Alteration in handwriting is one of the first signs of Alzheimer's disease. The handwriting gets shaky, due to loss of muscle control, confusion, and forgetfulness. The symptoms get progressively worse. It gets illegible and the phonological spelling mistakes become inevitable. In Chapter 18, the authors use a feature extraction technique to be used as a parameter for diagnosis. A variational auto encoder (VAE), a deep unsupervised learning technique, has been applied, which is used to compress the input data and then reconstruct it keeping the targeted output the same as the targeted input.

This edited volume on IDA gathers researchers, scientists, and practitioners interested in computational data analysis methods, aimed at narrowing the gap between extensive amounts of data stored in medical databases and the interpretation, understandable, and effective use of the stored data. The expected readers of this book are researchers, scientists, and practitioners interested in IDA, knowledge discovery, and decision support in databases, particularly those who are interested in using these technologies. This publication provides useful references for educational institutions, industry, academic researchers, professionals, developers, and practitioners to apply, evaluate, and reproduce the contributions to this book.

May 07, 2019
New Delhi, India *Deepak Gupta*
Bengaluru, India *Siddhartha Bhattacharyya*
New Delhi, India *Ashish Khanna*
Uttar Pradesh, India *Kalpna Sagar*

1

Intelligent Data Analysis: Black Box Versus White Box Modeling

Sarthak Gupta, Siddhant Bagga, and Deepak Kumar Sharma

Division of Information Technology, Netaji Subhas University of Technology, New Delhi, India,

1.1 Introduction

In the midst of all of the societal challenges of today's world, digital transformation is rapidly becoming a necessity. The number of internet users is growing at an unprecedented rate. New devices, sensors, and technologies are emerging every day. These factors have led to an exponential increase in the volume of data being generated. According to a recent research [1], users of the internet generate 2.5 quintillion bytes of data per day.

1.1.1 Intelligent Data Analysis

Data is only as good as what you make of it. The sheer amount of data being generated calls for methods to leverage its power. With the proper tools and methodologies, data analysis can improve decision making, lower the risks, and unearth hidden insights. Intelligent data analysis (IDA) is concerned with effective analysis of data [2, 3].

The process of IDA consists of three main steps (see Figure 1.1):

1. *Data collection and preparation*: This step involves acquiring data, and converting it into a format suitable for further analysis. This may involve storing the data as a table, taking care of empty or null values, etc.
2. *Exploration*: Before a thorough analysis can be performed on the data, certain characteristics are examined like number of data points, included variables, statistical features, etc. Data exploration allows analysts to get familiar with the dataset, and create prospective hypotheses. Visualization is extensively used in this step. Various visualization techniques will be discussed in depth later in this chapter.
3. *Analysis*: Various machine learning and deep learning algorithms are applied at this step. Data analysts build models that try to find the best possible fit to the data points. These models can be classified as white box or black box models.

A more comprehensive introduction to data analysis can be found in prior pieces of literature [4–6].

Intelligent Data Analysis: From Data Gathering to Data Comprehension,
First Edition. Edited by Deepak Gupta, Siddhartha Bhattacharyya, Ashish Khanna, and Kalpna Sagar.
© 2020 John Wiley & Sons Ltd. Published 2020 by John Wiley & Sons Ltd.

Figure 1.1 Data analysis process.

1.1.2 Applications of IDA and Machine Learning

IDA and machine learning can be applied to a multitude of products and services, since these models have the ability to make fast, data-driven decisions at scale. We're surrounded by live examples of machine learning in things we use in day-to-day life.

A primary example is web page ranking [7, 8]. Whenever we search for anything on a search engine, the results that we get are presented to us in the order of relevance. To achieve this, the search engine needs to "know" which pages are more relevant than others.

A related application is collaborative filtering [9, 10]. Collaborative filtering filters information based on recommendations of other people. It is based on the premise that people who agreed in their evaluation of certain items in the past are likely to agree again in the future.

Another application is automatic translation of documents from one language to another. Manually doing this is an extremely arduous task and would take a significant amount of time.

IDA and machine learning models are also being used for many other tasks [11, 12] like object classification, named entity recognition, object localization, stock prices prediction, etc.

1.1.3 White Box Models Versus Black Box Models

IDA aims to analyze the data to create predictive models. Suppose that we're given a dataset D(X,T), where X represents inputs and T represents target values (i.e., known correct values with respect to the input). The goal is to learn a function (or map) from inputs (X) to outputs (T). This is done by employing supervised machine learning algorithms [13]. A model refers to the artifact that is created by the training (or learning) process. Models are broadly categorized into two types:

1. *White box models*: The models whose predictions are easily explainable are called white box models. These models are extremely simple, and hence, not very effective. The accuracy of white box models is usually quite low. For example – simple decision trees, linear regression, logistic regression, etc.
2. *Black box models*: The models whose predictions are difficult to interpret or explain are called black box models. They are difficult to interpret because of their complexity. Since they are complex models, their accuracy is usually high. For example – large decision trees, random forests, neural networks, etc.

So, IDA and machine learning models suffer from accuracy-explainability trade-off. However, with advances in IDA, the explainability gap in black box models is reducing.

1.1.4 Model Interpretability

If black box models have better accuracy, why not use them all the time? The problem is that a single metric, such as classification accuracy, is an incomplete description of most real-world tasks [14, 15]. Sometimes in low-risk environments, where decisions don't have severe consequences, it might be sufficient to just know that the model performed well on some test dataset without the need for an explanation. However, machine learning models are being extensively used in high-risk environments like health care, finance, data security, etc. where the impact of decisions is huge. Therefore, it's extremely important to bridge the explainability gap in black box models, so that they can be used with confidence in place of white box models to provide better accuracy.

Interpretability models may be local or global. Global methods try to explain the model itself, thereby explaining all possible outcomes. On the other hand, local models try to explain why a particular decision was made.

As artificial intelligence (AI)-assisted decision making is becoming commonplace, the ability to generate simple explanations for black box systems is going to be extremely important, and is already an area of active research.

1.2 Interpretation of White Box Models

White box models are extremely easy to interpret, since interpretability is inherent in their nature. Let's talk a few white box models and how to interpret them.

1.2.1 Linear Regression

Linear regression [16, 17] attempts to model the relationship between input variables and output by fitting a linear equation to the observed data (see Figure 1.2). A linear regression equation is of the form:

$$y = w_0 + w_1 x_1 + w_2 x_2 + \cdots + w_p x_p \qquad (1.1)$$

where,

y is the output variable,
x1, x2,..., xp are "p" input variables,
w1, w2,..., wp are the weights associated with input variables, and
w0 makes sure that the regression line works even if the data is not centered around origin (along the output dimension).

The weights are calculated using techniques like ordinary least squares and gradient descent. The details of these techniques are beyond the scope of this chapter; we will focus more on the interpretability of these models.

The interpretation of the weights of a linear model is quite obvious. An increase by one unit in the feature xj results in a corresponding increase by wj in the output.

Another metric for interpreting linear models is **R^2** measurement [18]. R^2 value tells us about how much variance of target outcomes is explained by the model. R^2 value ranges

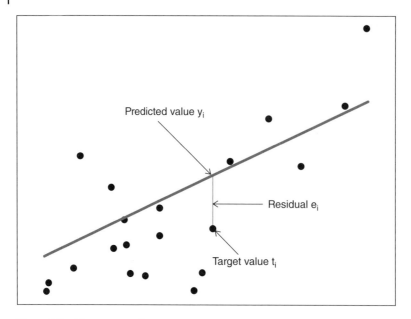

Figure 1.2 Linear regression.

from 0 to 1. Higher the R^2 value, better the model explains the data. R^2 is calculated as:

$$R^2 = 1 - SS_r/SS_t \tag{1.2}$$

where

SS_r is the squared sum of residuals, and
SS_t is the total sum of squares (proportional to variance of the data)

Residual e_i is defined as:

$$e_i = y_i - t_i \tag{1.3}$$

where

y_i is the model's predicted output, and
t_i is the target value in the dataset.

Hence, SS_r is calculated as:

$$SS_r = \sum_{i=1}^{n} (y_i - t_i)^2 = \sum_{i=1}^{n} e_i^2 \tag{1.4}$$

And SS_t is calculated as:

$$SS_t = \sum_{i=1}^{n} \left(t_i - \bar{t} \right)^2 \tag{1.5}$$

where $\bar{t}$ is the mean of all target values.

But, there is a problem with R^2 value. It increases with number of features, even if they carry no information about the target values. Hence, adjusted R^2 value $(\overline{R^2})$ is used, which takes into account the number of input features:

$$\overline{R^2} = 1 - (1 - R^2)\frac{n-1}{n-p-1} \tag{1.6}$$

Where

    $\overline{R^2}$ is the adjusted R^2 value,

    n is the number of data points, and

    p is the number of input features (or input variables)

1.2.2 Decision Tree

Decision trees [19] are classifiers – they classify a given data point by posing a series of questions about the features associated with the data item (see Figure 1.3).

Unlike linear regression, decision trees are able to model nonlinear data. In a decision tree, nodes represent features, each edge or link represents a decision, and leaf nodes represent outcomes.

The general algorithm for decision trees is given below:

1. Pick the best attribute/feature. Best feature is that which separates the data in the best possible way. The optimal split would be when all data points belonging to different classes are in separate subsets after the split.
2. For each value of the attribute, create a new child node of the current node.
3. Divide data into the new child nodes.
4. For each new child node:
   a. If all the data points in that node belong to the same class, then stop.
   b. Else, go to step 1 and repeat the process with current node as decision node.

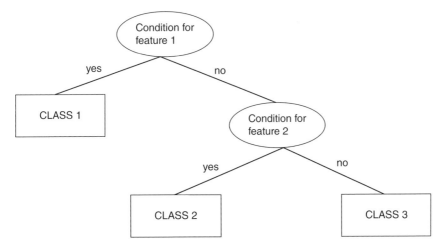

Figure 1.3 Decision tree.

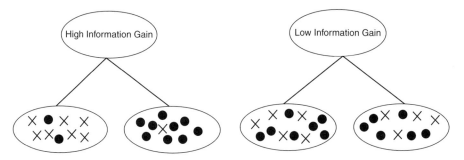

Figure 1.4 Distribution of points in case of high and low information gain.

Many algorithms like ID3 [20] and CART [21] can be used to find the best feature.

In ID3, information gain is calculated for each attribute, and the attribute with highest information gain is chosen as best attribute (see Figure 1.4).

For calculating information gain, we first calculate entropy (H) for each feature (F) over the set of data points in that node (S):

$$H_F(S) = \sum_{v \in values(F)} - p_v \times \log(p_v) \tag{1.7}$$

where *values*(F) denotes the values that feature F can take.

Note that the entropy is 0 if all data points in S lie in the same class. Information gain (G) is then calculated as:

$$Information\ Gain = Entropy(parent) - AverageEntropy(children)$$

$$G_F(S) = H_F(S) - \sum_{v \in values(F)} \frac{n_v}{n} \times H_F(S_v) \tag{1.8}$$

where

S_v denotes the subset of data points in S who have value v for feature F, and
n_v denotes the number of data points in S_v, and
n denotes the number of data points in S.

Then, the feature with maximum information gain is selected.

Another algorithm for selecting best feature is CART. It uses the Gini Index to decide which feature is the best. The Gini Index is calculated as:

$$Gini_F(S) = 1 - \sum_{v \in values(F)} p_v^2 \tag{1.9}$$

Minimum value of the Gini Index is 0 when all data points in S belong to the same class.

Interpreting decision trees is extremely simple. To explain any decision made by the decision tree, just start from the root node, and move downward according to the input features. Eventually, a leaf node will be reached, which the prediction is made by the decision tree.

Hence, it is clear that if a decision tree is small, then it may be considered white box. But for large decision trees, it's not possible to say which factor has how much effect on the final outcome. So, large decision trees are considered black box models.

There are a lot of other white box IDA algorithms like naive Bayes, decision rules, k-nearest neighbors, etc. whose predictions are easy to explain.

1.3   Interpretation of Black Box Models

In case of white box models, being interpretable is a property of the model itself. Hence, we studied a few white box models and learned how to interpret them. However, in case of black box models, special algorithms are needed for their interpretability. These algorithms are model-agnostic, i.e., they do not depend on a particular model. This has a huge advantage – flexibility.

We will discuss a few of these algorithms in this section.

1.3.1   Partial Dependence Plot

This technique was developed by Friedman [22]. Partial dependence plot (PDP) reveals the marginal effect of a feature on the predicted outcome of a model. The chosen features are varied, while the other features are fixed at their average value to visualize the type of relationship between the chosen features and the model's outcome. For example, in case of linear regression, PDP will always show a linear relationship because the outcome depends linearly on every feature.

Let x_s be a subset of the set of all features, and let x_c be the complement of x_s. Here, x_s is the set of selected features for which partial dependence function should be plotted, and x_c is the set of other features used in the model. Usually, there are one or two features in x_s. This is because visualizing and interpreting a two-dimensional or three-dimensional plot is easy. Let the model be represented by f. Then, partial dependence function of f on x_s is given by:

$$f_s = E_{x_c}\left[f(x_s, x_c)\right] = \int f(x_s, x_c)dP(x_c) \qquad (1.10)$$

Each subset of features x_s has its own partial dependence function f_s, which gives the average value of f at a certain fixed value of x_s, while x_c varies over its marginal distribution $dP(x_c)$. Partial dependence marginalizes the output predictions of the model f over the distribution of features x_c. Hence, the function f_s shows the relationship between x_s and the predicted outcome.

Since $dP(x_c)$ is not known, it is estimated by taking average over all data points in the training data:

$$f_s = \frac{1}{n}\sum_{i=1}^{n} f(x_s, x_{c_i}) \qquad (1.11)$$

where

> n is the number of training examples, and
> x_{c_i} is the set of features x_c in the ith training example.

If f_s is evaluated at all x_s observed in data, then we'll have n pairs of the type (x_s, f_s), which can be plotted to see how the machine learning model f varies with the set of features x_s.

Figure 1.5 [23] shows the partial dependence of house value for on various features – median income (MedInc), average occupants per household (AvgOccup), and median house age (HouseAge).

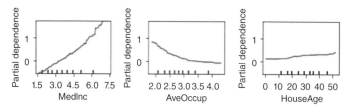

Figure 1.5 Partial dependence plots from a gradient boosting regressor trained on California housing dataset [23].

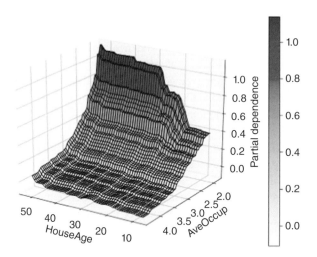

Figure 1.6 Partial dependence plot from a gradient boosting regressor trained on California housing dataset [23].

PDP can also be plotted for two features in x_s instead of one, as shown in Figure 1.6 [23] (partial dependence of house value on house age and average occupancy).

However, PDP sometimes gives incorrect results too. As an example [24], consider the following data generation process:

$$Y = 0.2 \times X_1 - 5 \times X_2 + 10 \times \mathbb{1}_{X_3 \geq 0}(X_2) + \varepsilon \qquad (1.12)$$

where

X_1, X_2, X_3 are random variables uniformly distributed over $(-1,1)$

$\mathbb{1}$ is indicator function, defined as $\mathbb{1}_{A(x)} = \begin{cases} 1 \ if \ x \in A \\ 0 \ if \ x \notin A \end{cases}$, and

ε is a random variable normally distributed over $(0,1)$.

1,000 observations were generated from this equation, and a stochastic gradient boosting model was fit to the generated data. Figure 1.7a [24] shows the scatter plot between X_2 and Y, and Figure 1.7b [24] shows the partial dependence of Y on X_2. The PDP shows that there is no meaningful relationship between the two variables, but we can clearly see from the scatter plot that this interpretation is wrong.

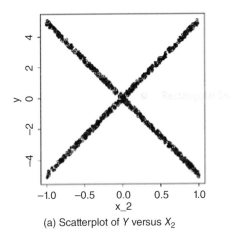

(a) Scatterplot of Y versus X_2

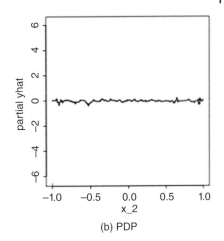

(b) PDP

Figure 1.7 Relationship between X_2 and Y [24].

These plots sometimes produce wrong results because of averaging, which is an inherent property of PDPs. Moreover, PDPs provide a useful summary only when the dependence of selected features on the remaining features is not too strong.

To overcome some of the issues with PDP, individual conditional expectation (ICE) plots are used.

1.3.2 Individual Conditional Expectation

ICE [24] plots basically disperse the output of PDPs. Instead of averaging over the other (nonselected) features, ICE plots make one line for each data point, representing how the prediction of that data point changes when the selected feature changes. So, a PDP is the average of all lines of the corresponding ICE plot.

More formally, for the observations $\{(x_{s_i}, x_{c_i})\}_{i=1}^{n}$, a curve f_s^i is plotted against x_{s_i}, while x_{c_i} is kept fixed. The individual plots taken collectively $\{f_s^i\}_{i=1}^{n}$ constitute ICE plot for a set of selected features x_s.

Consider the example given above. The ICE plot for the same is shown in Figure 1.8 [24]:

From Figures 1.7b and 1.8, we can see that ICE plot shows the relationship between X_2 and Y in a much better way than PDP.

However, ICE plots are not able to resolve one issue with PDPs. When the feature of interest is correlated with the other features, the joint feature distribution is not taken into account and as a result, the plots might provide incorrect interpretations. To overcome this problem, accumulated local effects (ALE) plots are used.

1.3.3 Accumulated Local Effects

ALE [25] describe how features affect the predictions of a machine learning model. They are unbiased to correlation between features, unlike a PDP.

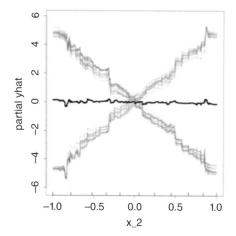

Figure 1.8 ICE plot between feature X_2 and Y [24].

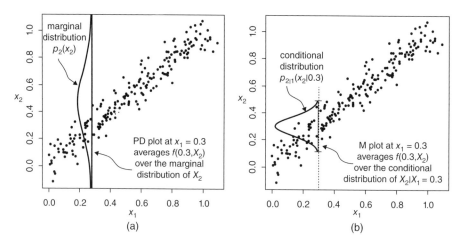

Figure 1.9 Calculation of PDP and M-plot [25].

The equation for calculating partial dependence of f on x_s is given by:

$$f_{s,PDP}(x_s) = E_{X_c}\left[f(x_s, X_c)\right] = \int_{x_c} f(x_s, x_c)P(x_c)dx_c \tag{1.13}$$

Where X represents a random variable and x represents observed values of X.

In Figure 1.9 [25], the two variables x_1 and x_2 are highly correlated.

Figure 1.9a shows how the partial dependence is calculated for $x_1 = 0.3$. According to the equation above, f_s is calculated as the integration of marginal distribution of x_2 over the entire vertical line $x = 0.3$ shown in the figure. This would require extrapolation much beyond the envelope of the training data. This problem occurs because marginal distribution is not concentrated around the data.

To overcome this problem, marginal plots (M-plots) [26] are used (see Figure 1.9b), which replaces conditional probability for marginal probability. In case of highly correlated variables, conditional probability is more or less concentrated around the data as shown in the figure. So, the formula for calculating M-plot is:

$$f_{s,M}(x_s) = E_{X_c|X_s}\left[f(X_s, X_c)|X_s = x_s\right] = \int_{x_c} f(x_s, x_c)P(x_c|x_s)dx_c \tag{1.14}$$

Integrating the conditional distribution has the advantage that excessive extrapolation is not required. However, averaging the local predictions leads to mixing the effects of both features, which is undesirable.

ALE addresses this problem by taking the differences, instead of averages over the conditional probability distribution. Ale is calculated as follows:

$$f_{s,ALE}(x_s) = \int_{z_{0,1}}^{x_s} E_{X_c|X_s}\left[f^s(X_s, X_c)|X_s = z_s\right] dz_s - constant \tag{1.15}$$

$$f_{s,ALE}(x_s) = \int_{z_{0,1}}^{x_s} \int_{x_c} f^s(z_s, x_c)P(x_c|z_s)dx_c dz_s - constant \tag{1.16}$$

where

$$f^s(x_s, x_c) = \frac{\delta f(x_s, x_c)}{\delta x_s} \tag{1.17}$$

There are quite a few changes as compared to M-plots. First, we average over the changes of predictions, not the predictions themselves (f^s instead of f). Second, there is an additional integration over z. $z_{0,1}$ is some value chosen just below the effective support of the probability distribution over the selected features x_s. The choice of $z_{0,1}$ is not important, since it only affects the vertical translation of ALE plot, and the constant is chosen to vertically center the plot.

ALE calculates prediction differences, and then integrates the differential (see Figure 1.10 [6]). The derivative effectively isolates the effect of the features of interest, and blocks the

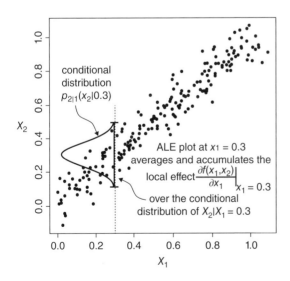

Figure 1.10 Calculation of ALE plot [25].

effect of correlated features, thereby overcoming the correlation problem of PDPs and ICE plots.

1.3.4 Global Surrogate Models

A global surrogate model is an interpretable model that is trained to approximate the predictions of a black box model. Interpretation of the black box model can then be done by simply interpreting the surrogate model.

Formally, we want the approximation g of a black box model f to be as close as possible to f under the constraint that model g is interpretable. So, any white box model can be used as the function g (like linear regression, decision tree, etc.).

R^2 measure can be used to determine how well the surrogate model replicates the black box model:

$$R^2 = 1 - \frac{SS_e}{SS_t} = 1 - \frac{\sum_{i=1}^{n} (y_i^* - y_i)^2}{\sum_{i=1}^{n} (y_i - \overline{y_i})^2} \tag{1.18}$$

where

y_i^* is the prediction of the surrogate model,
y_i is the prediction of the black box model, and
$\overline{y_i}$ is average of black box model predictions.

A major disadvantage of this method is that we can never be sure how well the surrogate model replicates the black box model. It may happen that the interpretable surrogate model is very close to the black box model for a subset of the data, but diverges for another subset.

1.3.5 Local Interpretable Model-Agnostic Explanations

Local interpretable model-agnostic explanations (LIME) [27] can be thought of as a local surrogate method. LIME says that to understand the model, we need to perturb the input values, and see how the predictions change. Instead of learning a global surrogate model, it focuses on fitting local surrogate models to explain why a single prediction was made.

To explain a single prediction, that particular input data point is perturbed over and over again, and a white box model is trained based on the predictions from the black box model. This provides a good local approximation of the model around that data point.

This provides an advantage over global surrogate models – the learned model should be a good approximation of the black box model locally, but it does not have to be so globally, thus overcoming a major shortcoming of global surrogate models.

1.3.6 Feature Importance

A feature's importance is calculated by the change in model's prediction error after permuting the feature [28]. If the model error increases, it means that the feature is important, because the model relied on it for its prediction. A feature is considered unimportant if permuting its value doesn't change the model's prediction error.

This method suffers from the same problem as PDP – correlated features. If features are correlated, then permuting the values creates unlikely data points and gives false interpretation results.

1.4   Issues and Further Challenges

Although a lot of work is being done to understand these white box and black box models, but there are still a lot of issues that need to be tackled.

A major issue is the distinction between correlation and causation. Figure 1.11 [29] makes it clear why correlation does not imply causation.

However, machine learning models cannot distinguish between the two. This is an issue with such models. Hence, interpretability is of paramount importance, so that false causations could be detected and removed from these models.

Another issue is that machine learning models operate in high dimensional vector spaces, and it's usually beyond human perception to visualize and interpret more than three dimensions. So, one option is try to explain only one or two features at a time, while the other option is dimensionality reduction, which leads to some loss in data. Internal states of machine learning models are extremely difficult to interpret; hence most of the interpretability algorithms try to explain the predictions instead of trying to understand the model.

Another challenge is the testing of these interpretability models, which as of now relies mostly on humans, since automating it is significantly hard.

1.5   Summary

In this chapter, the types of IDA models have been discussed, namely white box models and black box models. These models are caught in the explainability-accuracy trade-off. White box models have low accuracy, but are able to produce high-quality explanations of the decisions made by the model. Black box models, on the other hand, are more accurate models, but suffer from low explainability. To highlight the differences between the two models, various interpretability techniques have been reviewed.

White box models (like linear regression, decision trees, naive Bayes, etc.) are inherently interpretable. A few of these models have been discussed, along with ways to interpret their predictions.

Figure 1.11   Correlation does not imply causation [29].

However, to explain the decisions of black box models, special algorithms have been developed. A few of these "model interpretability algorithms" have been discussed in this chapter, namely, PDPs, ICE, ALE, global and local surrogate models, and feature importance.

Although researchers have made tremendous progress, a lot of challenges still remain untackled. A few shortcomings of machine learning models have also been presented.

References

1 Data Never Sleeps 5.0. www.domo.com/learn/data-never-sleeps-5 (accessed 02 November 2018).

2 Berthold, M.R. and Hand, D.J. (eds.) (2007). *Intelligent Data Analysis: An Introduction.* Berlin, Heidelberg: Springer.

3 Cohen, P. and Adams, N. (2009, August). Intelligent data analysis in the 21st century. In: *International Symposium on Intelligent Data Analysis* (eds. N.M. Adams, C. Robardet, A. Siebes and J.F. Boulicaut), 1–9. Berlin, Heidelberg: Springer.

4 Bendat, J.S. and Piersol, A.G. (2000). *Random Data: Analysis and Measurement Procedures*, vol. 729. Hoboken, NJ: Wiley.

5 Var, I. (1998). Multivariate data analysis. *Vectors* 8 (2): 125–136.

6 Mosteller, F. and Tukey, J.W. (1977). Data Analysis and Regression: A Second Course in Statistics. In: *Addison-Wesley Series in Behavioral Science: Quantitative Methods.* Addison-Wesley Publishing.

7 Page, L., Brin, S., Motwani, R., and Winograd, T. (1999). *The PageRank Citation Ranking: Bringing Order to the Web.* Stanford InfoLab.

8 Sharma, D.K. and Sharma, A.K. (2010). A comparative analysis of web page ranking algorithms. *International Journal on Computer Science and Engineering* 2 (08): 2670–2676.

9 Linden, G., Smith, B., and York, J. (2003). Amazon.com recommendations: item-to-item collaborative filtering. *IEEE Internet Computing* 7 (1): 76–80.

10 Breese, J.S., Heckerman, D., and Kadie, C. (1998). Empirical analysis of predictive algorithms for collaborative filtering. In: *Proceedings of the Fourteenth Conference on Uncertainty in Artificial Intelligence*, 43–52. Madison, WI: Morgan Kaufmann Publishers Inc.

11 Segre, A.M. (1992). Applications of machine learning. *IEEE Expert* 7 (3): 30–34.

12 Langley, P. and Simon, H.A. (1995). Applications of machine learning and rule induction. *Communications of the ACM* 38 (11): 54–64.

13 Kotsiantis, S.B., Zaharakis, I., and Pintelas, P. (2007). Supervised machine learning: a review of classification techniques. *Emerging Artificial Intelligence Applications in Computer Engineering* 160: 3–24.

14 Doshi-Velez, F., & Kim, B. (2017). Towards a Rigorous Science of Interpretable Machine Learning. arXiv:1702.08608.

15 Vellido, A., Martín-Guerrero, J.D., and Lisboa, P.J. (2012). Making machine learning models interpretable. In: *ESANN*, vol. 12, 163–172.

16 Seber, G.A. and Lee, A.J. (2012). *Linear Regression Analysis*, vol. 329. Hoboken, NJ: Wiley.

17 Montgomery, D.C., Peck, E.A., and Vining, G.G. (2012). *Introduction to Linear Regression Analysis*, vol. 821. Hoboken, NJ: Wiley.

18 Cameron, A.C. and Windmeijer, F.A. (1997). An R-squared measure of goodness of fit for some common nonlinear regression models. *Journal of Econometrics* 77 (2): 329–342.

19 Safavian, S.R. and Landgrebe, D. (1991). A survey of decision tree classifier methodology. *IEEE Transactions on Systems, Man, and Cybernetics* 21 (3): 660–674.

20 Quinlan, J.R. (1986). Induction of decision trees. *Machine Learning* 1 (1): 81–106.

21 Breiman, L. (2017). *Classification and Regression Trees*. New York: Routledge.

22 Friedman, J.H. (2001). Greedy function approximation: a gradient boosting machine. *Annals of Statistics* 29 (5): 1189–1232.

23 Partial Dependence Plots. http://scikit-learn.org/stable/auto_examples/ensemble/plot_partial_dependence.html (accessed 02 November 2018).

24 Goldstein, A., Kapelner, A., Bleich, J., and Pitkin, E. (2015). Peeking inside the black box: visualizing statistical learning with plots of individual conditional expectation. *Journal of Computational and Graphical Statistics* 24 (1): 44–65.

25 Apley, D. W. (2016). Visualizing the Effects of Predictor Variables in Black Box Supervised Learning Models. arXiv:1612.08468.

26 Cook, R.D. and Weisberg, S. (1997). Graphics for assessing the adequacy of regression models. *Journal of the American Statistical Association* 92 (438): 490–499.

27 Ribeiro, M.T., Singh, S., and Guestrin, C. (2016). Why should i trust you?: explaining the predictions of any classifier. In: *Proceedings of the 22nd ACM SIGKDD International Conference on Knowledge Discovery and Data Mining*, 1135–1144. ACM.

28 Breiman, L. (2001). Random forests. *Machine Learning* 45 (1): 5–32.

29 Spurious Correlations. http://tylervigen.com/spurious-correlations (accessed 02 November 2018).

2

Data: Its Nature and Modern Data Analytical Tools

Ravinder Ahuja, Shikhar Asthana, Ayush Ahuja, and Manu Agarwal

Computer Science and Engineering Department, Jaypee Institute of Information Technology, Noida, India

2.1 Introduction

Data is a collection of values that pertain to qualitative or quantitative variables. Data is a vital and irreplaceable asset to any form of an organization or an enterprise. From the smallest scale, up to a multinational corporation, data is always omnipresent. Whether it's internal to the four walls of the organization or outside, data is always there – just waiting to be organized and analyzed. Data without any processing or order, in its most primitive form, is called raw data [1]. In essence, raw data is just an unordered set of qualitative or quantitative variables with very little or no processing done. Even though raw data has a major perspective of becoming "information," it requires selective extraction, formatting, organizing, and analysis for transforming itself into it. For example, a broker transaction terminal in a busy open stock market collects huge volumes of raw data each day, but this data, even though having the capacity to, does not yield much information until it has undergone processing. Once processed, the data may indicate the specific units of stock that each buyer or seller has bought or sold, respectively. Hence, from the raw data, we get processed data, from which we get information, and from this information, we derive our end point, which is insight [2] as shown in Figure 2.1. That means that the information we obtain from processed data can derive insights that help us better understand and perceive the future.

Returning to our stock market example, once the processed data is acquired, we can obtain the general trend of the stock, which is right now being bought the most. This, in turn, will provide us with the insight pertaining to the probability of that stock's price being increased, which will be very high due to intense demand for that stock. And on the basis of this insight, one can plan whether he or she wants to invest in that stock or if selling off an already owned stock would serve them better. As more and more data is taken up for consideration, the information and insights derived from them start getting more accurate and serve as a better representative for the general trend. As the stakes and competition between various organizations and enterprises are rising day by day, these obtained insights are what make the difference between the leading organizations and others. Hence, the need of the hour is to better understand, manage, analyze, and process the humongous amounts of data available to obtain valuable insights. Depending on the field of operation and type of values

Intelligent Data Analysis: From Data Gathering to Data Comprehension,
First Edition. Edited by Deepak Gupta, Siddhartha Bhattacharyya, Ashish Khanna, and Kalpna Sagar.
© 2020 John Wiley & Sons Ltd. Published 2020 by John Wiley & Sons Ltd.

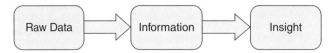

Figure 2.1   Various stages of data.

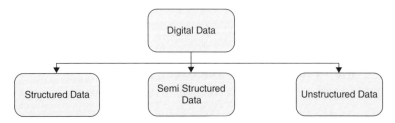

Figure 2.2   Classifications of digital data.

and variables, data can be divided into a lot of categories, but for the sake of this chapter, we would be focusing on the digital data and we will loosely use data and digital data interchangeably. Digital data, as shown in Figure 2.2, can be broadly classified into structured, semi-structured, and unstructured data. We will explore each of these categories one by one in the upcoming sections and understand their definitions, sources, and the ease or unease of working with and processing them.

The rest of the chapter is organized as follows: Section 2.2 contains various data types and different file formats, Section 2.3 contains an overview of big data, Section 2.4 contains various phases of data analytics, Section 2.5 contains various data analytics tools, Section 2.6 contains database management tools for big data, Section 2.7 contains challenges in big data analytics, and finally, Section 2.8 contains conclusion.

2.2   Data Types and Various File Formats

There are different data types and different file formats, which are described as follows:

2.2.1   Structured Data

Structured data is the data that is in an organized form. In simpler terms, structured data refers to information having a high degree of organization (recall that since the term used is information, it means that the data in question is already processed). We will be focusing mainly on structured data as it is the most used data in terms of data analytics. Structured data conforms to a predefined schema or structure, such that its integration into a relational database is smooth and effortlessly searchable by various search algorithms and operations. This ease of accessibility also augments the ability to analyze, update, modify, and delete a certain part or the whole of the data. Data stored in databases is an example of structured data. In structured data, there exist relationships between entities of data, such as classes and their instances. From an understanding point of view, structured data should be synonymous with data model – any model with the intent to provide a way to

Table 2.1 Schema of an employee table in a broker company.

Column name	Data type	Constraints
EmpNum	VARCHAR(10)	PRIMARY KEY
EmpName	VARCHAR(25)	
DeptNum	VARCHAR(5)	NOT NULL
BrokerID	VARCHAR(10)	UNIQUE
ContactNum	VARCHAR(10)	NOT NULL

store, access, analyze, and modify data. Let us discuss this in the context of a relational database management system (RDBMS) due to their vast use, availability, robustness, and efficiency attributes. Most of the structured data is generally stored in RDBMS, which is also very cost-effective. An RDBMS conforms to the relational data model wherein the data is stored in rows and columns and where storage, retrieval, and management of data has been tremendously simplified.

The first step of RDBMS is the design of a relational (a relation is generally implemented using a table), the fields and columns to store the data, the type of data that we would like to store in these fields (they can range from just numbers, characters to dates, and Boolean). The number of records/tuples in a relation is called the cardinality of the relation and the number of columns is called the degree of the relation. Often there is a need to implement constraints to which our data conforms to. These constraints include UNIQUE, which makes sure no two values are same in tuples; NOT NULL, which makes sure there is no null or empty value in a column; PRIMARY KEY, a combination of UNIQUE and NOT NULL constraints and there exists only one primary key in a relation. To better understand, let us design our own relation to store the details of an employee who's working in a stock brokering enterprise. Table 2.1 portrays a good example of a well-structured relation (in the form of a table), complete with all the required fields and constraints, which adheres to the relational data model.

The above-shown example is a perfect example of an ideal data model and the values inserted into such a data model would be a prime specimen for structured data. The data would be following an organized structure and would be adhering to a data model.

Various reasons [3], due to which working with structured data are so much easier and efficient, are as follows:

1. *Data manipulation*: It is very easy to manipulate data when the data is structured data. Operations like inserting, updating, and deleting, which come under data manipulation language (DML), provide the required ease in operations pertaining to storing, accessing, processing, etc.
2. *Scalability*: Structured data possess great potential to be scalable due to the ease of storage and processing attributes.
3. *Security*: Structured data can be better secured as applying encryption and decryption techniques upon a structured schema, which is easier for the organization, and through this, they can control access to information and data.

4. *Transaction processing*: Usually a structured data set has better transaction processing due to the adherence to a defined structure. Taking our previous example of RDBMS, we have support for atomicity, consistency, isolation, and durability (ACID) properties, which makes the transitioning of data and its processing easier and uniform.
5. *Retrieval and recovery*: Structured data is easier to retrieve from a database and, due to its conformity to a data model, it is also easier to create backups and provide back-up security to structured data.

2.2.2 Semi-Structured Data

Semi-structured data, also often referred to as self-describing data, is one level lower than structured data in terms that it does not conform to the data models that one typically associates with relational databases or any other such type. However, it is still more structured than unstructured data since it uses tags to segregate semantic elements and to administer different levels of records and fields within the data. However, in semi-structured data, instances belonging to the same class need not have the same attributes and are usually just clumped together. The above-stated facts pertain to the fact that in semi-structured data, there is an amalgamation of data and schema.

Semi-structured data has many sources, and nowadays, the use of semi-structured data has shown growth as compared to previous years. When we discuss semi-structured data, the first names that come to mind are extensible markup language (XML) and JavaScript object notation (JSON). These are among the most famous of the various file formats that are used and these, in particular, deal with a lot of semi-structured data. XML was made widely popular due to its heavy usage in web services. JSON also followed a similar path to popularity due to its wide applications and extensive use in implementing web services. Each of these files formats and many others is discussed in detail in the later sections.

2.2.3 Unstructured Data

Unstructured data, as the name suggests, does not conform to any predefined data model. To facilitate our understanding let us consider examples like audio files, free-form text, social media chats, etc. Each of these examples has unpredictable structures. Unstructured data is very hard to deal with in terms of processing and analysis and is usually used as sources from which we can obtain structured data after applying various techniques and methods. Unstructured data nowadays constitutes up to 80% of the total data that is being generated in any organization. As we can see, in Figure 2.3, this is a big amount of data and it cannot be ignored. Since the majority of the data being generated is unstructured data, the essence of handling unstructured data lies in our ability to convert this unstructured data into a more usable form. In order to accomplish this very task, we have at our disposal many techniques and methods. The most commonly used of them are listed below with a brief explanation:

1. *Data mining*: Data mining is the action of sorting across very large data sets to recognize and categorize patterns and institute relationships to solve and overcome various difficulties through data analysis [4]. Data mining tools allow us to make use of even unstructured data and utilize it so as to provide information, which would have been impossible

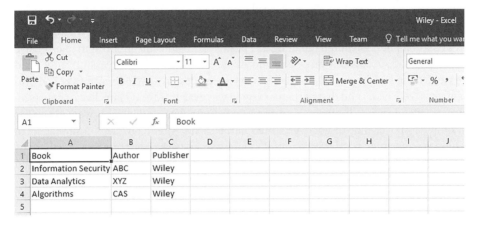

Figure 2.3 CSV file opened in Microsoft Excel.

to extract directly from unstructured data. The previously said patterns and relationship between variables are identified using cutting-edge methods, which have their roots in artificial intelligence, machine learning, deep learning, statistics, and database systems. Some popular data mining algorithms include associative rule mining, regression analysis, collaborative filtering, etc.

2. *Text mining*: Text data is largely unstructured, vague, and difficult to deal with when being under the target of an algorithm. Text mining is the art of extraction of high quality and meaningful information from a given text by methods, which analyze the text thoroughly by means of statistical pattern learning [5]. It includes sentimental analysis, text clustering, etc.

3. *Natural language processing*: Commonly referred to simply as natural language processing (NLP), is related to bridging the gap between human–computer communications. It is the implementation of various computational techniques on the scrutiny and fusion of natural language and speech [6]. In simpler terms, NLP is largely concerned with how to analyze large amounts of natural language data and make it better understandable by a computer and is a component of artificial intelligence.

4. *Part of speech tagging (POST)*: The process of reading and analyzing some text and then tagging each word in a sentence as belonging to a particular part of the speech such as "adjective," "Pronoun," etc. [7]

There are many other tools and methods available too, however, the main gist remains the same for them all and that is to use various mathematical and statistical tools to analyze and find patterns present in unstructured data in lieu of making it usable for data analytics.

2.2.4 Need for File Formats

As more and more data is being produced, it is becoming impossible to process and analyze them as soon as they are produced/found. Hence, there is a vital need to store these data in somewhere and process them later. Moreover, the need for storing data is not restricted to the raw data end but is also prominent for the end product. To better understand, let us take the scenario of data related to weather forecasts. Consider the situation where we had a set of natural phenomenon happening in a particular region and after resource-extensive methods, we had successfully processed and analyzed this data. Our successfully processed data yielded the insights that these phenomena are precast to a cyclone. Afterward, appropriate steps were taken and that particular region was evacuated and many lives were saved. However, after deriving our insights, all the calculations, data, and insights were not stored and simply lost over time or got overwritten in the cache. After a few days, a similar set of phenomena appeared in another region. As we can see, if we had our previous calculations and the derived insights, we could have benefited greatly on the resource part and also had gained a time advantage as the insights were already ready. What if this, in turn, had facilitated us to better implement precautions and broadcast warnings to the particular region? Thus, let us look into some commonly used file formats. Also, from a real-time perspective, the industry today is utilizing various file formats for different purposes and one hardly ever gets a neat tabular data. As a data scientist or data analyst, it is essential to be aware and up to date with each of these file formats and be armed with information on how to use them.

2.2.5 Various Types of File Formats

A file format is a standardized way in which some information may be encoded for storage. A typical file format infers from itself two major things, whether the said file is a binary or an ASCII file and secondly, it shows how the information will be stored in the said file. We can easily identify a file format by looking at the extension of a file. For example, a file named "Wiley," which is stored as "Wiley.csv," clearly shows that the file is a comma separated values (CSV) file and its data will be stored in a tabular form. Hence in a way, each new file format represents a different type of data. To further enhance our knowledge, we will undertake the most commonly used file formats and understand each one of them individually [8].

2.2.5.1 Comma Separated Values (CSV)

Comma separate values (CSV) file format is a subset of the spreadsheet file format, in which data is stored in cells. Each cell is organized in rows and columns. Each line in the CSV file represents an entry or commonly called a record. Each record contains multiple fields that are separated by a comma. Another type of file format that is essentially very similar to CSV file format is the tab separated values (TSV), in which as the name suggests, the multiple fields are separated using tab. Figure 2.3 shows a CSV file named "Wiley.csv," which has been opened using Microsoft Excel.

2.2.5.2 ZIP

ZIP format is an archive file format. In simpler terms, a file is said to be an archive file if it contains multiple files along with some metadata. Archive file formats are used for compressing files so that they occupy less space. ZIP file format is a lossless compression format, which means that the original file could be fully recovered after decompressing an already compressed ZIP file. Along with ZIP, there are many more commonly used archive file formats that include RAR, Tar, etc.

2.2.5.3 Plain Text (txt)

This is perhaps the most commonly known file format. In this file format, everything is written in plain text, which is usually in an unstructured format and has no associated metadata. Reading a plain text file format is very easy; however, interpretation of the same by a computer program is very difficult. Figure 2.4 shows a plain text file opened with the notepad.

2.2.5.4 JSON

JSON file format [9] is one of the front-runners in the data file format used for semi-structured data. JSON, which is an acronym for JavaScript object notation is a text-based open standard that has been designed with the idea of exchanging data over the web in mind. This format is commonly used for transmitting semi-structured and structured data over the web and is a language-independent data. Figure 2.5 shows a JSON file having the details of two meteorologists.

2.2.5.5 XML

Extensible markup language [10] also serves as a major file format in the context of semi-structured data and is a part of the markup languages. This file format has certain

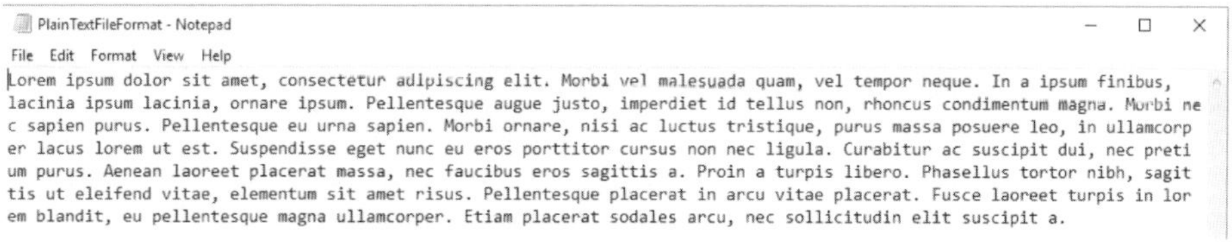

```
PlainTextFileFormat - Notepad                                                    —   □   ×
File  Edit  Format  View  Help
Lorem ipsum dolor sit amet, consectetur adipiscing elit. Morbi vel malesuada quam, vel tempor neque. In a ipsum finibus,
lacinia ipsum lacinia, ornare ipsum. Pellentesque augue justo, imperdiet id tellus non, rhoncus condimentum magna. Morbi ne
c sapien purus. Pellentesque eu urna sapien. Morbi ornare, nisi ac luctus tristique, purus massa posuere leo, in ullamcorp
er lacus lorem ut est. Suspendisse eget nunc eu eros porttitor cursus non nec ligula. Curabitur ac suscipit dui, nec preti
um purus. Aenean laoreet placerat massa, nec faucibus eros sagittis a. Proin a turpis libero. Phasellus tortor nibh, sagit
tis ut eleifend vitae, elementum sit amet risus. Pellentesque placerat in arcu vitae placerat. Fusce laoreet turpis in lor
em blandit, eu pellentesque magna ullamcorper. Etiam placerat sodales arcu, nec sollicitudin elit suscipit a.
```

Figure 2.4 Plain text file opened in Notepad.

```
                        Jason File Format

        {
                "Meteorologists" : [
                        {
                                "id" : "1",
                                "Name": "Shikhar",
                                "Salary": "230000",
                        },
                        {
                                "id" : "2",
                                "Name": "Manu",
                                "Salary": "230000",
                        }
                ]
        }
```

Figure 2.5 Sample JSON document.

```
XML -example format
<? Xml version= "1.0"?>
<contact-info>
<name>  Shikhar </name>
<company> HCL</company>
<phone> +91-9999999999</phone>
 </contact-info>
```

Figure 2.6 Sample XML code.

rules that it must follow while encoding the data. This type of file format is readable by both humans and machines. As discussed earlier, it plays a vital role in sending information over the internet and is a self-descriptive language, which is a bit similar to hypertext markup language (HTML). Figure 2.6 shows a sample XML document.

2.2.5.6 Image Files
Image files are among the most recent entries into the most used file formats for data analytics. With the boom in the utilization of computer vision applications, image clas-sifications, and various other aspects, images have started becoming a very frequently used file. The images that we are familiar with generally tend to be three-dimensional images having RGB values. However, there exists even two-dimensional images (usually called greyscale) and even four-dimensional in which there is metadata associated with the regu-lar three-dimensional pixels and RGB values. Every image consists of one or more frames and each frame consists of pixel values. Most commonly used file formats for image files are JPG, png, etc.

2.2.5.7 HTML
HTML is also a subset of the markup language of which XML is a part. It is the standard and the most widely used markup language for webpage creation. HTML uses predefined tags and hence can easily be identified. Some example tags that are used in HTML are the <p> tag, which represents the tag used for a paragraph, and the <br> tag, which is used

```
HTML Example File
<!DOCTYPE html>
<html>
<head>
<title> Page Title </title>
</head>
<body>
<h1> Wiley</h1>
<p> This is the first para.</p>
</body>
</html>
```

Figure 2.7 Sample HTML file.

for a line break. Many tags in HTML need to be closed while some do not have that need associated with them. Figure 2.7 shows a sample code for an HTML file.

We have discussed many of the most commonly used file formats (and data types), however, there exist even more such file formats, which one might see, including but not limited to PDF, DOCX, MP3, MP4, etc. These are fairly easier to understand and have a certain affinity for them in terms of understanding as we have used them in our day-to-day life. As we can see, each file format has a different style of encoding the data and thus in a way, each file type represents a different type of data. Based on the method of storing data, each file format has its own advantages and disadvantages. Every organization understands and analyzes which file format will give them the best results and then takes that specific file format into use. From a data analytics point of view, each file format and data type must be viable for processing and analysis.

2.3 Overview of Big Data

Big data is characterized by features, known as *7V's,* which are: *volume, variety, velocity, variability, veracity, visualization, and value* [11, 12] as shown in Figure 2.8. Data associated with big data need not necessarily to be structured, it contains structured, unstructured, and semi-structured data [11]. Around 85% of the data generated in organization is unstructured, and the data generated by individuals in the form of e-mail, messages, etc., are also unstructured in nature. Traditional databases are not able to store data of different forms like text, video, audio, etc., and that's why there is need for special databases that can store such data and data analytics tool also. Brief description of different characteristics is given below:

1. *Magnitude/volume*: The magnitude, as we know can be defined by the number of bits, which comprise the data. We know the magnitude grows from bits, and from bytes to petabytes and exabytes. Table 2.2 paints a comprehensive image as to how the magnitude grows with the increase of number of bits.

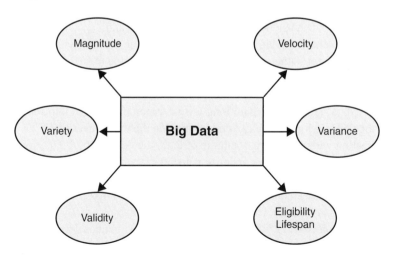

Figure 2.8 Characteristics of big data.

Table 2.2 Data storage measurements.

Units	Value
Byte	8 Bits
1 KB	1024 bytes
1 MB	1024 KB
1 GB	1024 MB
1 TB	1024 GB
1 PB	1024 TB
1 EB	1024 PB

2. *Velocity*: Velocity here refers to the ability to store the vast amounts of data and pre-pare it for processing and then finally analyzing and deriving information and enhanced insights, which would facilitate better decision making. We can measure the velocity by the amount of time it takes to do the abovementioned tasks. Earlier, batch processing was the way to go but with advancements in technology, we've upgraded to periodic, and then came near real-time, and currently, we are in the days of real-time processing.

3. *Variety*: As discussed earlier, the variety aspect of big data deals with how wide the range of the data types included in the big data sets are and what are their sources. The sources can range anywhere from traditional transaction processing systems and ratio-nal database management system (RDMS) (which produce structured data) to HTML pages (which produce semi-structured data) and unstructured emails or text documents (which produce unstructured data).

4. *Validity*: Although not a decisive characteristic in determining big data, validity serves as a must-have characteristic for the big data to be of any use. Validity associates to the accu-racy and correctness of the data in question. There are many ways in which the validity of the data can be affected. For example, due to the introduction of biases, abnormality, and noise into the data, the data can be rendered corrupted or invalid.

5. *Eligibility lifespan*: While analyzing big data, it is very important to know whether the data is actually eligible or not. Otherwise, the insights obtained due to such vast mag-nitudes of ineligible data can be disastrous in terms of decision making based on the insights. A data's eligible lifespan pertains to the time aspect of how long any data will be valid. Sometimes, data can be required for long-term decision making but if the eligi-bility lifespan of the data is a lot less, we can end up making the wrong decision.

6. *Variance*: When dealing with such a high volume of data, the collection of data must be kept under check for strict variance standards as its effect can snowball into wrong insights and decision making.

7. *Visualization*: After data has been processed it should be represented in readable format. There may be parameters in big data, which cannot be represented by usually available formats. Thus visualization is also a big problem for big data. Nan cubes developed by AT&T laboratories are used for visualization nowadays.

2.3.1   Sources of Big Data

Moreover, from a big data analytics point of view, the source of the data will play a vital role in determining and verifying the characteristics of big data and in turn will affect the analytical techniques that are to be applied and the insights that will be derived. In order to achieve success with big data, it is important that one develops the know-how to sift between the plethora of data sources available and accordingly select and use those resources that yield best usability and relevance. Let us explore some of the common big data sources available.

2.3.1.1   Media

Media is the most popular source of big data, which is available with much ease and provides the fastest way for organizations to get an in-depth overview of a targeted group of people according to their preference. Media data not only provides valuable data on consumer preferences but also encapsulates in itself the essence of the changing trends. Media includes social platforms like Instagram, Facebook, YouTube, as well as generic media like images, PDF, DOCX, etc., which provide comprehensive insights into every aspect of the target population.

2.3.1.2   The Web

The public web, which is made up of tons of HTML web pages, acts as the perfect picking grounds for extracting huge amounts of structured data. The internet is commonly available to everyone and hence this source serves as one of the most easily available sources of big data. Data sourced from the web ensures variety characteristic of the big data simply due to the enormity of the data available on the web. The web provides a premade big data infrastructure and repositories and can serve as beneficial sources to entrepreneurs and organizations who are on a budget.

2.3.1.3   Cloud

With the boom in cloud computing, companies have started integrating and shifting their data onto the cloud. Cloud storage serves as a storehouse of structured and unstructured data and can provide access to them in real time. The main aspect of taking the cloud as a source is the flexibility, availability, and scalability it offers. Not only that, but the cloud also provides an efficient and economical big data source.

2.3.1.4   Internet of Things

With the internet of things (IoT) expanding its applications, machine-generated content or data from the IoT has started becoming a valuable and sought-after source of big data. The benefit of IoT as a source for big data is that the data it brings to the table is not restricted to the data generated by computers or smartphones but also incorporates devices such as sensors, vehicular processes, meters, video games, etc.

2.3.1.5   Databases

One of the oldest sources of big data. However, with time, organizations have started utilizing a hybrid approach. Databases are used along with another big data source so as to utilize the advantages of both the sources. This strategy lays down the groundwork for a hybrid data model and requires low investment and IT infrastructural costs. Most popular databases include SQL, Oracle, DB2, Amazon Simple, etc.

2.3.1.6 Archives

Archives of scanned documents, paper archives, customer correspondence records, patients' health records, students' admission records, and so on, can also serve as a viable but crude source of big data.

2.3.2 Big Data Analytics

Big data is the process of exploring data to get useful and hidden information and patterns lying in the data [13]. The more data we have as input for our analysis, the better will be the analytical accuracy. This, in turn, would increase the reliability of the insights derived and decisions made, which are based on our analytical findings. All these would result in a positive change in terms of enhancing efficiency, reducing cost and time, and fuel us to innovate new products and services while enhancing and optimizing already existing ones. Hence, big data analytics is paving the way for an advanced and better tomorrow. Big data analytics applications are broadly classified as prescriptive, predictive, and descriptive as shown in Figure 2.9.

2.3.2.1 Descriptive Analytics

This is the first step of analytics, which utilizes the raw data of the company like sales, financial operations, inventory, etc., to find patterns like change in sales of a product, buying preferences of the customers [14]. This analytic involves simple techniques such as regression to find out how variables are related to each other and draw the charts and visualize data in a meaningful way. Dow Chemical's applied descriptive analysis to identify underutilized space in its offices and labs. As a result, they were able to increase space utilization by 20% and save approximately $4 million annually [15].

2.3.2.2 Predictive Analytics

This analysis is used to predict future trends and by combining data from multiple sources. Industries use this technique to evaluate the life of a machine by using sensor data [16].

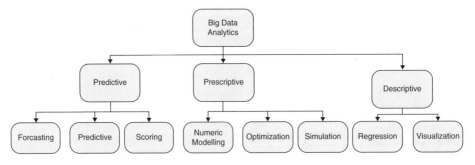

Figure 2.9 Different types of big data analytics.

Organization is able to predict their sales trends [17]. Predictive analytics is also used in academics [18], Purdue University applied this technique to predict that a student success in a course and give labels to risk of success (green means success probability is high, yellow means some problems, and red specifies high risk of failure).

2.3.2.3 Prescriptive Analytics

We learned in the previous two techniques how they help to predict the past and future. Next is how we can execute it, and for this, we use prescriptive analysis, which assists professionals in knowing the impact of various possible decisions. As per Garter [19], only 3% of organizations use this approach, because this analysis involves technique of simulation, numeric modeling, and optimization. Health care applications use this analysis to recommend diagnoses and treatment by analyzing a patient's history. Aurora Health Care Center saves millions of USdollars by utilizing big data analytics [20]. The oil and gas exploration industry utilizes prescriptive analytics to identify the best drilling point[21], by utilizing the various parameters like temperature, soil type, depth, chemical composition, seismic activity, machine data, etc., to find out the best locations to drill [22, 23].

2.4 Data Analytics Phases

The analytical process depends on the users' utility and experience. The users can be from different domains, like engineers, end users, scientists, medical field professionals, and business analysts, which uses the information generated from data. The operations on the data can vary from simple data queries to data mining, algorithmic processing, text retrieval, and data annotation [24].There are mainly five phases of data analytics as shown in Figure 2.10.

A. *Preparation phase*: In this process of analytics, users collects data through surveys or uses existing data for further processing, data is collected according to the application for which we have to work upon.
B. *Preprocessing phase*: The data collected in the first phase may have some noise issues, having nil entries, etc. Due to these issues, data is not in a form for further processing. So we need to remove these from data, and this is done by transformation, cleaning, scaling, filtering, and completion of data.
C. *Analysis phase*: In this phase of analysis, various operations like classification, clustering, regression, and visualization is applied to get useful information from the data.
D. *Post-processing phase*: In this phase of analytics, data is being interpreted, and a document is created based on the output of the analysis phase, and the evaluation of data is done.

Figure 2.10 Various phases of data analytics.

2.5 Data Analytical Tools

There are number of data analytics tools available in the market. In this chapter, the authors will discuss Microsoft Excel, Apache Spark, Open Refine, R programming, Tableau, and Hadoop. The comparison among these tools are given in Table 2.3 and their description is as follows:

2.5.1 Microsoft Excel

The capability to analyze data is a powerful skill that helps a company or an organization make knowledgeable decisions. Microsoft Excel is one of the leading tools for data analysis and offers a plethora of inbuilt functionalities, which help to make data analysis easy and efficient without having to lose the validity of the analysis [25]. Microsoft Excel is a spreadsheet program included in the Microsoft Office suite of applications (including MS Word, MS PowerPoint, etc.) which can be used on Windows, macOS, Android, and iOS. Installing MS Excel is fairly easy. As mentioned above, it is a part of the Microsoft Office suite and can be purchased from the official website of Microsoft. After the purchase, a setup file download will be provided. After downloading, basic installing steps will ensure the successful installation of the software. This section will illustrate the powerful features that MS Excel has to offer, as well as its advantages and the various disadvantages that one might face while choosing MS Excel as the data analysis tool.

Microsoft Excel sports a wide variety of powerful functionalities for data analysis. From creating flexible data aggregations using pivot tables, representing data visually using pivot charts, to calculating margins and other common ratios using inbuilt calculation-friendly functionalities, MS Excel is the de facto standard for data analysis. Let us dig into some of these features and functionalities:

1. *Sort*: MS Excel has the ability to sort our collected data on one column or multiple columns and offers a varying flexibility on the type of sort applied. The most common sorts include ascending sort (in which the data is sorted from small to large manner) and the descending sort (in which the data is sorted from large to small manner). Sorting data is often one of the basic steps in data analysis as it often increases the efficiency of other analytical algorithms.
2. *Filter*: In many cases, often we have a very large data source in which we have to select and analyze only that data, which is relevant or which satisfies certain conditions and constraints. MS Excel can help filter and will only display data records, which meet the input constraints
3. *Conditional formatting*: Depending on the value of certain data, one might often need to format it differently. This not only helps understand vast amounts of data easily but also helps in better readability and also sometimes makes applications of computer-based data analysis algorithms easier. A very common example can be coloring all the profit-based final sales as green in color and the loss based on final sales as red.
4. *Charts and graphs*: Diagrammatic representation of any data helps increase its understanding, handling, and presentation properties among various other benefits. Even a simple MS Excel chart can say more than a page full of numbers. MS Excel offers various

Table 2.3 Comparison of different data analytic tools.

Features	MS Excel	Apache Spark	Open Refine	Hadoop	NoSQL	Tableau	R Programming
Flexibility	It offers great flexibility to structured and semi-structured data.	Apache Spark is a great tool for managing any sort of data irrespective of its type.	It operates on the data resembling to the relational database tables.	Hadoop manages data whether structured or unstructured, encoded or formatted, or any other type of data	There are all sorts of information out there, and at some point you will have to go through tokenizing, parsing, and natural-language processing	Tableau does not provide the feature of automatic updates. Always a manual work is required whenever user changes the backend.	This tool can incorporate all of the standard statistical tests, models and analyses as well as provides for an effective language so as to manage and manipulate data.
Scalable	MS Excel has decent scalability both horizontally and vertically	Apace Spark boats a scalable model is a preferred tool as it offers all the features of Hadoop plus the added speed advantage.	Scalability for Open Refine has not been one of its strong points, however, it accommodates this drawback with the ability to transform data from one format to another.	This is a huge feature of Hadoop. It is an open source platform and runs on industry-standard hardware	Horizontally Scalable Increasing load can be manage by increasing servers etc.	Tableau is very easy to learn as compared to any other visualization software. Users can incorporate Python and R to implement tableau	when it comes to scalability, R has some limitations in terms of available functions to handle big data efficiently and a lack of knowledge about the appropriate computing environments to scale R scripts from single-node to elastic and distributed cloud services.

(Continued)

Table 2.3 (Continued)

Features	MS Excel	Apache Spark	Open Refine	Hadoop	NoSQL	Tableau	R Programming
Cost Effective	MS Excel basic version is free for all. However, professional packs are costly.	Apache Spark is an open source software and hence is free and easily available.	It is an open source software and has easy availability and offers a nice solution to messy data.	ApacheHadoop being open source software is easily available and runs on low-cost hardware platform.	NoSQL being open source software is easily available and runs on low-cost hardware platform.	Much costlier when compared to other Business intelligent tools providing approximately same functionality.	R is open-source language. It can be used anywhere in any organization without investing money for purchasing license.
Data visu-alization	Pivot tables, graphs, and pie charts are among the many tools available for visualizing data in MS Excel	Spark supports many APIs through which data visualization can be done which include open source visualization tools such as, D3, Matplotlib, and ggplot, to very large data.	Since it allows linking to websites and other online APIs, it supports external visualization tools. However, it also has some built in reconciliation tools as well.	Stores data in form data nodes. Client Server Architecture.	Visualizes data in form of Document database, Key-value pair database, Wide column database, Node based database	Tableau is purely dedicated for data visualization	R visualizes the data in the form of graphical representation, this attribute of R is extremely exemplary and this is the reason why it is able to surpass most of the other statistical and graphical packages with great ease.

charts and graphs, which include pie chart, bar graph, histogram, etc. MS Excel offers a very user-friendly way to make these charts and graphs.

5. *Pivot tables*: One of the most powerful and widely used features of MS Excel, pivot tables allow us to extract the significance from a large, detailed data set. Excel pivots are able to summarize data in flexible ways, enabling quick exploration of data and producing valuable insights from the accumulated data.

6. *Scenario analysis*: In data analysis, we are often faced with a lot of scenarios for analysis and are constantly dealing with the question of "what if." This what-if analysis can be done in MS Excel and it allows us to try out different values (scenarios) for various formulas.

7. *Solver*: Data analysts have to deal with a lot of decision problems while analyzing large sets of data. MS Excel includes a tool called solver that uses techniques from the operations research to find optimal solutions for all kinds of decision problems.

8. *Analysis ToolPak*: This add-in program provides data analysis tools for various applications like statistical, engineering, and financial data analysis.

As we can see, MS Excel as a tool for data analysis has a lot of functionalities. One can see many advantages in using this tool. Some of these advantages include easy and effective comparisons, powerful analysis of large amounts of data, ability to segregate and arrange data, and many more. However, like everything in life, there are many disadvantages associated with MS Excel as well. These disadvantages vary from a relatively higher learning curve for full utilization of MS Excel, costly services to time-consuming data entries, problems with data that is large in volume, and many times just simple calculation errors. Thus, MS Excel offers a holistic tool, which serves a variety of data analysis needs but at the same time, there is still a lot of room for improvement and it fails in very specific analysis with scaling data.

2.5.2 Apache Spark

Spark was launched by the Apache Software Foundation and developed at UC Berkeley [26] for speeding up the Hadoop computational computing software process. Nowadays, industries are relying heavily on the Hadoop tool to analyze their data sets. The reason for this is that the Hadoop framework is based on the MapReduce programming model, which facilitates a computing solution that is scalable, flexible, and fault tolerant. However, one of the major drawbacks was maintaining speed in processing large data sets in terms of waiting time between queries and waiting time to run the program. This was solved by Spark. Apache Spark is an open-source distributed lightning-fast cluster computing technology, designed for fast computation. Spark provides the user a method for programming entire clusters with fault tolerance and implicit data parallelism. Since it is an open-source project, Spark can easily be downloaded and installed from the internet and even from the official Apache website. Some of the features of Spark is described as below (shown in Figure 2.11):

1. *Swift processing*: As mentioned earlier, Spark overcomes one of the most major drawbacks of Hadoop – which was its slow speed. Spark reduces the number of read-write to disk, which results in achieving high data processing speed of about 10× faster on the disk and 100× faster in memory.

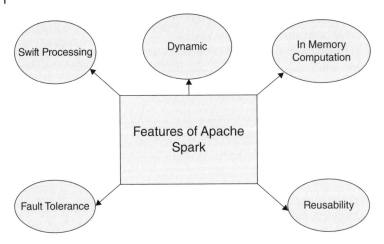

Figure 2.11 Features of Apache Spark.

2. *Dynamic*: Sporting implicit parallel processing, Spark is dynamic in nature. It provides over 80 high-level operators, which when paired with its processing, can help to develop a parallel application.
3. *In-memory computation*: Another feature of Spark that helps save time is its ability to have a cache. That means we overcome the time required to fetch the data from disk every time. This increases processing speed.
4. *Fault tolerance*: The resilient distributed data set (RDD), which is the fundamental data structure of Apache Spark, provides the ability of fault tolerance to Spark. It is designed to handle the failure of any worker node in the cluster, which ensures that the loss of data is reduced to zero.
5. *Reusability*: The Spark code can be reused for various processes, which include running batch-processing, joining stream against historical data, or running ad-hoc queries on stream state.

Even though Spark overcomes a major drawback of one of the most extensively used tools, it is not free of its own drawbacks. Spark's in-memory capabilities can form a bottleneck when we want cost-efficient processing of big data as keeping data in memory is quite expensive. Also, the latency of Spark is a bit high as compared to its fellow software Apache Flink. Apache Spark also lacks its own file management system and hence relies on some other platforms like cloud-based platforms. Spark has proved to be 20× faster than Hadoop for iterative applications, and was shown to speed up a real-world data analytics report by 40×, and has been used interactively to scan a 1 TB data set with 57 seconds latency [27].

2.5.3 Open Refine

Open Refine [28] is a sophisticated tool that is really helpful in working with big data especially when the data set is quite messy. It operates on the data that looks very similar to the relational database tables (having data in the form of rows and columns).

It performs various tasks like cleaning of data, transforming it from one format to another, and it can also perform the task of extending it to the web services and external data. Open refine provides the feature of exploring data, which helps to work with the huge and messy data set with ease and also a reconcile and match feature, which extends the data with the web.

Open Refine is available in multiple languages including English, Japanese, Russian, Spanish, German, French, Portuguese, Italian, Hungarian, Hebrew, Filipino, and Cebuano, etc., and it is also supported by the Google News Initiative.

Steps for an Open Refine Project

1. **Importing the Data Set**
   The software allows the importation only when the data is present in these formats: TSV, CSV, Excel spreadsheet, JSON, XML, and XHTML. The data file must either be present on the system or as an online file. To import the file we have to first launch OpenRefine and click on Create Project on the top right of the screen and select Get Data from the Computer and then locate your file. After importing the data is present in the form of a table, which has rows, and the corresponding columns with a title.

2. **Cleaning and Transforming**
   After successfully importing the data, Open Refine has multiple tools and features, which can be used for the transformation. The actions that are going to be performed are recorded so that at any point we can go to the previous required state.
   - *Row-based operation*: We cannot add a row to the data. The only operations supported are either marking or deleting the rows. By combining filters and facets the rows can be selected and the operations are performed.
   - *Column-based operation*: Unlike Rows, we can add new columns. There are multiple operations that are supported while dealing with columns.
   - *Cell-based operation*: We can just modify the value present in the cell. We use GREL (Google Refine Expression Language) statements for modifying the data on the selected cell.

3. **Exporting Results**
   Once the data has been transformed, it can be exported to different formats like CSV, JSON, TSV, HTML, and Excel spreadsheets, which are supported by OpenRefine. By default, the format selected is JSON, which can be easily changed to other formats.

Uses of OpenRefine

1. Cleaning of the data set that is huge and messy.
2. Transforming the data set.
3. Parsing data from the web.

Limitations of OpenRefine

1. To perform simple operations, complicated steps are needed to be followed
2. Sometimes the tool degrades and returns false results, which only has the solution of restarting the software or sometimes even the project itself.

2.5.4 R Programming

R is a programming language for statistical computation, representation of graphics, and reporting [29]. R was invented by Robert Gentleman and Ross Ihaka and is based on the language S designed at Bell Lab for statistical analysis in 1990s. It is a free, open-source language available at [30] and maintained by the R project. R can be used in command line mode as well as many graphic user interfaces (GUIs), like RStudio, R Commander, etc. Some of the features of R are:

1. It supports both procedural- and object-oriented programming language.
2. It provides printing options for the reports of the analysis performed on the data in hard copy as well on the screen.
3. Large number of operators is available in R for handling arrays and matrices.
4. R is most popular language because it consists of more than 10 000 packages.
5. R is used in many domains like business, education, soil science, ecology, and in remote sensing.

2.5.4.1 Advantages of R

1. R is an open-source and platform-independent language. It can be used anywhere in any organization without investing any money for purchasing license.
2. R can be integrated to other languages such as C/C++, Java, and Python. It also allows you to interact with other data sources/packages like SPSS, SAS, Stata, Access, and Excel.
3. Parallelism can be easily performed in R. There are so many packages that explore multiple cores on the same machine or over the network.
4. R has very large community of people. This means there are so many QA forums, which helps you in case you become stuck somewhere.
5. R is extensible. New packages can be easily added by programmers/users into R.

2.5.4.2 Disadvantages of R

1. R is relatively slower than its competitive languages such as Python and Julia.
2. R is not scalable as compared to its competitive languages.
3. R can handle data sets that can fit into the memory of the machine. The most expensive machines also may not have a big enough memory size that can accommodate a big enterprise data set.

2.5.5 Tableau

Tableau [31] is business visualization software used to visualize large volumes of data in form of graphs, charts, figures, etc. Tableau is mostly used to visualize spreadsheets (Excel files). Since it is difficult to analysis numbers, text, and their interdependencies, Tableau helps to analyze and understand how the business is going. Hence, Tableau is also known as business intelligent software.

2.5.5.1 How TableauWorks
Tableau at the backend runs SQL queries to fetch data from the database and represents it in interactive and impressive visual forms such as graphs, pie charts, etc. Its drag and drop GUI helps to facilitate easy designing and exploration. It can work on multiple databases, combining their results into a single visualized form.

2.5.5.2 Tableau Feature

1. *Easy and user-friendly*: Tableau provides easy installation and does not use any high-level programming language. Its drag-and-drop functionality provides a user-friendly interface.
2. *Variety*: Tableau can visualize data in numerous forms such as graphs, pie charts, bar graph, line graph, etc., and in numerous colors and trends.
3. *Platform independent*: Tableau is platform independent, i.e., it can work on any hardware (Mac, PC, etc.) or software (MacOS, Windows, etc.).

2.5.5.3 Advantages

1. *Data visualization*: Tableau is purely dedicated to data visualization. Tableau visualizes data in many interactive forms such as bar graphs, pie charts, and histograms. It is used to visualize and compare company stocks, success, market shares, etc.
2. *Implementation*: Tableau provides visuals in its tableau gallery. The drag-and-drop functionality of tableau allows easy implementation and we can create visualizations in minutes.
3. *Easy implementation*: Tableau is very easy to learn as compared to any other visualization software. Users can incorporate Python and R to implement tableau.
4. *Handles large amounts of data*: Tableau can deal with millions of rows allowing for various visualizations.

2.5.5.4 Disadvantages

1. *Flexibility*: Tableau does not provide the feature of automatic updates. Manual work is always required whenever the user changes the backend.
2. *Cost*: Much costlier when compared to other business intelligent tools providing approximately the same functionality.
3. *Screen resolution*: If the developer screen resolution is different from the user screen resolution, then the resolution of the dashboard might get disturbed.

2.5.6 Hadoop

The Hadoop is an open-source Apache Software utility framework library used for big data that allows it to store and process large data sets in parallel and distributed fashion. The core Apache Hadoop utility is basically divided into two parts: the storage part – well known as HDFS (Hadoop distributed file system) and the processing part – Map reduce, as shown in Figure 2.12.

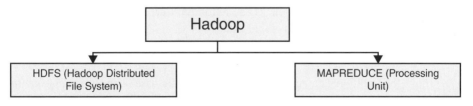

Figure 2.12 Components of Hadoop.

2.5.6.1 Basic Components of Hadoop

A. *Common*: this module of Apache Hadoop includes libraries and utilities needed by other Hadoop modules.
B. *Distributed file system (HDFS)*: a distributed file system that provides very high bandwidth across the cluster.
C. *YARN*: this module was introduced in 2012 as a platform responsible for the management and computing of resources as clusters that are utilized for scheduling users' applications.
D. *Map reduce*: this is the processing part of Hadoop used for large-scale data processing.

2.5.6.2 Benefits

A. *Scalability and performance*: Operates a very large data set over many inexpensive parallel servers making it highly scalable. Hadoop can store thousands of terabytes of data.
B. *Reliable*: Hadoop provides fault tolerance. It creates a copy of original data in the cluster, which it uses in the event of failure.
C. *Flexibility*: Structured format is not required before storing data. Data can be stored in any format (semi-structured or unstructured).
D. *Cost effective*: Apache Hadoop is open-source software that is easily available and runs on a low-cost hardware platform.

2.6 Database Management System for Big Data Analytics

There is a special type of data storage required for huge data generated. In this section, two such types of data storage system for big data analysis will be described.

2.6.1 Hadoop Distributed File System

HDFS [31] is a product of Hadoop and works closely with it. This is a distributed file system that is used for reliable running of software and scalability on hardware. This is highly tolerant because it divides the data into smaller parts and replicates it to several machines in the cluster. It can scale up to 4,500 machines and 200 PB data in a single cluster. HDFC works in batch mode. It is based on the master/slave architecture with single NameNode and large number of DataNode. NameNode is responsible for handling meta data and DataNode is used to carry actual data. Data is divided into chunks of 64 MB and stored in at least three different locations for fault tolerance. NameNode contains all the information about replication of data and file distribution. Every DataNode sends a signal to NameNode; if it does not send, then it is considered as failed. HDFC has an inbuilt function that maintains uniform distribution of data and stores copies of said data.

2.6.2 NoSql

NoSql, unlike the traditional database, is a new set of the file-based database management system independent of RDMS. Big companies like Amazon and Google use NoSql to manage their databases. This was designed to overcome two main drawbacks of traditional

Table 2.4 Comparison between SQL and NoSQL.

	SQL	NoSQL
Relational or distributed	Rational Database(RDBMS)	Nonrational Database(or) Distributed Database
Schema and syntax	Structured data Table based Predefined syntax	Unstructured data Document-based, key-value pair, graphs etc.
Complex queries	Good to fit complex queries	Not good
Scalability	Vertically Scalable	Horizontally Scalable
	Increasing load can be managed by increasing RAM, CPU, etc.	Increasing load can be managed by increasing servers, etc.
operties and theorem	Follows on ACID properties	Follows CAP theorem
Example	MySql, Oracle	MongoDB, Neo4j

databases system: high operational speed and flexibility in storing data. The basic difference between SQL and NoSQL is that the SQL works on predefined schema and all the data are entered accordingly. The field defined as an integer can store only the integer value, whereas in NoSql there is no predefined structure and data is stored in a different format such as JSON object, etc.

2.6.2.1 Categories of NoSql

A. *Document database*: Data is stored in the form of document in an encoded format such as a JSON object, or XML. Documents are identified via a unique key. Each key is paired with a complex data structure called a document, e.g., Mongo DB, Apache Couch DB, and Arango DB.
B. *Key-value pair database*: This data is stored in the form of key and value pair that are called columns. For every data value, there must exist a key for reference. The column name is string and can contain any data type or size. The schema of the database is dynamic and can be easily changed by just adding columns. It is the same as an associative array, e.g., Aerospike, Apache Ignite, and Amazon DynamoDb.
C. *Wide column database*: In this, the data is stored in form of rows and columns. But unlike a traditional database, the data type of each row can vary, e.g., HBase and Cassandra.
D. *Node-based database*: This type of database is also known as a graph-based database. Here the data is stored in nodes that are connected to each other. This type of data set is used where there is high connectivity among the nodes, e.g., Allegro Graph and Arango DB (Table 2.4).

2.7 Challenges in Big Data Analytics

It seems to be particularly helpful to classify the problems that analysts have when we talk about sorting big data into three different categories, since these hierarchical classifications are found to be really useful and versatile. We should not be confused by this problem and the problems faced by IT, since IT has problems like storing data, accessing data, and putting it somewhere.

The four categories are:

2.7.1 Storage of Data

Big data, as the name suggests, is really hard and uneconomical to work with. Working with massive data that is out of the scope of your system's ability to hold it in its RAM isn't going to work. Hence we need to change our approach in the analysis project. First, before starting to work on big data we always need to subset the data, which involves extracting a data set so that we can work with some smaller data and develop models; after building such a model, we can come up with the entire data.

2.7.2 Synchronization of Data

In big data analytics, we need a lot of data. This data is collected from different sources and needs to be transformed into a specific analytical platform to get the output as required by the business professionals. It is a really big challenge faced by the professionals as the task of collecting and synchronizing is enormous and time-consuming.

2.7.3 Security of Data

As with increasing data and its opportunities, the risk of data being stolen by others increases. Big data analytical tools that are used for analyzing and data mining uses disparate data sources that also increase the potential risk as our data gets more exposed to the big data world. These security threats create a lot of tension.

2.7.4 Fewer Professionals

As the demand for big data analytics is increasing, the problem of acute shortage of big data scientists and big data analysts is faced by organizations. There are very few professionals who completely understand big data and are able to get some meaningful insights out of it.

2.8 Conclusion

The data generated on the internet is being doubled every year and the type of data is mainly unstructured in nature. The size of data is too large, which cannot be processed with traditional programming languages, visualization tools, and analytical tools. In this book chapter, the authors explained various types of data, i.e., structured, unstructured, and semi-structured, as well as their various resources and various file formats in which data is stored. Further, analytical process of big data is categorized into descriptive, prescriptive, and predictive along with real-world applications. Then the authors described various analytical tools like R, Microsoft Excel, Apache Spark, Open Refine, Tableau, and two storage management tools, HDFS and NoSql. In the last section, various challenges and applications of big data were presented.

References

1 https://en.wikipedia.org/wiki/Raw_data

2 www.ngdata.com/what-is-data-analysis

3 Castellanos, A., Castillo, A., Lukyanenko, R., and Tremblay, M.C. (2017). Understanding benefits and limitations of unstructured data collection for repurposing organizational data. In: *Euro Symposium on Systems Analysis and Design*. New York, Cham: Springer.

4 Liao, S.-H., Chu, P.-H., and Hsiao, P.-Y. (2012). Data mining techniques and applications–a decade review from 2000 to 2011. *Expert Systems with Applications* 39 (12): 11303–11311.

5 Tan, A.-H. (1999). Text mining: the state of the art and the challenges. In: *Proceedings of the PAKDD 1999 Workshop on Knowledge Discovery from Advanced Databases*, vol. 8. sn, 65–70.

6 Gelbukh, A. (2005). Natural language processing. In: *Fifth International Conference on Hybrid Intelligent Systems (HIS'05)*, 1. Rio de Janeiro, Brazil: IEEE https://doi.org/10 .1109/ICHIS.2005.79.

7 Jahangiri, N., Kahani, M., Ahamdi, R., and Sazvar, M. (2011). A study on part of speech tagging. *Review Literature and Arts of the Americas*.

8 www.analyticsvidhya.com/blog/2017/03/read-commonly-used-formats-using-python/

9 www.json.org

10 www.w3.org/standards/xml/core

11 Assunção, M.D., Calheiros, R.N., Bianchi, S. et al. (2015). Big data computing and clouds: trends and future directions. *Journal of Parallel and Distributed Computing* 79: 3–15.

12 Gudivada, V.N., Baeza-Yates, R., and Raghavan, V.V. (2015). Big data: promises and problems. *Computer* 48 (3): 20–23.

13 Zaharia, M., Chowdhury, M., Das, T. et al. (2012). Resilient distributed datasets: a fault-tolerant abstraction for in-memory cluster computing. In: *Proceedings of the 9th USENIX Conference on Networked Systems Design and Implementation*, 2–2. USENIX Association.

14 Russom, P. (2011). Big data analytics. *TDWI Best Practices Report, Fourth Quarter* 19 (4): 1–34.

15 Wolpin, S. (2006). An exploratory study of an intranet dashboard in a multi-state health-care system. *Studies in Health Technology and Informatics* 122: 75.

16 Delen, D. and Demirkan, H. (2013). Data, information and analytics as services. *Decision Support Systems* 55: 359–363.

17 IBM netfinity predictive failure analysis. http://ps-2.kev009.com/pccbbs/pc_servers/pfaf .pdf (accessed 20 November 2018).

18 Lohr, S. (2012). The age of big data. *New York Times* 11 (2012).

19 Purdue university achieves remarkable results with big data. https://datafloq.com/read/ purdueuniversity-achieves-remarkable-results-with/489 (accessed 28 February 2015).

20 Gartner taps predictive analytics as next big business intelligence trend. www. http:// enterpriseappstoday.com/business-intelligence/gartner-taps-predictive-analytics-as-nextbig-business-intelligence-trend.html (accessed 28 February 2015).

21 The future of big data? Three use cases of prescriptive analytics. https://datafloq.com/read/future-big-data-use-cases-prescriptive-analytics/668(accessed 02 March 2015).

22 Farris, A. (2012). How big data is changing the oil & gas industry. *Analytics Magazine.*

23 The oil & gas industry looks to prescriptive analytics to improve exploration and production. www.exelisvis.com/Home/NewsUpdates/TabId/170/ArtMID/735/ArticleID/14254/The-Oil--Gas-Industry-Looks-to-Prescriptive-Analytics-To-Improve-Exploration-and-Production.aspx (accessed 28 February 2015).

24 Kantere, V. and Filatov, M. (2015). A framework for big data analytics. In: *Proceedings of the Eighth International C* Conference on Computer Science & Software Engineering* (eds. M. Toyama, C. Bipin and B.C. Desai), 125–132. New York: ACM.

25 Levine, D.M., Berenson, M.L., Stephan, D., and Lysell, D. (1999). *Statistics for Managers Using Microsoft Excel*, vol. 660. Upper Saddle River, NJ: Prentice Hall.

26 Ham, K. (2013). OpenRefine (version 2.5). http://openrefine.org. Free, open-source tool for cleaning and transforming data. *Journal of the Medical Library Association* 101 (3): 233.

27 Zaharia, M., Chowdhury, M., Franklin, M.J. et al. (2010). Spark: cluster computing with working sets. In: *Proceedings of the 2nd USENIX Conference on Hot Topics in Cloud Computing, HotCloud'10*, 10–15. New York: ACM.

28 Ihaka, R. and Gentleman, R. (1996). R: a language for data analysis and graphics. *Journal of Computational and Graphical Statistics* 5 (3): 299–314.

29 www.tableau.com/trial/data-visualization

30 www.r-project.org/about.html

31 http://hadoop.apache.org

3

Statistical Methods for Intelligent Data Analysis: Introduction and Various Concepts

Shubham Kumaram, Samarth Chugh, and Deepak Kumar Sharma

Division of Information Technology, Netaji Subhas University of Technology (Formerly Netaji Subhas Institute of Technology), Delhi, India

3.1 Introduction

In recent years, the widespread use of computers and the internet has led to generations of data on an unprecedented scale [1–4]. To make an effective use of this data it is necessary that this data must be collected and analyzed so that inferences can be made to improve various products and services. Statistics deals with collection, organization, and analysis of data. Organization and description of data is studied under descriptive statistics, whereas analysis of data, and making predictions based on it is dealt with in inferential statistics.

3.2 Probability

3.2.1 Definitions

Before we delve deeper into the understanding of statistical methods and its applications, it is imperative that we review some definitions and go through concepts that will be used all throughout the chapter [5].

3.2.1.1 Random Experiments
A random experiment is defined as follows:

1. All outcomes of an experiment that are possible must be known;
2. The outcome of each trial in an experiment must not be known before it takes place; and
3. Multiple trials, as per requirement, can be repeated without any change in the chances of attaining the outcome.

Outcome space, Ω, of an experiment refers to all possible values that can be achieved as the outcome of a trial. For example, while tossing a single coin, outcome space encompasses the outcomes, head and tail. A related concept is that of an *event*. An event can be simple such as "showing six on a die roll" or complex such as "showing numbers below four on a die roll." It is a subset of the outcome space, which is dependent on the outcome.

Intelligent Data Analysis: From Data Gathering to Data Comprehension,
First Edition. Edited by Deepak Gupta, Siddhartha Bhattacharyya, Ashish Khanna, and Kalpna Sagar.
© 2020 John Wiley & Sons Ltd. Published 2020 by John Wiley & Sons Ltd.

3.2.1.2 Probability

In any experiment, probability is defined with reference to a particular event. It is measured as the chances of occurrence of that event among all the possible events. In other words, it can be defined as the number of desired outcomes divided by the total number of possible outcomes of an experiment.

3.2.1.3 Probability Axioms

Any experiment for which we may calculate the probability of an event, must satisfy the following axioms:

1. $Probability(Outcome) > = 0$
2. If Outcome A and Outcome B are mutually exclusive i.e., $A \cap B = \varnothing$ then $Probability(A \cup B) = Probability(A) + Probability(B)$
3. $Probability(Outcome\ space) = 1$

3.2.1.4 Conditional Probability

The probability of an event is subject to change based on the values of probabilities of other events of the same experiment. Conditional probability of outcome A given outcome B is denoted as follows:

$$P(A|B) = \frac{P(A \cap B)}{P(B)} \tag{3.1}$$

where $P(B) > 0$.

3.2.1.5 Independence

It follows that two events may not always be interrelated and cause any influence over each other. Outcome A is said to be independent of outcome B when occurrence of B does not affect the probability of outcome A.

$$P(A|B) = P(A) \tag{3.2}$$

While the reverse is true, the following expression is also true and is derivable

$$P(A \cap B) = P(A)P(B) \tag{3.3}$$

3.2.1.6 Random Variable

Random variable X is a function defined from the domain of the outcome space and it is the value obtained from different outcomes for a particular event that is being sought. For instance, when tossing two coins, a random variable can be the quantity of heads obtained shown as follows:

$$X(H, T) = X(T, H) = 1$$
$$X(H, H) = 2$$
$$X(T, T) = 0 \tag{3.4}$$

3.2.1.7 Probability Distribution

The assignment of a value of probability to each possible outcome of a random variable X is done through a probability function P such that $P(X = x) = p(x)$ where x is the particular outcome in concern and $p(x)$ is its likeliness of occurrence.

While this concept is valid for discrete values, continuous variables require the representation through area under a curve. This is represented by a function f, which is the probability density of X.

$$P(a < X \leq b) = \int_b^a f(x)dx \qquad (3.5)$$

The probability distribution function, F, is defined for both types of variables that provides a value for an outcome x at most equal to it

$$F(x) = P(X \leq x) \qquad (3.6)$$

3.2.1.8 Expectation

Expectation is similar to probability distribution in the sense that it has different definitions for different nature of variables, discrete and continuous. In the case of discrete variables, the expected value or *mean* μ is calculated as:

$$E(X) = \sum_x xp(x), \text{ and } E(h(X)) = \sum_x h(x)p(x) \qquad (3.7)$$

Whereas for continuous variables, the summation is simply replaced with an integration:

$$E(X) = \int_{-\infty}^{\infty} x f(x)dx, \text{ and } E(h(X)) = \int_{-\infty}^{\infty} h(x)f(x)dx \qquad (3.8)$$

3.2.1.9 Variance and Standard Deviation

Variance σ^2 and standard deviation $\sigma = \sigma^2$ tell us about the spread of data about the mean value. Variance is obtained by squaring the difference of each outcome x with the mean μ, i.e.,

$$\sigma^2 = V(X) = E(X - \mu)^2 \qquad (3.9)$$

3.2.2 Bayes' Rule

Bayes' rule brings an interesting phenomenon to probability where given a particular evidence, we try to calculate the probability of the event that led to it. Let B_i represent different partitions of Ω then according to Bayes' rule, we have

$$P(B_i|A) = \frac{P(A|B_i)P(B_i)}{\sum_j P(A|B_j)P(B_j)} \qquad (3.10)$$

Here we have B_i an event whose probability has to be found out given that we have an evidence of the outcome A. B_i must be influenced by A to have an effect on its value. An extremely famous example to aid the understanding of the Bayes' rule is that of a false- or true-positive test for a clinical trial. Let us have a trail to detect cancer in patients. Now the trial could give us a true-positive result in the case where the patient has cancer and the trial indicates so. However, there are cases where a false-positive is obtained when the patient does not have cancer but the trial reports otherwise. Similarly, there are cases possible for negative test results and for people who do not have cancer. Bayes' rule allows us to answer questions of the kind: given a positive test result, what is the probability of a patient actually having cancer?

3.3 Descriptive Statistics

Descriptive statistics, as its name implies, refers to the branch of statistics that deals with collection, presentation, and interpretation of data. Raw data, by itself, is of little use. Descriptive statistical methods can be used to transform raw data in a form such that its gist can be easily found out and it can be interpreted using established methods. Descriptive statistical methods are of the following types:

1. Picture representation
2. Measure of central tendency
3. Measure of variability
4. Distributions

3.3.1 Picture Representation

In this type of representation, the complete data set is represented in a way that its related properties can easily be found. In essence, it presents a "picture" of the entire data set. However, it should be noted that picture representations are not limited to just graphical methods, but also encompass tabular methods for representing data.

3.3.1.1 Frequency Distribution
Frequency distribution is the simplest type of data representation. It just shows the number of times a given element appeared in a data set. It can be of two types – simple and grouped.

3.3.1.2 Simple Frequency Distribution
In this type of distribution, the leftmost column of the table lists the various categories, and the corresponding entries in the right columns depict the number of times elements in the data set correspond to the given category. Here, each single category is given a row of its own, hence it is not really suited for large data sets.

3.3.1.3 Grouped Frequency Distribution
This distribution is applicable where multiple categories exhibit similar properties that can be grouped together, especially in case of numerical values where the independent variables can be grouped to form ranges. Here, the cumulative frequency values for the whole group occupy a single row. This achieves a compact representation of data, but suffers from loss of intragroup information.

3.3.1.4 Stem and Leaf Display
Frequency distributions force us to choose between compactness and lossless representation of data. Stem and leaf display are a special type of grouped frequency distribution,

which also preserves intragroup information. Here, class intervals are shown on the left of a vertical line. Each class interval is represented by a *stem*, which show the lowest value in the interval. The column to the right of stem is the *leaves* column, which contains a series of numbers, corresponding to the last digit of values for each value in the interval. The value of the dependent variable for any entry is given by:

$$\text{Final Value} = \text{Stem value} \times \text{Stem width} + \text{leaf} \tag{3.11}$$

where "stem width" is the number of values a stem can represent.

3.3.1.5 Histogram and Bar Chart

Histogram and bar chart are similar ways of representing the given data graphically. In these representations, the abscissas stands for the independent variable, whereas the length of the bars along the ordinate axis is proportional to the values of the corresponding dependent variable. While histograms and bar charts appear almost the same, there are subtle differences between the two. In histograms, the dependent variable must be numerical, whereas in a bar chart, it can be qualitative as well as quantitative. The ordering of the variable on the abscissa is important and follows some logical directive, whereas in a bar chart, the order of variables is not important and can be arbitrary.

3.3.2 Measures of Central Tendency

Measures of central tendency give us a view of what an average element of the data set is likely to look like. There are a multitude of ways to define an average, such a mean, median, and mode.

3.3.2.1 Mean

Mean of a distribution is the sum of all measurements divided by the number of observations.

$$\bar{x} = \frac{1}{n} \sum_{i=1}^{n} x_i \tag{3.12}$$

where $x_1, x_2,..., x_n$ are a set of values of n observations. In case the distribution is continuous, the mean is calculated by:

$$\bar{f} = \frac{1}{b-a} \int_{a}^{b} f(x)dx \tag{3.13}$$

where $f(x)$ is the distribution function, $\bar{f}$ is the mean of the function, $[a,b]$ is the domain of the function and $x \in$ R.

3.3.2.2 Median

Median of a data set is the value that separates the higher half from the lower half.

3.3.2.3 Mode

Mode is defined as the most frequent value in a data set. There can be one or multiple modes for a given set of values.

3.3.3 Measures of Variability

In most of the practical scenarios, measures of central tendency are not enough to sufficiently describe a data set. Such methods do not show whether the values are mostly consistent or pertain to extremities. Hence, measures of variability such as standard deviation and variance, along with average values, are required to succinctly describe the data without losing much information.

3.3.3.1 Range

Range is defined as the difference between the highest and the lowest value in a data set. While it's a pretty simple measure, it informs us about the distribution of data in the data set. Simple range, though, is affected by outliers that can expand it abnormally, hence it has been supplanted by other measures such as *interquartile range* and *semi-interquartile range*.

Interquartile range is the difference between the highest and the lowest values when only values in Q2 and Q3 are considered. Semi-interquartile range is the interquartile range divided by two. These measures mitigate the effects of outliers and provide a better view of the data than the simple range.

3.3.3.2 Box Plot

Box plot is a graphical way of depicting interquartile range. A rectangle is drawn on a coordinate plane, where the lower edge corresponds to the boundary between Q1 and Q2, and the upper edge corresponds to the boundary between Q3 and Q4. The median is shown as a line segment within the rectangle. Two lines emanate from the upper and lower ends of the box, called *whiskers*, whose ends denote the maximum and minimum values.

3.3.3.3 Variance and Standard Deviation

Variance is defined as the average of squared difference from the mean.

$$\sigma^2 = \frac{1}{n} \sum_{i=1}^{n} (x_i - \bar{x})^2 \tag{3.14}$$

where σ^2 is the variance, n is the number of data instances, x_i are the independent variables, and $\bar{x}$ is the mean. Standard deviation is the measure of dispersion in a data set. A low variance implies that the values in the data set lie close to the mean, while a high variance means that the data is spread out. It is calculated as the square root of variance.

3.3.4 Skewness and Kurtosis

Skewness is the measure of asymmetry of the probability distribution of a random variable about its mean. It can be positive, negative, or undefined. In a positive skew, also known as the right skew, the right tail is longer than the left and hence the mass of the resulting distribution leans heavily to the left, whereas in a negative skew, or the left skew, the left tail is longer and most of the mass is concentrated toward the right part of the curve. Many textbooks say that in a right skew, the mean is to the right of median, and it is to the left of median in the left skew, but this assertion has been shown to be unreliable [6]. Formally, skewness is defined as:

$$\gamma = E\left[\left(\frac{X - \bar{x}}{\sigma}\right)^3\right] \tag{3.15}$$

where γ is Pearson's moment coefficient of skewness, E is the expectation function, X is the random variable, $\bar{x}$ is the mean, and σ is the standard deviation.

Kurtosis is the measure of thickness of tails in a distribution. With regard to normal distribution, a distribution whose tail width is similar to that of the standard normal distribution is called mesokurtic, one whose tail is wider is known as platykurtic and the distributions with narrower tails are called leptokurtic. Kurtosis is defined as:

$$\text{Kurt}[X] = E\left[\left(\frac{X - \bar{x}}{\sigma}\right)^4\right] \tag{3.16}$$

where the variables mean the same as in Eq. (3.15).

3.4 Inferential Statistics

In inferential statistics [7], generalized conclusions are drawn from a sample of a population of data that is available. Here population refers to the entire group of subjects that are of interest while gathering the data for a particular observation. Sample is a small subset of that group that is mostly selected in an unbiased manner such that each subject had an equal probability of being part of the subset. Since the sample only represents a part of the population, any inferences drawn upon it are only an estimation of the correct possible conclusion subject to the error of generalization.

The difference between inferential and descriptive statistics and consequently the meaning of inferential can also be understood by looking from a viewpoint of empirical and theoretical distributions. An empirical distribution is one wherein the values of concern of a particular population can be observed. For example, the loss and victories of a football club in their home stadium. Whereas theoretical distributions are based on predictions and formulations on the premise of basic facts about the data or some assumptions. The amalgamation of both empirical and theoretical distributions gives rise to inferential statistics. It is based on proving the correctness of a hypothesis that begins with an empirical consideration and concludes with a theoretical standpoint.

The need of inferential statistics, in the light of descriptive statistics and the risk of producing generalization errors, is to figure out a general trend and elements about the population. This has to be performed through sampling of data since the entirety of data is not available or cannot be harnessed in a usable manner. Hence, the aim is to reasonably determine facts that can be concluded through the utilization of the sample.

3.4.1 Frequentist Inference

In this method of predictive approach, we look at the changes in behavior based upon the number of repetitions while keeping the conditions constant [8]. The two different approaches that will be discussed are testing and estimation. Testing is based on deciding whether a particular hypothesis regarding the population stands accepted or not with respect to the patterns observed in the sample data. The other approach of estimation can further be divided into two types depending upon whether the estimation is of a particular value or a range of values, i.e., point estimation or interval estimation.

3.4.1.1 Point Estimation

Point estimation aims to predict a particular value for the population, θ with a single *point* estimate. Let G stand for the predictor of θ, which leads to prediction error being the difference between the two. For a better prediction, the value of $G - \theta$ should be minimal. Another related value that can be calculated is the bias of the estimator. Here, the multiple repetitions of the sample is considered while calculating the expectation value, $E_\theta(G)$, which when equal to θ makes the estimator G unbiased. The bias, B_θ, is calculated as

$$B_\theta = E_\theta(G) - \theta \tag{3.17}$$

In addition to bias, variance of an estimator, V_θ, is also an essential quality measure that shows how much the individual estimates g fluctuate from the $E_\theta(G)$.

$$V_\theta(G) = E_\theta(G - E_\theta(G))^2 \tag{3.18}$$

In fact, mean squared error, $M_\theta(G)$ is an umbrella term for the two values calculated above. It is defined as the sum of the squared bias and variance.

$$M_\theta(G) = B_\theta{}^2(G) + V_\theta(G) \tag{3.19}$$

3.4.1.2 Interval Estimation

An interval estimation of a particular θ is defined using (G_L, G_U). The expected width of the interval

$$E_\theta(G_U - G_L), \tag{3.20}$$

and the probability of actually having the value of population in that interval

$$P_\theta(G_L < \theta < G_U), \tag{3.21}$$

are important quality measures. An obvious trade-off between a higher probability of having the value in interval and width of the interval can be seen. Thus, according to convention, a value of $(1 - \alpha)$ is used as a confidence level for the estimator

$$P_\theta(G_L < \theta < G_U) \geq 1 - \alpha, \tag{3.22}$$

for all possible θ. An individual instance (g_U, g_L) of this interval is called a $100(1 - \alpha)\%$ confidence interval.

Comprehensive study of both point estimation and interval estimation is beyond the scope of this book.

3.4.2 Hypothesis Testing

In this type of prediction, two hypotheses are formed about the value for a particular θ and then a choice between these hypotheses is made to allocate a value. The important hypothesis to be noted is the *null hypothesis*, H_0, which is accepted to be true unless the sample data provides strong reasons not to do so. Moreover, the null hypothesis is rejected only in the case where an individual estimate g of a predictor G is in the *critical region*, C. Following these rules, the only scope for error is left when a false-positive or false-negative case arises.

Type I Error: Reject H_0 when it is true.
Type II Error: Accept H_0 when it is false.

Type I errors are considered to be a grave as compared to type II errors. It is more serious to distort the predictions made on populations through a false negative since the population data would also be rejected based on the sample data. G is usually accepted as a point estimator for θ. For example, if a hypothesis is based on the mean μ, then X would be the appropriate choice for G.

$$H_0 : \theta \geq \theta_0, H_a : \theta < \theta_0. \tag{3.23}$$

Critical region mentioned earlier is governed by the maximum value that still gets rejected by H_0. All values from $-\infty$ to c_u, which is the critical value bounding the region are rejected. Hence, this is also known as left rejection.

The measure of quality of a test is the test's power β

$$\beta(\theta) = P_\theta(Reject H_0) = P_\theta(G \in C). \tag{3.24}$$

Since H_0 is the domain in which we want most of our g to fall as it falls under type II errors, we wish to have a higher value of $\beta(\theta)$ for $\theta \in H_a$. Type I errors are avoided by having a small value of $\beta(\theta)$ for $\theta \in H_0$. Additionally, we try to restrict the type I errors to a maximum value called *significance level* α of the test

$$\max_{\theta \in H_0} \beta(\theta) \leq \alpha. \tag{3.25}$$

3.4.3 Statistical Significance

A conclusion drawn about the population from a sample can be said to be reasonable enough when the inferences made in the sample of data also hold true for the entire data set, i.e., population. Since the sample is a considerably smaller section of the data, it may so happen by chance that correlations may be found out within the limited instances of variables. However, when looking at the larger picture, no correlation may be present. For example, when two coins are tossed together 10 times, there is a possibility of both the coins showing a heads or tails together a significant number of times or even in all situations. However, we are aware of the absence of any correlation between these independent events. Statistical significance refers to the calculation of confidence levels beyond which we can accept a correlation between different variables to be not just a coincidence but an actual dependency.

When statistical significance [9] is discussed in the context of correlation, we come across two values, r and ρ. r refers to the degree of correlation between pair of instances in the sample. ρ on the other hand refers to the degree of correlation that exists in the actual entirety of the population. We seek to eliminate the possibilities, through the use of statistical significance, of having a positive or negative value of r (for positive or negative correlation) while the value of ρ equals zero. This is a bigger concern when the size of the sample is small since such inferences can be easily formed. It has been conventionally accepted that 5% of statistical significance is the benchmark for accepting that an inference drawn on a sample is not arrived at by chance or luck and may hold true for the entire population.

3.5 Statistical Methods

The basic concepts of statistics are used to develop various standardized methods for intelligent analysis of data. While on a basic level, these methods can tell us if the given data has any statistical significance, advanced methods can be used to extract principal features from the data set, or to make predictions based on existing data. The different statistical methods are described in this section.

3.5.1 Regression

Regression defines the relationship between the response and the regressors. Here, the independent variables are called "regressors," whereas the dependent variable, also known as the output, is called the "response." Regression can be linear as well as nonlinear.

3.5.1.1 Linear Model

In the linear model of regression, the response is defined as a linear function of the regressors, as shown in Eq. (3.26).

$$E(y) = \beta_0 + \sum_{i=1}^{n} \beta_i x_i \tag{3.26}$$

Here, y is the response, $E(y)$ is the expected value of y, n is the number of data instances, β_i are the weights and x_i are the regressors.

While the linear model is not accurate for several classes of data, it can provide a close approximation of the underlying function. Moreover, a large class of nonlinear functions, such as polynomials, can be easily converted into the linear model [10]. Linear models can be easily solved using simple numerical methods, and hence it is the most widely used model for data analysis. Moreover, linear models can be used to perform statistical inference, especially hypothesis testing and interval estimations for weights and predictions.

3.5.1.2 Nonlinear Models

Any regression function in which the response is not a linear function of the regressors is known as nonlinear regression. Examples of such models are the **allometric model** [11] and **Mitscherlich model** [12]. The allometric model is given as:

$$y_j = \beta_0 x_{1j}^{B_1} + \varepsilon_j \quad j = 1, \ 2, \ldots, m. \tag{3.27}$$

It is used to describe the relationship between the weight of a part of a plant and the weight of the whole plant.

The Mitscherlich model used to predict crop yield [13] with respect to amount of fertilizer applied is expressed as:

$$y_j = \beta_0(1 - e^{-\beta_1(x_{1j}+\beta_2)}) + \varepsilon_j \quad j = 1, \; 2, \dots, m. \tag{3.28}$$

While nonlinear methods can model real-world phenomena more accurately, there exists no algebraic method to estimate the least square estimators for βs. Moreover, the statistical properties of the responses and regressors are unknown, hence no statistical inference can be performed. Due to these shortcomings, nonlinear models are employed only in cases where the relationships between the regressors and the responses are known, and the goal is to find the weights. If output is to be predicted, an approximate linear model should be used.

3.5.1.3 Generalized Linear Models

The linear models are very flexible, and are applicable in a large number of cases, but there still remain some classes of data that cannot be modeled accurately by purely linear models. Nonlinear models are one such way of handling such data, but they come with their own shortcomings, the biggest being that they cannot be reliably used for making predictions. Generalized linear models are another such way of dealing with such data. The generalization is as follows:

1. The distribution of the data can be any of the distributions in the exponential family. Hence, such models can be used to analyze data following distributions such as binomial, Poisson, or gamma, as well as the normal.
2. The expected value of the response is not a linear function of the regressors, but is given by:

$$g(E(y_j)) = \beta_0 + \sum_{i=1}^{n} \beta_i x_{ij}. \tag{3.29}$$

Here, the variables are the same as in Eq. (3.26). $g(.)$ represents a differentiable monotone function, known as *link function*.

A general algorithm has been proposed in [14], which though iterative, has natural starting values and uses repeated least square estimations. It also gives us a common method for statistical inference.

3.5.1.4 Analysis of Variance

Analysis of variance (ANOVA) is used to identify if the βs in the linear model (described in Section 3.5.1) are zero. In other words, ANOVA is used to eliminate variables from the regressor vector that do not have an effect on the final output.

As stated in Eq. (3.26), the general equation of a linear model is

$$y_j = \beta_0 + \sum_{i=1}^{n} \beta_i x_{ij}. \tag{3.30}$$

Estimated values for weights and responses are found using least square estimators. The estimated weights are written as $\hat{\beta}_i$, and hence the *fitted values* are calculated as:

$$\hat{y}_j = \hat{\beta}_0 + \sum_{i=1}^{n} \hat{\beta}_i x_{ij}. \tag{3.31}$$

The *fitted values* are the predication made by the regression according to the estimated weights. A quantity *residual* is defined as the difference between the observed value and the fitted value for a given data point:

$$r_j = y_j - \hat{y}_j. \tag{3.32}$$

The residuals are related to the variance σ^2, which can be estimated from the residuals by:

$$S^2 = \frac{\sum_{j=1}^{m}(y_j - \hat{y}_j)^2}{m - (n+1)}, \tag{3.33}$$

where the numerator is called the *residual sum of squares(R.S·S)* and the denominator is called the *residual degrees of freedom(v)*. If the fitted model is adequate or redundant, i.e., it contains at least the required nonzero weights, S^2 is a good estimate of σ^2; whereas if the fitted model is deficient, S^2 is larger than σ^2.

We can use the concepts defined above to test whether some of the inputs in a model are extraneous. We can fit a base model Ω_1, which is known to be either adequate or redundant. For simplicity, Ω_1 can contain all the possible inputs. Another model, Ω_0, is fitted for the data in which the weights and inputs under investigation are absent.

The residual sum of squares (RSS_1) and degrees of freedom(v_1) are calculated for Ω_1. The estimate of variance according to the base model is given by

$$S_1^2 = \frac{RSS_1}{v_1}. \tag{3.34}$$

The residual sum of squares (RSS_0) and degrees of freedom(v_0) are calculated for Ω_0. The estimate of variance according to the test model is given by

$$s_0^2 = \frac{RSS_0}{v_0} \tag{3.35}$$

If Ω_1 is also an adequate model, the values of S_1^2 and S_0^2 should be fairly close. For better accuracy, *extra sum of squares* is calculated as

$$ESS = RSS_0 - RSS_1 \tag{3.36}$$

and extra degrees of freedom as

$$v_E = v_0 - v_1 \tag{3.37}$$

A third estimate of variance is given by

$$S_E^2 = \frac{ESS}{v_E} \tag{3.38}$$

S_E^2 is a good estimate of variance if Ω_0 is adequate, but is more sensitive than S_1^2. F-statistic, defined as the ratio between S_E^2 and S_1^2 is calculated as shown in Eq. (3.39).

$$F = \frac{S_E^2}{S_1^2} \tag{3.39}$$

If F is approximately equal to 1, we can state that Ω_0 is an adequate model for the given data. An F-statistic value above 4 leads to rejection of Ω_0 at 5% level of significance.

3.5.1.5 Multivariate Analysis of Variance

In some cases, the output of a data set is a vector, instead of a scalar quantity. Here, we use a variation of the method described in Section 3.5.1. Such data is analyzed using a multivariate linear model, defined as:

$$\begin{matrix} y_i \\ (c \times 1) \end{matrix} = \beta_0 + \beta_1 \, x_{1j} + \beta_2 \, x_{2j} + \ldots + \beta_n \, x_{nj} + \varepsilon_j \quad j = 1, \ 2, \ldots, m, \tag{3.40}$$

where the ε_j are independently distributed over $N^c(\mathbf{0}, \Sigma)$, m is the number of data instances and $\mathbf{y}$ and βs are vectors of dimension $(c \times 1)$.

The above model can be fitted in a manner similar to that in Section 3.5.1, to give the fitted values

$$\hat{\mathbf{y}}_{\mathbf{j}} = \hat{\beta}_0 + \sum_{i=1}^{n} \hat{\beta}_{\mathbf{i}} x_{ij} \quad j = 1, \ 2, \ldots, m. \tag{3.41}$$

The residuals are calculated in a similar manner.

$$\mathbf{y}_{\mathbf{j}} - \hat{\mathbf{y}}_{\mathbf{j}} \, j = 1, 2, \ldots, m \tag{3.42}$$

A matrix of residual sum of squares and products is calculated, which is analogous to the residual sum of squares in ANOVA. The matrix is given as:

$$R = \sum_{i=1}^{m} (\mathbf{y}_{\mathbf{j}} - \hat{\mathbf{y}}_{\mathbf{j}})(\mathbf{y}_{\mathbf{j}} - \hat{\mathbf{y}}_{\mathbf{j}})^T. \tag{3.43}$$

The residual degrees of freedom (v) is same as that in ANOVA for the same set of inputs. Σ is then given by R/v. The extra sum of squares and products matrix can be calculated in the same manner by which we calculated the extra sum of products in the univariate model, that is, by taking a difference for the matrices for the two models under comparison.

For the test statistic, there are four commonly used methods – *Wilks' lambda, Lawley-Hotelling trace, Pillai trace, and Roy's greatest root*. If the dimensionality of the output is 1, all these test statistics are the same as F-statistic. Once the superfluous βs have been eliminated, the relationship between the elements of the output and the βs is found by *canonical variate analysis*.

3.5.1.6 Log-Linear Models

Log-linear models are used to identify relationships between categorical variables, i.e., variables that cannot be represented as a numeric value. Categorical variables are of two types – *nominal*, which are those variables in which order is of no significance; and *ordinal*, which are variables which have certain natural ordering, represented as *none-mild-moderate-severe*. The log-linear model is analogous to ANOVA, the difference being that ANOVA only works with numerical variables, whereas these models deal with categorical inputs.

Log-linear model is a type of generalized linear model, where the output Y_i follows a Poisson distribution, with an expected value μ_i. The log of μ_i is taken to be linear, hence the name. Such models are represented by:

$$y_j \sim Pois(\mu_j) \text{ and } \log(\mu_j) = \beta_0 + \beta_1 x_{1j} + \beta_2 x_{2j} + \ldots + \beta_n x_{nj}. \tag{3.44}$$

Since the variables are categorical, they can either be present, represented by 0; or be absent, represented by 1. Hence, the input variables can only be binary.

Log-linear model is used to find the associations between different variables. These associations, also called interactions, are represented by β in Eq. (3.44). Hence, the problem boils down to finding which of the β s are zero, same as that in ANOVA.

For log-linear analysis, we first need to calculate *deviance* of the fitted model. Deviance in the log-linear model is analogous to *Residual Sum of Squares* in ANOVA. The deviance can be calculated for a data set using any of the popular statistical software packages available, to get a deviance table for the data set. The change in deviance after including the interaction between a pair of variables is similar to *Extra Sum of Squares* in ANOVA. If the interaction is redundant, the change in deviance has a distribution that is chi-squared with one degree of freedom. If the change is much higher than expected, the interaction has a significant effect on the output, and must be included in the model. This analysis is performed for each effect and their interactions, and the superfluous ones are weeded out. One thing is to be remembered is that change in deviance depends on the order in which the terms are included in the deviance table, hence a careful analysis of data is required.

After the deviance analysis, the model is represented by the significant effects and interactions. A conditional independence graph is then prepared. In this graph, each effect is represented as nodes, and the interactions in the model are represented as edges between them. If, for a given node N, we know the values of its neighbor nodes, then any information about any other nodes does not tell us anything new about N.

3.5.1.7 Logistic Regression

Logistic regression is appropriate to use in situations where the output variable is dichotomous, i.e., it is either fully present or absent. The regression function is defined as:

$$\text{logit}(P_j) = \log \frac{P_j}{1 - P_j} = \beta_0 + \beta_1 x_{1j} + \beta_2 x_{2j} + \ldots + \beta_n x_{nj} \quad j = 1, \ 2, \ldots, m, \tag{3.45}$$

where $P(y_j = 1) = p_j$ and $P(y_j = 0) = 1 - p_j$ and m is the number of data instances. For a good model, the variables need to be independent, with a correlation coefficients no greater than 0.9 [15].

3.5.1.8 Random Effects Model

In most practical studies, there are certain hidden attributes affecting the output that cannot be easily measured or identified. For example, in a batch of patients under study, the innate immunity varies from individual to individual. These unknown, random variables are also represented in the random effects model.

There are two types of random effects, which are as follows:

3.5.1.9 Overdispersion

The data is said to be overdispersed when the observed variability in output is much greater than what is predicted from the fitted model. It can happen if there's a failure in identifying a hidden variable which should have been included. Another reason may be that the output is not entirely decided by the input, but also on certain intrinsic variations between individuals. Hence, to model such effects, an extra variable U_j is added to the linear part of the model, which is distributed over $N(0,\sigma_U{}^2)$. A logistic regression with random effects can be expressed as:

$$\text{logit}(p_j) = \beta_0 + \beta_1 x_{1j} + \ldots + \beta_n x_{nj} + U_j. \tag{3.46}$$

Overdispersion has been studied in [16].

3.5.1.10 Hierarchical Models

Hierarchical models, also known as mixed models, are suitable when a treatment is applied to a group of individuals, instead of a unique treatment to each individual under study. In such models, the subjects of the study are divided into groups, known as *blocks*, by a process known as *blocking*.

It is assumed that members of the same group share certain properties with each other, which are not shared with members outside the group. The group properties become a source for random effects in the model. The random effects can be present for each level in the hierarchy. These effects are incorporated in the model using random coefficients or random intercepts. For a thorough discussion, see [14].

3.5.2 Analysis of Survival Data

Analysis of survival data is the study of given data to find out the time taken for an event to happen. For example, it can be performed on patients with fatal illnesses to find out the average time to death; or it can be performed on a batch of industrial equipment to find the optimal time before it should be replaced. Though it does not fall under the domain of generalized linear functions, the methods used are quite similar.

The data set for such analyses differ a lot with those used in regression methods. The reasons for such differences are:

Censoring The time period for which the study is conducted is limited, and it is very likely that many of the subjects would survive beyond the time it has ended. For such subjects, it is known that they had survived for the time period of the study, but exact time of death, or failure, is not known. Hence, we have some information about the subject, but it is not conclusive. Such observations are known as *censored observations*.

Time-Dependent Inputs The studies might be run for a long time, during which some of the inputs may change. For example, a machine can be services, or a patient may start doing more physical activity. Hence, the inputs are not constant, but rather a function of time. So, it is not always clear which value of the input should be used.

The *survivor function* describes the probability that the survival time is greater than a time t, and it is defined as:

$$S(t) = P(y \geq t) = 1 - F(t), \tag{3.47}$$

where $F(t)$ is the cumulative distribution function of the output. The *hazard function* is the probability density of the output at time t with respect to survival till time t:

$$h(t) = \frac{f(t)}{S(t)}, \tag{3.48}$$

where $f(t)$ is the probability density function of the output. In most cases, the hazard increases with time. This is because though the probability distribution of the output decreases, the number of survivors declines at a greater rate.

For most models, a *proportional hazard function* is used, which is defined as:

$$h(t) = \lambda(t)\exp.\{G(x,\beta)\} \tag{3.49}$$

where $\lambda(t)$ is *baseline hazard function*, a hazard function on its own, and $G(x,\beta)$ is any arbitrary known function. Conventionally, $G(x,\beta)$ is taken as a linear function, similar to that in the linear model. A linear G implies that if the baseline hazard is known, the logarithm of likelihood is a generalized linear model with Poisson distribution output and a log link function, which is the log-linear model.

In industrial studies, the baseline hazard can be known by performing controlled experiments, but the same cannot be done for human patients due to ethical, moral, and legal concerns. Hence, *Cox proportional hazard model* [17] is used in medical studies that can work with arbitrary baseline hazard. It loses information related to the baseline, and the resulting likelihood is independent of it. It uses the information about the order in which the failures take place, instead of the survival time. This model, too, can be fitted using a log-linear model for further analysis.

3.5.3 Principal Component Analysis

In multivariate statistics, increasing number of variables increases the complexity of analysis. If some of the variables are correlated, they can be collapsed into a single component. The method of reducing a set of variables to a set of principal components is known as *Principal Component Analysis* [18].

Consider a random vector $\mathbf{X}$ with dimensions $(p \times 1)$. It's variance-covariance matrix is given by:

$$\Sigma = \begin{pmatrix} \sigma_1^2 & \sigma_{12} & \cdots & \sigma_{1p} \\ \sigma_{21}^2 & \sigma_{22} & \cdots & \sigma_{2p} \\ \vdots & \vdots & \ddots & \vdots \\ \sigma_{p1} & \sigma_{p2} & \cdots & \sigma_p^2 \end{pmatrix}. \tag{3.50}$$

The *first principal component* (Y_1) is defined as a linear combination of x-variables with maximum variance among all possible linear combinations. We need to find $\mathbf{e_{11}, e_{12}, ..., e_{1p}}$ for the linear expression

$$Y_1 = e_{11}X_{11} + e_{12}X_{12} + ... + e_{1p}X_{1p} \tag{3.51}$$

that maximized variance of Y_1,

$$var(\mathbf{Y_1}) = \sum_{k=1}^{p} \sum_{l=1}^{p} e_{1k}e_{1l}\sigma_{kl} = \mathbf{e_1'} \sum \mathbf{e_1} \tag{3.52}$$

subject to the constraint that

$$\mathbf{e}_1'\mathbf{e}_1 = \sum_{j=1}^{p} e_{1j}^2 = 1. \tag{3.53}$$

The constraint makes sure that a unique value is obtained.

The *second principal component* (Y_2) is a linear combination of x-variables that accounts for the maximum amount of remaining variance, with the constraint that there is no correlation between the first and the second component. Similarly, the ith *principal component* (Y_i) is the linear combination of x-variables that maximizes

$$var(\mathbf{Y}_i) = \sum_{k=1}^{p} \sum_{l=1}^{p} e_{ik} e_{il} \sigma_{kl} = \mathbf{e}_i' \sum \mathbf{e}_i \tag{3.54}$$

subject to the following constraints:

$$\mathbf{e}_i'\mathbf{e}_i = \sum_{j=1}^{p} e_{ij}^2 = 1 \tag{3.55}$$

$$\text{cov}(Y_1, \ Y_i) = \sum_{k=1}^{p} \sum_{l=1}^{p} \mathbf{e}_{1k} \mathbf{e}_{il} \sigma_{kl} = \mathbf{e}_1' \Sigma \mathbf{e}_i = 0, \tag{3.56}$$

$$\text{cov}(Y_2, \ Y_i) = \sum_{k=1}^{p} \sum_{l=1}^{p} \mathbf{e}_{2k} \mathbf{e}_{il} \sigma_{kl} = \mathbf{e}_2' \Sigma \mathbf{e}_i = 0, \tag{3.57}$$

$$\text{cov}(\mathbf{Y}_{i-1}, \mathbf{Y}_i) = \sum_{k=1}^{p} \sum_{l=1}^{p} \mathbf{e}_{i-k,k} \mathbf{e}_{il} \sigma_{kl} = \mathbf{e}_{i-1}' \sum \mathbf{e}_i = 0 \tag{3.58}$$

To calculate the coefficients for the principal components, the eigenvalues for the variance-covariance matrix are found are arranged in a decreasing order. The corresponding eigenvectors for these eigenvalues are also found. The coefficients of the principal components are same as the values of the elements of the eigenvectors, while the eigenvalues give the variances of the components. We should find k principal components such that

$$\frac{\sum_{i=1}^{k} \lambda_i}{\sum_{j=1}^{p} \lambda_j} \sim 1, \tag{3.59}$$

where λ_i is the ith eigenvalue.

3.6 Errors

We had read in Section 3.4 that the aim of inferential statistics was to correctly or at least reasonably make predictions about the population from the given sample without making an error of generalization. That is, we tend to make predictions about values based on the training sample. If these predictions result in a numeric value, then they are categorized as regression, otherwise if the values fall into discrete unordered sets, they are categorized as classification.

The basis for prediction of values is the assumption that a random variable Y, is dependent in some way on the values of X_i , which are the outcomes of the given sample. This dependency can be any function such as a simple linear regression model. However, since the prediction is based on assumption, if the true relationship between the random variable and the sampled data is not linear, it can lead to prediction error [19]. If on the other extreme end, no assumptions are made then the random variable might overfit the data. Overfitting refers to a phenomenon in which the parameters drawn from a sample of data to define a relationship between the random variable and the sampled data become too sensitive to a particular sample. In such a case, the predictions made about the relationship can be found out to be wrong even on a new sample of data. Thus, overfitting more often than not, contributes to prediction error.

3.6.1 Error in Regression

Let us illustrate the concept of prediction errors and the way to handle them through an example. Consider a true linear relationship between samples and the random variable. We will try to model this relationship through three different classes of models: linear, quadratic, and cubic. It should be noted that the cubic model encompasses the quadratic model if the cubic coefficient becomes zero. Similarly, the cubic and quadratic model both encompass the linear model through appropriate values of the coefficients. Let the sample size be 10 and a total of 1000 different samples be taken. Ideally, all possible variations of a fixed size of sample should be taken but that would require a lot of computation. From these 1000 samples, values of random variable are to be predicted which will help in the calculation of bias and variance.

Bias is the difference between the true value of the random variable and its average estimate drawn from the 1000 different samples. Variance refers to the expected squared difference between an estimate of a single sample and the average of that over the 1000 different samples. On evaluation of the results of the experiment, it would be obvious that there is no evident bias since all of the models encompass the correct model. However, an interesting observation is the change in values of variance over the different models. The linear model displays the minimal variance while the cubic model has the largest variance.

When the size of the sample is fixed and small, variance tends to play a larger role in contributing to the prediction error. However, as we will see in our next example, as sample size increases, bias tends to overtake the contribution toward error. Highly flexible models tend to suffer from a high variance as the model adapts to randomness. On the other hand, incorrect models may present a lower variance (linear model) but present larger deviations from the nominal value, hence, having a larger bias. This trade-off between bias and variance needs to be handled appropriately for different sample sizes.

Let us now consider a true quadratic relationship between the samples and the random variable. This time, we will consider 1000 different samples each of size 10, 100, and 1000. In this case, it is evident that linear model will be biased while the quadratic and cubic models remain unbiased. To illustrate how the different sample sizes play a role in prediction error, we will compute the mean squared errorm which is the sum of squared bias and variance. It is observed that for sample sizes 10 and 100, the variance of linear model is smaller as compared to that of quadratic and cubic. However, the mean squared error for sizes 100 and 1000 is significantly smaller for the latter two models. This arises from the contribution of bias that is almost negligible for the two models and the gradual decrease in variance as size of sample increases.

3.6.2 Error in Classification

In the case of classification, every value of the sample is linked to a class label. Let us consider the case where there are two class labels, 0 and 1. The rule for categorization that takes error rate into consideration is known as the Bayes rule. The rule states that a class should be assigned to an input point that shows the highest probability of a particular class.

There are two different ways to assign a point to a class. The first requires the averaging of class probabilities of the individual classifier and assigning the input point to the class with the greatest average probability. The second requires assignment of input point to a class based on majority voting of individual classifiers. Here also we encounter bias and variance that contribute toward the classification prediction error. However, the difference between classification and regression is that a wrong estimate of a predicted value does not necessarily result in classification into the wrong category. This protection is offered through the Bayes rule.

According to the Bayes allocation rule, there are two distinct regions that can be drawn in a classifier based on the true classifications and the predicted ones. A positive boundary bias [20] refers to the regions where the classifier is on average on the wrong side of correct allocations. A negative boundary bias [20] refers to regions that are unbiased. An increase in bias with constant variance actually leads to a decrease in error rate in the negative boundary bias, whereas in the positive boundary bias, an increase in bias is followed by an increase in error rate. The opposite relationship between variance and error is observed when bias is kept constant in these two regions of boundary bias. The interaction between bias and variance is more complex than in the case of regression. As long as a negative boundary bias is maintained, the minimization of variance reduces error rates.

3.7 Conclusion

An introductory view of the basics of statistics and the statistical methods used for intelligent data analysis was presented in this chapter. The focus was on building an intuitive understanding of the concepts. The methods presented in this chapter are not a comprehensive list, and the reader is encouraged to audit the literature to discover how these methods are used in real world scenarios. The concepts presented in the chapter form the basics of data analysis, and it is expected that the reader now has a firm footing in the statistical concepts that can then be used to develop novel state-of-the-art methods for specific use cases.

References

1 Berthold, M.R., Borgelt, C., Höppner, F., and Klawonn, F. (2010). *Guide to Intelligent Data Analysis: How to Intelligently Make Sense of Real Data*. Berlin: Springer Science & Business Media.

2 Lavrac, N., Keravnou, E., and Zupan, B. (2000). Intelligent data analysis in medicine. *Encyclopedia of Computer Science and Technology* 42 (9): 113–157.

3 Liu, X. (2005). Intelligent data analysis. In: *Encyclopedia of Data Warehousing and Mining*, 634–638. IGI Global.

4 John E. Seem. Method of intelligent data analysis to detect abnormal use of utilities in buildings, New York, NY: US Patent 6,816,811, 9 November 2004.

5 DeGroot, M.H. and Schervish, M.J. (2012). *Probability and Statistics*. Boston, MA: Pearson Education.

6 Von Hippel, P.T. (2005). Mean, median, and skew: correcting a textbook rule. *Journal of Statistics Education* 13 (2), 1–13.

7 Richard Lowry. (2014). Concepts and applications of inferential statistics.

8 Berthold, M.R. and Hand, D.J. (2007). *Intelligent Data Analysis: An Introduction*. Springer.

9 Huck, S.W., Cormier, W.H., and Bounds, W.G. (1974). *Reading Statistics and Research*. New York, NY: Harper & Row.

10 Draper, N.R. and Smith, H. (2014). *Applied Regression Analysis*, vol. 326. Hoboken, NJ: Wiley.

11 Niklas, K.J. (1994). *Plant Allometry: The Scaling of Form and Process*. Chicago, IL: University of Chicago Press.

12 Mitscherlich, E.A. (1909). Des gesetz des minimums und das gesetz des abnehmended bodenertrages. *Landwirsch Jahrb* 3: 537–552.

13 Ware, G.O., Ohki, K., and Moon, L.C. (1982). The mitscherlich plant growth model for determining critical nutrient deficiency levels 1. *Agronomy Journal* 74 (1): 88–91.

14 McCullagh, P. and Nelder, J.A. (1989). Generalized linear models, vol. 37 of monographs on statistics and applied probability. London, UK: Chapman and Hall, Second edition.

15 Tabachnick, B.G. and Fidell, L.S. (2007). *Using Multivariate Statistics*. New York: Allyn & Bacon/Pearson Education.

16 Collett, D. (1991). *Modelling binary data*, vol. 380. London, UK: Chapman & Hall.

17 Cox D.R. (1992). Regression models and life-tables. In: Kotz, S. and Johnson, N.L. (eds). *Breakthroughs in Statistics*. New York, NY: Springer.

18 Jolliffe, I. (2011). Principal component analysis. In: *International Encyclopedia of Statistical Science*, 1094–1096. Boston, MA: Springer.

19 Salkever, D.S. (1976). The use of dummy variables to compute predictions, prediction errors, and confidence intervals. *Journal of Econometrics* 4 (4): 393–397.

20 Friedman, J.H. (1997). On bias, variance, 0/1—loss, and the curse-ofdimensionality. *Data Mining and Knowledge Discovery* 1 (1): 55–77.

4

Intelligent Data Analysis with Data Mining: Theory and Applications

Shivam Bachhety, Ramneek Singhal, and Rachna Jain

Department of Computer Science and Engineering, Bharati Vidyapeeth's College of Engineering, New Delhi, India

Objective

Having explained various aspects of intelligent data analysis (IDA) in previous chapters, we are now going to see it along with the essence of data mining. In this chapter we are going to base our discussion on the following major points:

- What data mining is? What are the similarities and differences between data and knowledge?
- Process of knowledge discovery in data. Along with it, various mining methods, including classification, clustering, and decision tree will be discussed.
- Issues related to data mining and its evaluation.
- In addition to that, the chapter will be concluded with a view of data visualization and probability concepts for intelligent data analysis (IDA).

4.1 Introduction to Data Mining

Data mining has been defined as:

> ... the process of discovering meaningful new correlations, patterns and trends by sifting through large amounts of data stored in repositories, using pattern recognition technologies as well as statistical and mathematical techniques [1].

Initially, the term "data fishing" was used, which meant analyzing data without a prior hypothesis. The term was used in a negative way and continued to be used until Michael Lovell published an article in the *Review of Economic Studies* in 1983. He used the term "data mining" but the semantics of the term was still negative. Later on, it was in the 1990s when "data mining" was seen as a positive and beneficial practice. First, the International Conference on Data Mining and Knowledge Discovery (KDD-95) was started in 1995 for research purposes. It was in Montreal under AAAI sponsorship. Usama Fayyad and Ramasamy Uthurusamy chaired it. The journal *Data Mining and*

Intelligent Data Analysis: From Data Gathering to Data Comprehension,
First Edition. Edited by Deepak Gupta, Siddhartha Bhattacharyya, Ashish Khanna, and Kalpna Sagar.
© 2020 John Wiley & Sons Ltd. Published 2020 by John Wiley & Sons Ltd.

Knowledge Discovery wasstarted by Usama Fayyad in 1996 as a primary research journal in data science.

With computerization there has been an explosive increase in the amount of data that is being produced on a daily basis. An enormous amount of data is produced every single day in the world. Some of it is structured, as in the case of data that is stored in relational databases and in data warehouses. This type of data is easy to understand and applying various data mining algorithms on it is comparatively easier. But still a majority of data is unstructured. Video data from YouTube and Facebook is a great example of unstructured data. The idea about the enormity of data generated can be understood from the fact that YouTube users upload 48 hours of new video each minute and Facebook users generate over 4 million posts per minute. Availability of low-cost hardware has increased our ability for storing this enormous volume of data, which is then to be analyzed.

In order to gain useful insights from this data, it is necessary to be analyzed; however, the volume of data has surpassed the human ability to analyze it. Therefore, we need powerful tools and the concept of data mining to extract useful information and hidden patterns from the data. Data mining is known as knowledge discovery from data (KDD) because it is the process of manipulating data and analyzing it in order to convert it into knowledge.

Thus, data mining can be defined as the process of analyzing large volume of data to identify patterns and knowledge to solve problems through data analysis. As a result, data mining has found its usage in various fields including health care, business, education, manufacturing, etc., which are explained in brief below.

4.1.1 Importance of Intelligent Data Analytics in Business

An enormous amount of data is generated each day in any business from a variety of sources like orders, trade transactions, and consumer feedback, and this data is cleaned, organized, and stored in data warehouse databases. The primary goals of dat a mining include the following:

- Discovering unknown strategic business information
- Performing market analysis to predict the chances of success of new products
- Designing marketing campaigns
- Predicting consumer loyalty
- Finding factors that cause manufacturing problems
- Categorizing and classifying items that consumers buy together
- Predicting risks and return of investment

All of these things aim to reduce risks and increase profitability in the business. Companies like Amazon analyze a trend of customers buying specific items together so that in future when customers buy an item, they then recommend the items that previous customers have brought together. Amazon has reported a 29% sales increase as a result of this recommendation system. The stock market is another field in which data mining is extensively used in order to predict stock prices and get better returns on investment.

4.1.2 Importance of Intelligent Data Analytics in Health Care

Data in the field of health care can be of great help in improving life expectancy if utilized properly. In this aspect, data science is of great importance in order to improve the methods of diagnosing, preventing, and treating diseases. In human genetics, sequence mining helps to find out how the changes in an individual's DNA sequence affect the risks of developing common diseases such as cancer. Sharing of information about patients, diseases, and medicines with other doctors around the world also helps better detection of diseases and better courses of treatment.

4.2 Data and Knowledge

Data is unprocessed facts and figures without any context or analysis. Data alone has no certain meaning. On being cleaned and analyzed, it gets converted to information and then to knowledge, as shown in Figure 4.1.

Data over the years has grown not only in volume but also in variety. In order to handle this variety of data, various kinds of databases have been evolved over the period of time. Initially, in the 1960s, a primitive file system was used to store data. In this system, the generated data was stored in simple files on hard disk. It was difficult to comprehend and analyze data using such a system. After the 1970s, relational databases came into the picture. Data was now stored in tables that were related to each other. It is still widely used in companies as a way of storing structured data. Since the entire data is stored in the form of rows and columns, it is easier to find patterns using data query languages like structured query language (SQL) and others. As time passed, storing the data as multiple tables became a problem. This gave rise to the need of developing a system where humongous amount of data can be stored. A data warehouse is one such environment. In order to obtain strategic information, data warehouses were put to use. Data was first gathered from various sources and it was cleaned. Online analytical processing (OLAP) can be viewed as the beginning

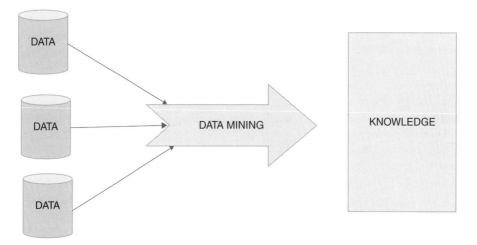

Figure 4.1 From data to knowledge.

Table 4.1 Dissimilarities between data and knowledge.

Data	Knowledge
It is collection of raw and unorganized facts.	It is basically possession of information and using it for the profit of business.
It is not specific or context based.	It is very specific and context based.
It is based on observations and records.	It is obtained from information.
It may be organized or unorganized.	It is always organized.
It is sometimes not useful.	It is always useful.
Data doesn't depend on knowledge.	Without data, knowledge cannot be generated.

of data mining, but it had some crucial differences. The analysts using OLAP needed some prior knowledge about the things they were looking for, whereas in data mining, the analysts have no prior knowledge of the expected results. Moreover, OLAP gives the answers to questions on the basis of past performance, but it cannot uncover patterns and relationships that predict the future. Table 4.1 represents some of the peculiar differences between data and knowledge.

Not only is the data huge in volume, it also comes from a variety of sources. Some of which are shown in Figure 4.2 below.

4.3 Discovering Knowledge in Data Mining

Data mining is aimed at extracting hidden patterns and relationships within data and to reveal strategic information that can be beneficial for the business. But data mining is more than just applying data mining tools on collected data. It begins from understanding the

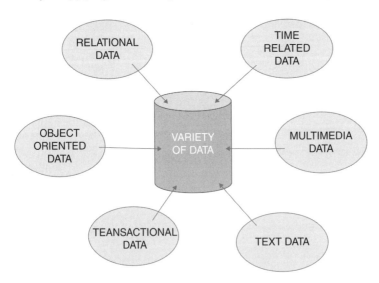

Figure 4.2 Variety of data in data mining.

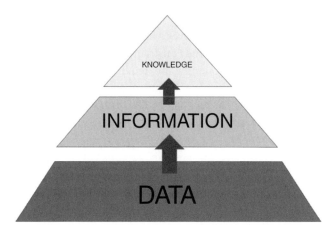

Figure 4.3 Knowledge tree for intelligent data mining.

requirements and business expectations from a particular data mining project and ends with applying the gained knowledge in business in order to benefit it. Data mining is used to extract this information from any given data as shown by knowledge tree in Figure 4.3.

4.3.1 Process Mining

Process mining is the study and analysis of event logs that help to bring insights into the company's business operations and help us to deepen our understanding of it and what are the improvements that are needed to make it optimal. This is done in order to find out the deep working of a process and shortcomings and bottlenecks that are present in any process of a business.

There are two possible cases under process mining:

1. *No prior model*: In this case there is no prior model for the process that is to be analyzed, so on the basis of event data, a flowchart of a process is drawn and it is optimized.
2. *With prior model*: If there exists a prior model, then the event logs can be compared to the model in order to find out if the workers are performing a specific process in the optimized way. Such a prior model is also useful if you want to optimize the model itself with new information such as processing time, cycle time, waiting time, cost, etc.

4.3.2 Process of Knowledge Discovery

The entire process of conversion of data to knowledge can be summarized in the following steps shown in Figure 4.4.

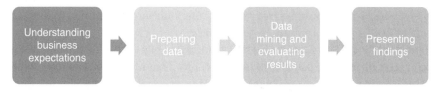

Figure 4.4 Knowledge discovery process.

1. **Understanding business expectations:**

   The first and foremost step is to understand the objectives, needs, and domain of a business. Ideas about the current situation, resources, and constraints that are present with a particular business is also important. After obtaining any required prior knowledge about the business and current situation, achievable and optimal objectives and targets for the data mining system should be set. Objectives can be like improving marketing campaigns, predicting risk and return of investment, predicting customer's loyalty, classifying items that customers buy together, etc.

2. **Preparing data:**

   Next step is collecting and preparing the data that has to be analyzed. There can be two scenarios: one in which the data to be collected is taken from a data warehouse and the other where the required data is not in any data warehouse and has to be collected by other means.

   If the required data is in any data warehouse then it is expected that the data is already cleaned and transformed. We just need to select the required data from the data warehouse and perform data mining on it. Appropriate metadata should also be included in order to have a better understanding about the data before it is analyzed.

   If the data is not in any warehouse then it has to be retrieved by a process involving the following steps: extraction, cleaning, transformation, and selection.

   Extraction: Data that will be used for data mining can be stored in variety of locations. It may be the internal or external data of an organization or archived data or data from web. This data has to be extracted and stored for further processing.

   Cleaning: The extracted data may contain a number of anomalies that can cause hindrances when the data mining is performed. In order to make the data appropriate for data mining it has to be cleaned first. Cleaning involves providing default values for missing data, eliminating the duplicate data, correcting the misspellings, etc.

   Transformation: Once the data is cleaned, it should be sorted and merged with other data in order to obtain the final clean, summarized, and standardized data that can be used for analysis, and upon on which data mining algorithms can be applied.

   Selection: This step is similar to the scenario in which we obtain data from a data warehouse. In this step, the required data has to be selected from the vast data that we have summarized in the step above. Different data mining algorithms require different data as input, thus while selecting the data mining algorithm and tools, this must be kept in mind.

3. **Performing data mining and evaluating results:**

   In this step the selected data mining algorithm is applied on the selected and prepared data. The result after each iteration is evaluated and matched with the objective of analysis. Sometimes, multiple iterations of data mining algorithm are required in order to obtain the desired result. The patterns or relationships found are evaluated and analyzed

4. **Presenting findings and applying it to business:**

   Any pattern or relationship that is found by data mining is of no use if it is not implied in the business. In this final step, the knowledge and information gained are applied in the business to benefit from it and to meet the objectives that were set initially, thus fulfilling the purpose of data mining.

4.4 Data Analysis and Data Mining

Data analytics is the super set of data mining as shown in Figure 4.5. Data mining is a type of analysis that focuses on modeling and discovery of knowledge in huge data sets while business intelligence, also a type of data analysis, focuses mainly on the business information that can be used to make the business more profitable.

Data analytics is a process by which the data is examined to draw insightful conclusions to help a business make decisions that will be profitable. It can be either exploratory, focusing on finding new features in data, or confirmatory, focusing on testing the already existent hypothesis.

On the other hand, data mining works on mathematical models and highly detailed data sets in order to extract hidden relationships and patterns.

Relationships: Data mining enables us to find hidden relationships between two or more objects or between the attributes of the same object. Evident use of data mining to extract relationships can be seen in recommendation systems. For example, when we click to buy a particular book from an online shopping website, another book that is similar to the book that we have ordered is shown in recommendation. A similar thing happens when we go to supermarket and buy a particular item. After keen analysis through data mining and identifying relationship between different items, the items in the marketplace are arranged in groups that customers tend to buy together, thus enabling better sales.

Pattern: Data mining also helps us to identify patterns that exist in a particular data, and from these patterns, it is possible to predict the future. Using past data of customer transactions, a pattern can be identified that tells the bank which customers are likely to commit fraud or which customers are unlikely to repay a loan. The bank can then use this for any new customer in order to judge whether the person is likely to repay the loan or not. This helps the banks reduce the burden of unpaid loans.

4.5 Data Mining: Issues

Data mining is still a new and relatively complex field. There are a number of issues that are related to it as displayed in Figure 4.6. Some of the issues are related to the availability of data, the performance of mythology, or handling the various needs of the users.

Figure 4.5 Relationship between data analysis and data mining.

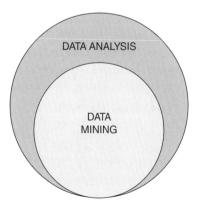

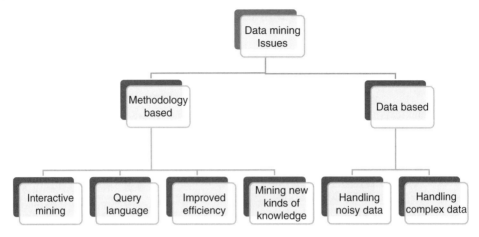

Figure 4.6   Issues in data mining.

Some of the issues related to mining methodology:

1. *Interactive mining*: At the present time, mining is highly noninteractive in nature. Most of the times, users don't even know what sort of knowledge is going to be obtained at the end of mining process, but in order to make the mining work more efficiently it should be made interactive. The user should be able to redirect the mining process in order to obtain desired results and knowledge, i.e., the mining should be able to be changed dynamically.

2. *Data mining query languages*: At present, data mining is highly nonflexible. We select the data on which a data mining algorithm has to be run and we get a pattern or relationship as knowledge, but in order to be more efficient, mining should be more flexible. To achieve this high level, data mining query languages should be developed. In such languages users should be able to give queries similar to those in SQL. This will enable the users to obtain the specific knowledge that they are looking for.

3. *Improved efficiency*: With time, the size of data sets is increasing exponentially, so the mining algorithms should be able to work on them. They must provide efficient results even on huge data sets and must provide this result in real time with high efficiency so that they are actually helpful in the real world.

4. *Mining new kinds of knowledge and at multiple levels*: Data mining already uses various techniques such as classification, regression, clustering, trend analysis, etc. With increasing complexity of data there is always scope to find some new kind of knowledge. Moreover, we should be able to obtain knowledge at multiple granular levels. Thus, different kinds of knowledge at multiple levels will motivate the end users and analysts to turn to data mining for their various objectives in respective business.

Some of the issues related to data to be mined:

1. *Handling noisy data*: Data that is collected from data warehouses is expected to be cleaned and noise free but still it may contain some empty columns, missing data, or inconsistent data. In addition to this, sometimes the outlining data is deliberately included to observe deviating patterns. The mining algorithms and tools should be able to handle any noisy data. Not only this will increase efficiency but it will also optimize the process and may give new types of knowledge in certain cases.

2. *Handling complex data*: Complexity of data can be seen either as complexity in the variety of data that is being generated or as a complexity that large amounts of data that is stored in different places has to be analyzed by the mining algorithm simultaneously and in real time. Variety of data include web data, multimedia data, time stamp data, hypertext data, sensor data, etc. A single mining system cannot produce optimal results for all kind of data. Thus specific systems should be made to handle them efficiently. As for distributed data, mining algorithms that can run on a distributed network efficiently can be the solution.

4.6 Data Mining: Systems and Query Language

4.6.1 Data Mining Systems

Data mining systems are an important aspect in the field of data mining to perform data analysis to identify important patterns contributing to research, business strategies, and knowledge bases.

The data mining systems are classified into the following categories as shown in Figure 4.7:

- Database technology
- Statistics
- Machine learning
- Information science
- Visualization
- Other disciplines

Figure 4.7 Various systems in data mining.

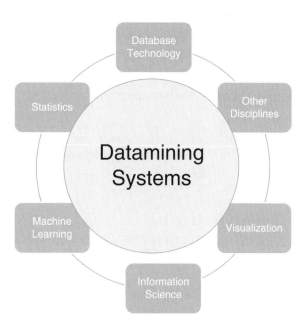

Other classification criteria for systems in data mining include:

Based on mining of databases: Database systems can be categorized based on data types, models, etc.

Based on mining of knowledge: Knowledge mining can be classified based on functionalities such as characterization, discrimination, association, classification, prediction, clustering, etc.

Based on utilization of techniques: The techniques can be classified based on the degree of interactivity of users involved in analysis of methods employed.

Based on adaptation of application: They can be classified based on applications in areas of finance, telecommunications, stock analysis, etc.

4.6.2   Data Mining Query Language

First proposed by Han, Fu, and Wang for data mining systems the data mining query language (DMQL) is based on the SQL. It is considered to maintain ad hoc and collaborative data mining. The DMQL commands specify primitives to work on databases and data warehouses. It is efficiently used in data mining tasks too.

It is very useful in the following cases:

- Applying model to a new data and provide parameters as input.
- Extract patterns or regression rules to get cases that fit the problem.
- Obtain statistical summary from the training data.
- Perform cross-prediction on adding new data.

The following considerations play an important role in designing the data mining language:

- Specification of data set in data mining request applicable to a data mining task.
- Specification of knowledge kind in data mining request to be revealed during the process.
- Availability of background knowledge available for data mining process.
- Expressing data mining results as generalized or multi-level concepts.
- Ability to filter out less interesting knowledge based on threshold values.

Syntax of DMQL for data mining:

```
⟨DMQL⟩ ::=
    use database ⟨database_name⟩
    {use hierarchy ⟨hierarchy_name⟩ for ⟨attribute⟩}
    ⟨rule_spec⟩
    related to ⟨attr_or_agg_list⟩
    from ⟨relation(s)⟩
    [where ⟨condition⟩]
    [order by ⟨order_list⟩]
    {with [⟨kinds_of⟩] threshold = ⟨threshold_value⟩
        [for ⟨attribute(s)⟩)]}
```

[Source: http://hanj.cs.illinois.edu/pdf/dmql96.pdf]

The different mining rules present in DMQL are:

- **Characterization:** The rule contains to state a characteristic that will be satisfied by maximum examples present in the target class.
  Syntax:

  ```
  mine characteristics [as pattern_name]
  analyze {measure(s) }
  The analyze clause, specifies aggregate measures,
  such as count, sum, or count%
  ```

- **Discrimination:** It means to discriminate a concept of target class from all other contrasting classes.
  Syntax:

  ```
  mine comparison [as {pattern_name]}
  For {target_class }where {t arget_condition}
  {versus {contrast_class_i}
  where {contrast_condition_i}}
  analyze {measure(s)}
  ```

- **Classification:** This is used to classify the set of data using certain rules to find a favored classification scheme.
  Syntax:

  ```
  mine classification [as pattern_name]
  analyze classifying_attribute_or_dimension
  ```

- **Association:** It looks for patterns present in the data set to associate elements.
  Syntax:

  ```
  mine associations [ as {pattern_name}]
  {matching {metapattern}}
  ```

- **Prediction:** It helps to predict values upon the existing values in the data set.
  Syntax:

  ```
  mine prediction_[as pattern_name]
  analyze predictian_attribute_or_dimension
  {set {attribute_or_dimension_i= value_i}}
  ```

4.7 Data Mining Methods

Data mining methods involves the techniques that are used to generate the results in the knowledge discovery process. These methods consist of separate parameters and are specific to certain problems and data sets.

The common components presents in data mining methods:

- **Model representation** consists the language to find the discoverable patterns in the knowledge process.

- **Model evaluation** involves the steps to fit the model to estimate a particular pattern and meet the conditions of the knowledge discovery of data (KDD) estimates how well a particular pattern (a model and its parameters) meet the criteria of the KDD process. It should meet both logical and statistical benchmarks.
- **Search methods** contain parameter search and model search to optimize the model calculation and further loops over the parameters in the search method.

The various available data mining methods are:

1. Classification
2. Cluster analysis
3. Association
4. Decision tree induction

4.7.1 Classification

Classification is a data mining technique used to categorize each item in a set of data into one of an already defined set of classes or collections. It is used to reveal important information about data and metadata. It involves building the classification model and then using the classifier for classification as shown in Figure 4.8.

The other classification methods include:

- *Genetic algorithm*: Centered on the idea of the survival of the fittest, the formation of the new population is dependent on the rules set by the current population and offspring values also. Crossover and mutation are the genetic parameters used to form offspring. The fitness rules are decided by the training data set and evaluated by its accuracy.
- *Rough-set approach*: This method is functional on discrete valued attributes, and sometimes requires conversion of continuous valued attributes. It is established by creation of equivalence classes using training set. The tuples forming the equivalence class are undetectable. This indicates that the samples are similar to the attributes that describe the data, thus, they can't be distinguished easily.
- *Fuzzy-set approach*: Also called possibility theory, it works at an extraordinary level of generalization. This method is an advantage when dealing with inexact data. The fuzzy theory is a solution to vague facts where finding relations in the data set is complex. There is no direct measure to classify in the data set.

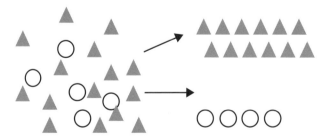

Figure 4.8 Diagrammatic concept of classification.

The most common algorithms include logistic regression, naïve Bayes, and support vector machine, etc.

Classification of e-mail messages as spam or authentic based upon the content present in them using large data set of e-mails.

4.7.2 Cluster Analysis

Identifying separate collections or subcategories within data set is known as cluster detection as shown in Figure 4.9. It is quite useful in data exploration and estimating natural groupings. Members of the same cluster have synonymous properties with each other than other members. It has wide applications such as pattern recognition, data analysis, and image processing. It is a vital tool to view insights into the distribution of data for each cluster. It is adaptable to changes and thus, most suitable for prediction purposes.

The different capabilities of clustering in data mining include:

- Scalability: Ability to deal with large data sets
- Works with both low as well as high dimensional data
- Ability to deal with different cluster shapes.
- Ability to deal with missing or noisy data.
- Clustering is comprehensible and able to deal with different datatypes such as interval data, categorical, and binary data too.

The common algorithms include: k-means, expectation maximization, and orthogonal partitioning.

Credit card fraud detection and deriving plant and animal taxonomies using gene similarities.

4.7.3 Association

It is used to discover relation between two or more items in the same transaction as shown in Figure 4.10. It is often useful to find hidden patterns in data, and also called relation

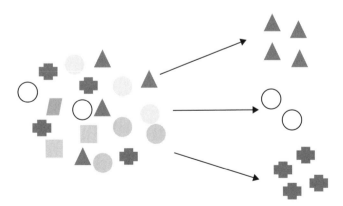

Figure 4.9 Diagrammatic concept of clustering.

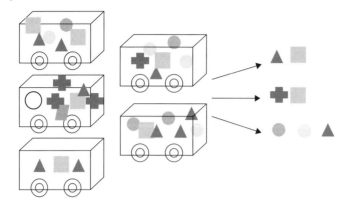

Figure 4.10 Diagrammatic concept of classification.

technique. It is often used to find rules associated with frequently occurring elements. It is the base for root cause analysis and market basket analysis.

The Apriori algorithm is widely used in market industry for association rule mining.

It is used in product bundling to research about customer buying habits based on historical customer transactions and defect analysis also.

4.7.4 Decision Tree Induction

Decision tree consists of a root node with its branches and leaf nodes. A decision tree is a structure that includes a root node, branches, and leaf nodes. The internal node denotes test and the branches lists the outcomes with a label.

The benefits include:

- Fast learning and easy classification process.
- Requires no domain knowledge.
- It is easy to understand.

Tree pruning: It is done to remove anomalies and reduce complexity of the tree.

It makes the tree smaller by removing extra branches aroused due to outliers. Its complexity is measured using the leaves present in the tree and error rate of the tree.

A sample instance with rules in decision tree is shown in Figure 4.11.

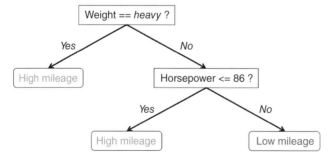

Figure 4.11 Specimen for decision tree induction.

4.8   Data Exploration

Data exploration is defined as the process to understand the characteristics present in the data set better. It includes listing down the characteristics using an analytics tool or a more innovative statistical software. It mainly focuses on summarizing the contents present in the data set knowing the variables present, any missing values, and a general hypothesis.

The various steps involved in data exploration are:

1. *Variable identification*: This involves deciding the role of the variables in the data set mainly as predictor and target variables. Also, identification of their data type and their category.
2. *Univariate analysis*: This involves analyzing variables one by one and used to find missing values and outliers. It consists of two categories:
   a. *Continuous variables*: It gives understanding about the central tendency and spread of the variables. Common visualizations include building histograms and boxplots for individual variable. They focus on the statistical parameters.
   b. *Categorical variables*: Count and count% are used to understand the distribution via frequency table. Bar chart is used for visualization purpose for categorical values.
3. *Bivariate analysis*: It determines the relation between two variables. It can be performed on the combination of continuous and categorical variables to search for association or disassociation. The combination can be:
   a. *Categorical and categorical*: The relation between two categorical variables can be discovered via following methods:
      – *Two-way table*: It is a table showing count and count% for each combination of observations in rows and columns.
      – *Stacked column chart*: It is a visual representation of two-way table showing columns as stacks shown in Figure 4.12.

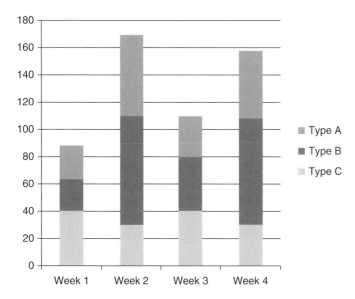

Figure 4.12   Sample representation for stacked column chart.

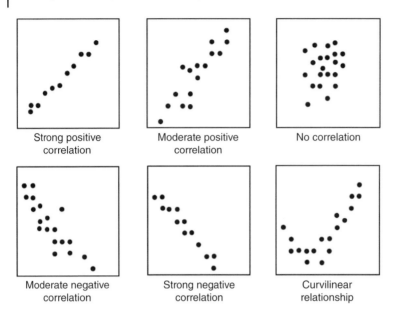

Figure 4.13 Different relationships shown by scatter plots for bivariate analysis.

- *Chi-square test*: It derives the statistical relation between two or more categories via difference in expected and observed frequencies. The formula is shown as below:

$$X^2 = \sum \frac{(\text{observed} - \text{expected})^2}{\text{expected}}$$

b. *Categorical and continuous:* Box plots can be drawn to display relation between categorical values and we can discover statistical significance using:
   - **Z-Test/T-Test:** This test is used to access if mean of two categories are statistically different or not. If the probability of Z is small then the difference of two means is more substantial. The T-test used when number of observation for both groups is less than 30.
   - **ANOVA:** Also called analysis of variance, it is used to analyze difference of means of two categories in a sample.
c. *Continuous and continuous*: In bivariate analysis, scatter plot is a nifty way when looking for relation between two continuous values as shown in Figure 4.13. The relation can be linear or nonlinear. To determine the strength of relation, correlation is used. Its value lies between −1 and 1, with 0 showing no correlation. The formula for correlation is as follows:

$$Correlation = \frac{Cov(x, y)}{\sigma x * \sigma y}$$

where,
Cov (x, y) is the covariance between the two variables
σx = Standard deviation of variable x
σy = Standard deviation of variable y

4. *Missing values treatment:* Missing values in a data set can reduce its fit in the model and can make the model biased, and can wrongly predict the relation among the variables. Missing values can occur at:

   - *Data extraction*: There may be errors with data extraction, which may lead to missing values. Sometimes the data source may also have inconsistencies or typographical errors. They can be corrected easily.
   - *Data collection*: They are harder to find and occur at data collection. These include missing at random, unobserved predictors, or no-response rate.

   The several methods for treating missing values:

   a. **Deletion:** It consists of two methods:

      List-wise deletion: In list-wise deletion, an observation is deleted is any variable seems missing. It is simple but the data set size gets reduced.

      Pair-wise deletion: In pair-wise deletion: It uses analysis of different cases before deletion of any observation, thus the size varies for different cases.

   b. **Mean/mode/median imputation:** This involves filling the missing values with the estimated ones. The three statistical parameters, i.e., mean, mode, and median denoting central tendency are a perfect example of this. The parameter is decided based on the data set observations.

   c. **Prediction model:** It is a refined method for treating missing data. This involves estimating the missing values via building a model that suits to the datatype and the predictor variables. The missing values are derived from the present values via techniques such as regression, ANOVA, logistic regression, etc. The disadvantage of this method includes the truth that the predicted values are sometimes more accurate and suited than the original missing value.

   d. **KNN imputation:** The k-nearest neighbor method involves prediction via a distance factor such as Euclidean distance that is the value of k. Its more nearest neighbors resemble the values in the model and are more closely related.

   Some of its advantages include:

   - It can predict both quantitative and qualitative attributes.
   - It is a robust method for predicting missing values.
   - It takes into account the correlation factor and model is easy to build.

   Some of its disadvantages include:

   - It is time-consuming when working with large databases
   - Selection of k-value is a critical task. It shows deviation at both higher and lower values of k. Hence, it must be decided suitably.

5. *Outlier treatment*: An outlier is an observation that can deviated from the general pattern in the data set. Outliers makes the data skewed and decreases accurateness. They can be univariate or multivariate depending upon the one and n-dimensional space respectively.

   The different causes of outliers include:

   - Data entry errors
   - Measurement error
   - Experimental error
   - Data processing error
   - Sampling error

Boxplots, histograms, and scatter plots are an easy way to detect outliers. We can remove outliers by following methods:

- Deleting observations
- Transforming and binning values
- Assigning values via mean, mode, and median based on the data set
- Treat outliers separately in the model

6. *Feature engineering:* It is the science of extracting more new information from the data set without adding any additional data. It involves two steps of variable creation (via derived variables and dummy variables) and transformation to make it more systematic.

4.9 Data Visualization

Data visualization is defined as producing graphs, charts, plots, or any graphical representation of data for better understanding and to extract meaningful information. It is a comprehensive way to analyze data and get statistics from the complete data set without reading the data. Patterns or trends are easily detectable using a visualization software that might get missed in text-based data. The more sophisticated software allow more complex visualizations or even with options such as demographics, geographic distribution, heat maps, dials, and gauges, etc.

Some of its advantages are:

- It can deal with very large databases containing nonhomogeneous and noisy data too.
- It requires no knowledge of statistics or algorithms for understanding.
- It provides a qualitative view for quantitative analysis.
- It can identify factors and critical areas for improvement.
- It is a comprehensive way to show results, recent trends, and to communicate to others too.

Some of the data visualization techniques are shown in Figure 4.14 and different visualization samples in Figure 4.15.

Data visualization is also used in business intelligence to get insights of the business model and other analytics.

- **Geometric techniques:** scatterplots matrices, Hyperslice, parallel coordinates

- **Pixel-oriented techniques:** simple line-by-line, spiral and circle segments

- **Hierarchical techniques:** Treemap, cone trees

- **Graph-based techniques:** 2D and 3D graph

- **Distortion techniques:** hyperbolic tree, fisheye view, perspective wall

- **User interaction:** brushing, linking, dynamic projections and rotations, dynamic queries

Figure 4.14   Different techniques used for data visualization.

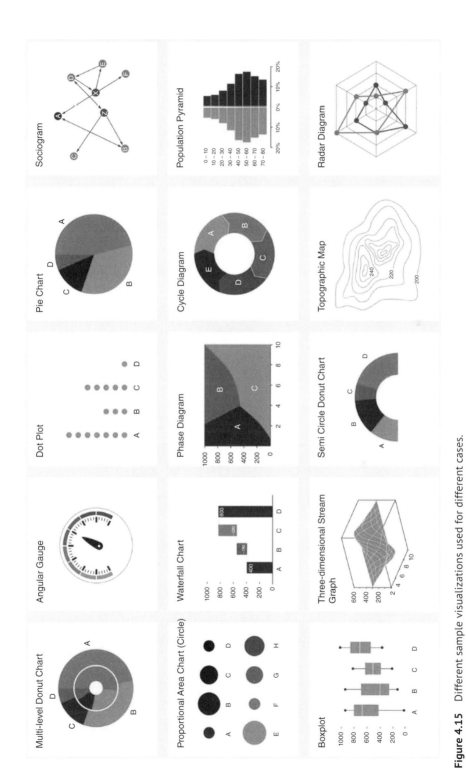

Figure 4.15 Different sample visualizations used for different cases.

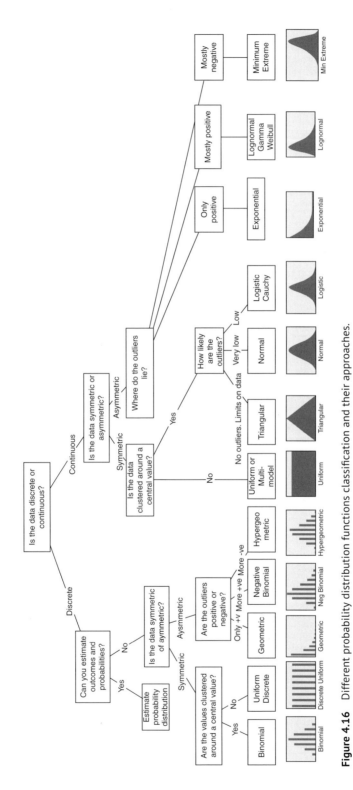

Figure 4.16 Different probability distribution functions classification and their approaches.

A data visualization can be in the form of a dashboard and can track live data from a source, showing variations using a metric like progress rate, performance monitoring, alert systems, etc. Tableau and Qlik are the major market vendors today in data visualization domain.

It is a useful environment for data engineers and scientists in the data exploratory process for detailed analysis.

4.10 Probability Concepts for Intelligent Data Analysis (IDA)

Probability is the chance of occurrence of an event. In similar cases to a data set, improbability, and uncertainty occur in many aspects in different scenarios. Thus, learning about probability help make firm decisions on what is likely to happen, based on an arrangement of data collected previously or by estimation. In intelligent data analysis, probability distributions are an important parameters to study variations and deviation in the trend.

Conditional probability: It is the measure of probability of an event when one event has already occurred. The formula is shown as below:

$$P(B \mid A) = \frac{P(A \text{ and } B)}{P(A)}$$

"It states the probability of occurrence of event B given event A equals the probability of occurrence of event A and event B divided by the probability of occurrence of event A."

This is used as the basis of naïve Bayes algorithm based on Bayes theorem as:

$$P(A \mid B) = \frac{P(B \mid A)P(A)}{P(B)}$$

"It states that how often event A occurs if given that event B occurs, written as P(A|B). We are given the happening probabilities of event B when event A had happened written as P(B|A) and occurrences of event A and B alone written as P(A) and P(B) respectively."

Probability distribution function: The *probability distribution* defined for a random variable is the distribution of probabilities over the random variables.

A random variable is a variable whose value is unknown that assigns value to the outcome of an experiment.

- In case of a *discrete random variable*, x, the probability distribution is defined by a **probability mass function**, and is denoted by $f(x)$. It gives information about the probability for each value of the random variable.
- In case of a *continuous random variable*, due to presence of an infinite number of values in the given interval, so the probability that a continuous random variable will lie inside a given interval is measured. So here, the probability distribution is defined by **probability density function**, and is also denoted by f(x).

The Figure 4.16 shows the difference probability functions with their approach.

Reference

1 Larose, D.T. (2005). *Discovering Knowledge In Data*, xi. Hoboken, NJ: Wiley.

5

Intelligent Data Analysis: Deep Learning and Visualization

Than D. Le[1] and Huy V. Pham[2]

[1]*University of Bordeaux, Labri, Bordeaux, France*
[2]*Ton Duc Thang University, Faculty of Information Technology, Ho Chi Minh, Vietnam*

5.1 Introduction

Deep learning [1] and deep reinforcement learning [2] are currently the resolution in applied artificial intelligence for many applications based on upgrading excellent performances. Basically, deep learning [1] is the best model for data representation by understanding and reinforcement learning [2] is a modern approach to solving the decision making. These are essential to represent the basic things forming the intelligent autonomous systems [3] that can be enabled to solve the basic level based on simultaneous localization and mapping (SLAM), which is based on interacting with the unknown environment. Deep learning is sometimes called hierarchical learning or deep structured learning. The history of deep learning and neural networks is not new. Indeed, the first mathematical model of neural networks was introduced by Walter Pitts and Warren McCulloch in 1943 [4]. However, it grew up just a few years ago by upgrading to graphic processing units (GPUs), which are increasing more opportunities for many applications in artificial intelligence. Figure 5.1 shows us the wide range of deep learning.

Let's look at the progression of fields of deep learning. In machine learning, there commonly exist three types of learning: supervised learning, unsupervised learning, and reinforcement learning. Most of the recent successes in applications are computer vision and image processing, language and speech processing, which includes natural language processing and speech recognition, and robotics. In the past decades, simple machine learning algorithms pointed directly to supervised learning. Supervised learning algorithms [5] are a mapping function from example data sets, where they will be labeled as targets of features, and data sets will be given large examples to classify, e.g., learning tasks learn to predict the output. There are no different previous decades which were tried for extracting the feature mapping, it is necessary to visualization in preprocessing data, called big data. And another challenge is how to visualize the network model, called tools to understand vectorization or deep learning frameworks.

In this article, the first section will cover deep learning and visualization in context. Next, the information of data processing and visualization is addressed to understand specific

Intelligent Data Analysis: From Data Gathering to Data Comprehension,
First Edition. Edited by Deepak Gupta, Siddhartha Bhattacharyya, Ashish Khanna, and Kalpna Sagar.
© 2020 John Wiley & Sons Ltd. Published 2020 by John Wiley & Sons Ltd.

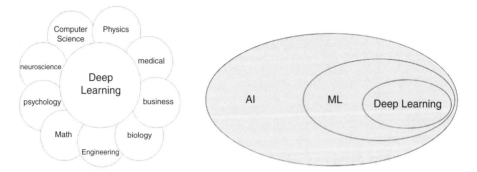

Figure 5.1   Left: overview of neural network and deep learning; Right: branch of deep learning in AI, artificial intelligence; ML, machine learning [1].

methods. Finally, several experiments are fully conducted with relevant applications in the real world, including data visualizations.

5.2   Deep Learning and Visualization

This chapter provides the basic virtualization concepts and system of mathematical notation, which is the first step for any foundation of intelligent data analysis (IDA). The content of this chapter is partially extracted from [1, 2, 6] where further understanding can be obtained.

5.2.1   Linear and Logistic Regression and Visualization

Let's look at an example in Figure 5.2a as a data-driven approach. Given (x_i, y_i) as the input and W as a weight accordingly, and score function can be defined by $f(x_i, W)$, and L as a loss function

The equation representing the linear model can be written as Figure 5.2b and Figure 5.3 by:

$$\hat{y} = \{(x_i, y_i)\}_{i=1}^{N} = x_i * w + b \tag{5.1}$$

in short, given x_i is the image input, and y_i is the integer label, it writes $\hat{y} = x * w$ and the loss function will be defined by summing of loss over examples:

$$loss = \frac{1}{N}\sum_i (\hat{y}_i - y_i)^2 = \frac{1}{N}\sum_i (x_i * w - y_i)^2. \tag{5.2}$$

(a) (b)

Figure 5.2   (a) Overview of visualization: score function, data loss, and regularization and (b) data-driven approach in linear model.

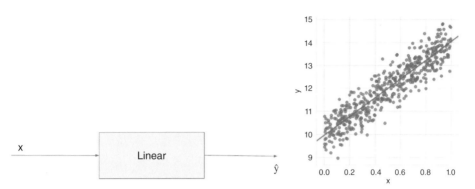

Figure 5.3 Linear model and sample data visualization: left: a simple linear module; right: visualization based on linear module.

And based on the scores vector, $s = f(x_i, W)$, we can write the support vector machine (SVM) loss as formulation:

$$L_i = \sum_{j \neq y_i} \begin{cases} 0 & \text{if } s_{y_i} \geq s_j + 1 \\ s_j & \text{otherwise} \end{cases}$$

$$= \sum_{j \neq y_i} \max(0, s_j - s_{y_i} + 1), \tag{5.3}$$

the average loss can be defined by:

$$L_i = \frac{1}{N} \sum_{j \neq y_i} L_i. \tag{5.4}$$

If we use the mean, then it will become:

$$L_i = \frac{1}{N} \sum_{j \neq y_i} \max(0, s_j - s_{y_i} + 1)^2. \tag{5.5}$$

From (5.3)

$$L = \frac{1}{N} \sum_{i=1}^{N} \sum_{j \neq y_i} \max(0, f(x_i; W)_j - f(x_i; W)_{y_i} + 1). \tag{5.6}$$

Regularization forms:

$$L(W) = \frac{1}{N} \sum_{i=1}^{N} (f(x_i, W), y_i) + \lambda R(W) \tag{5.7}$$

where λ is the regularization strength (hyperparameter). Data loss is the first part of Eq. (5.7) and it models predictions by matching the training data. The second part represents the regularization based on the training data, and is very useful on practices.

According to $f = f(g)$ and $g = g(x)$, chain rule:

$$\frac{df}{dx} = \frac{df}{dg} \frac{dg}{dx}. \tag{5.8}$$

In visualization context, the function (5.8) is used to define the feed forward and back propagation in neural networks. Especially, optimization is one of the most key research in deep learning. For instance, Figure 5.4 illustrates the loss function:

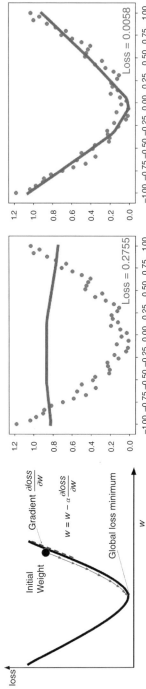

Figure 5.4 Gradient descent is the excellent to visualization in deep learning. left: describe the gradient descent concepts. center, right: Examples: Selected initial and final weights regarding the gradient descent.

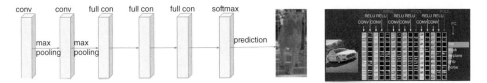

Figure 5.5 left: Design the model with simplify blocks regarding dog detection; right: Example shows interests in visualization feature layers [Fei-Fei Li].

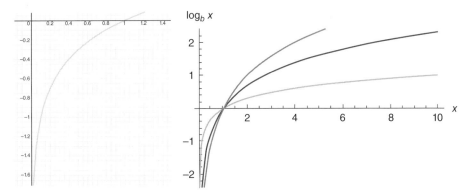

Figure 5.6 The loss of entropy.

There are three steps for defining the linear regression:

- Firstly, design the model using the class, shown in Figure 5.5.
- Next, it is constructed as the loss and optimizer function. In particles, it usually selects from API.
- Finally, it is often called training cycle by illustrating the forward, backward, and update.

Cross-entropy loss from Figure 5.6 shows the loss function based on entropy function:

$$\sigma(z) = \frac{1}{1 + e^{-z}} \tag{5.9}$$

From Eq. (5.2), we can rewrite the Eq. (5.1) as:

$$\hat{y} = \sigma(x * w + b). \tag{5.10}$$

Therefore, the loss function will become:

$$loss = -\frac{1}{N} \sum_{n=1}^{1} y_n \log \hat{y}_n + (1 - y_n) \log(1 - \hat{y}_n). \tag{5.11}$$

5.2.2 CNN Architecture

In $x \in \mathfrak{R}^{N \times 2}$, $w \in \mathfrak{R}^{2 \times 1}$, and $y \in \mathfrak{R}^{N \times 1}$, we can define a linear function as equation:

$$\begin{bmatrix} a_1 & b_1 \\ a_2 & b_2 \\ \vdots & \vdots \\ a_n & b_n \end{bmatrix} \begin{bmatrix} w_1 \\ w_2 \end{bmatrix} = \begin{bmatrix} w_1 \\ w_2 \\ \vdots \\ w_2 \end{bmatrix}. \tag{5.12}$$

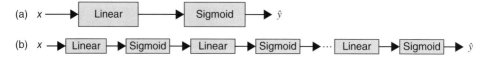

Figure 5.7 (a) Matrix multiplication for deep learning using linear model. (b) Visualized the linear function using sigmoid function extending the deep integration.

From Figure 5.7a, we can show the Eq. (5.2):

$$XW = \hat{Y}. \tag{5.13}$$

For this equation, the input defaults to $X = x_1, x_2, \ldots, x_n$. It can extend for the simplified structure, the deeper in-network structure, by considering adding more pairs of both linear and sigmoid-like functions (see Figure 5.7b). Similarly, we can also extend to a wider array of inputs and add more rows in the network; as a result, we will have a deep network.

5.2.2.1 Vanishing Gradient Problem

This is a problem when we train our data sets during training in practices. The performance of accuracy (loss or object detection and segmentation) using define by the equation chain rules: If there is a decomposition reduce to zero our parameters of networks will be learned nothings. In other hand, the network will be stable in case turning of regularization term.

Function	Formalization	Derivative	Plot
Binary Step	$f(x) = \begin{cases} 0, & \text{for } x \le 0 \\ x, & \text{for } x > 0 \end{cases}$	$f'(x) = \begin{cases} 0, & \text{for } x \ne 0 \\ ?, & \text{for } x \ge 0 \end{cases}$	
Piecewise Linear	$f(x) = \begin{cases} 0, & \text{for } x < x_{min} \\ wx + b, & \text{for } x_{min} \ge x \le x_{max} \\ 1, & \text{for } x > x_{max} \end{cases}$	$f'(x) = \begin{cases} 0, & \text{for } x < x_{min} \\ w, & \text{for } x_{min} \le x \le x_{max} \\ 0, & \text{for } x > x_{max} \end{cases}$	
Bipolar	$f(x) = \begin{cases} -1, & \text{for } x \le 0 \\ 1, & \text{for } x > 0 \end{cases}$	$f'(x) = \begin{cases} 0, & \text{for } x \ne 0 \\ ?, & \text{for } x = 0 \end{cases}$	
Sigmoid	$f(x) = \dfrac{1}{1 + e^{-x}}$	$f'(x) = f(x)(1 - f(x))$	
Bipolar Sigmoid	$f(x) = \dfrac{1 - e^{-x}}{1 + e^{-x}}$	$f'(x) = \dfrac{2e^x}{(e^x + 1)}$	

Function	Formalization	Derivative	Plot
Hyperbolic Tangent, TanH	$f(x) = \dfrac{2}{1 + e^{-2x}} - 1$	$f'(x) = 1 - f(x)^2$	
Arctangent, ArcTan	$f(x) = tan^{-1}(x)$	$f'(x) = \dfrac{1}{1 + x^2}$	
Rectified Linear Units, Leaky ReLU	$f(x) = \begin{cases} 0, & \text{for } x \leq 0 \\ x, & \text{for } x > 0 \end{cases}$	$f'(x) = \begin{cases} 0, & \text{for } x \leq 0 \\ 1, & \text{for } x > 0 \end{cases}$	
Leaky Rectified Linear Units, Leaky ReLU	$f(x) = \begin{cases} ax, & \text{for } x \leq 0 \\ x, & \text{for } x > 0 \end{cases}$	$f'(x) = \begin{cases} a, & \text{for } x \leq 0 \\ 1, & \text{for } x > 0 \end{cases}$	
Exponential Linear Units, ELU	$f(x) = \begin{cases} a(e^x - 1), & \text{for } x \leq 0 \\ x, & \text{for } x > 0 \end{cases}$	$f'(x) = \begin{cases} f(x) + a, & \text{for } x \leq 0 \\ 1, & \text{for } x > 0 \end{cases}$	
Softplus	$f(x) = ln(1 + e^x)$	$f'(x) = \dfrac{1}{1 + e^{-x}}$	

Optimizer: In Figure 5.8, it shows that Adam (adaptive learning rate optimization) [7] is one of the best performances.

5.2.2.2 Convolutional Neural Networks (CNNs)

Convolutional neural networks (CNNs) is the key idea to weight sharing with the large-scale data in the input layer. To visualize the CNN, there are generally three layers, involving input, feature extraction, and classification, as shown in Figure 5.9. Input layer represents the size image like 32×32. Then, the convolution layer is to define three components such as depth, width, and height. More information, convolution in action [8], illustrates the convolution animations with excellent demonstrations.

5.2.3 Reinforcement Learning

Let's consider that the Q-learning [9] (Figure 5.10) is model-free reinforcement learning and will be defined as action values:

$$Q(S_t, A_t) \leftarrow Q(S_t, A_t) + \alpha(R_{t+1} + \gamma \max_a Q(S_{t+1}, a) - Q(S_t, A_t)), \tag{5.14}$$

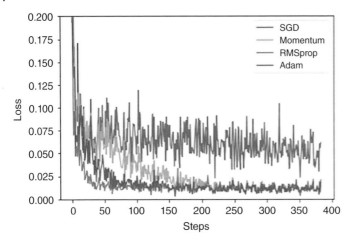

Figure 5.8 Optimizer [16]: Adam works and others shows.

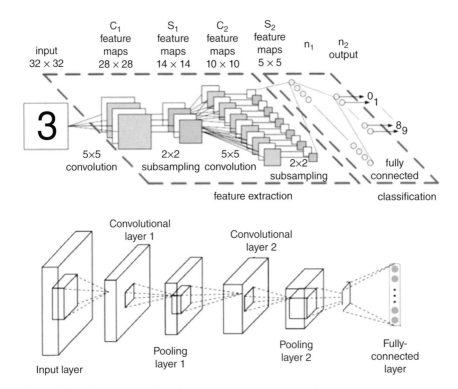

Figure 5.9 Left: example of block box most uses to visualize the complex networks.

where R_{t+1} is a value that an agent receives after taking action A in state S_t. Next, both parameters are also defined, noted *discount rate* γ and *learning rate* $\alpha(0<\alpha<1)$.

Solving for the optimal policy using Q-learning, if given set both γ and α to 1:

$$Q(s, a) \leftarrow R_{t+1} + \max_a Q(S_{t+1}, a) \tag{5.15}$$

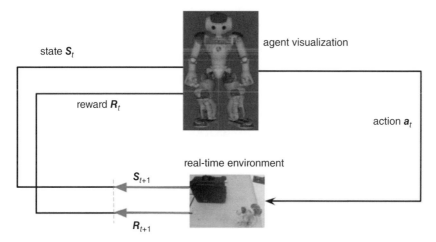

Figure 5.10 Overview of reinforcement learning model [9]: an agent is visualized with the real-time environment based on avoiding the obstacles experience.

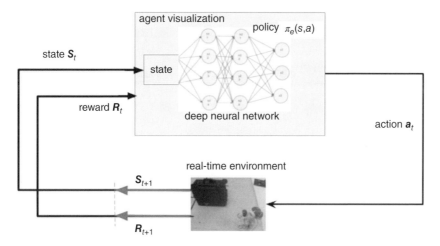

Figure 5.11 Deep reinforcement learning.

In the formulation, the Markov decision process is used to formalize the reinforcement learning problems. Our approach here is to visualize the meaning of Figures 5.11 and 5.12 and how it works in visualization contexts.

5.2.4 Inception and ResNet Networks

It will perform a series of operations on network modules (illustrated in Figures 5.13 and 5.14) to have the much better possible result in predictions based on classification. Each layer has each abstraction by teaching itself learning features in the first layer such as edge detection. Then it can be able to detect the shape detection in the middle layer by increasingly more abstractions until the end. Almost the last few layers there are the highest level detections for whole object classification.

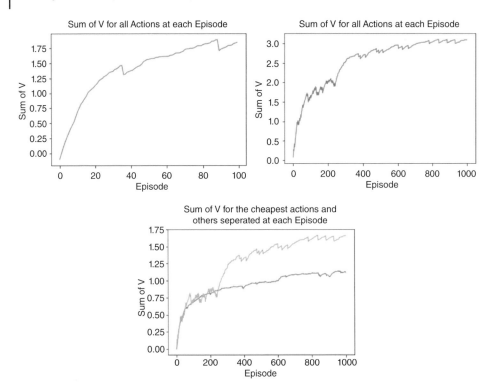

Figure 5.12 Reinforcement learning and visualization.

For transfer learning, it is basically to retrain that last layer on feature representation as objects.

5.2.5 Softmax

The formalization equation of Softmax:

$$\sigma(z)_j = \frac{e^{z_j}}{\sum_{k=1}^{K} e^{z_k}},$$ (5.16)

where $j = 1, \ldots, K$.

The course function can be defined by using cross-entropy (Figure 5.15). Firstly, it is needs to define L is loss, and (x_i, y_i) is the training set. Probably, the loss can define:

$$L = \frac{1}{N} \sum_i D(s(Wx_i + b), y_i),$$ (5.17)

and

$$D(\widehat{Y}, Y) = -Y \log \widehat{Y}.$$ (5.18)

Weight histogram is very useful for representing the debugging and visualization. For instance, the interpretation of distribution is to understand the training parameters such as weights, biases, loss of matrices, or statics parameters; we need to visualize the weights based on layers representation. Figure 5.16 is an example of this.

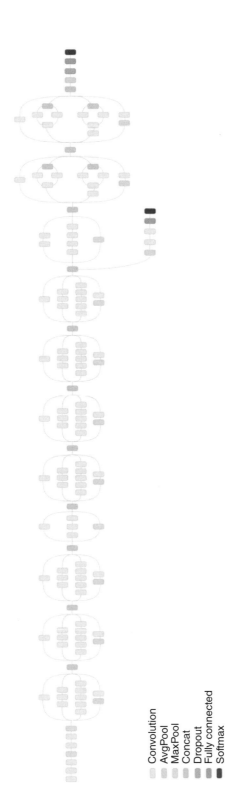

Figure 5.13 Inception v3 module: it was the powerful for visualizing the deep network model.

Convolution
AvgPool
MaxPool
Concat
Dropout
Fully connected
Softmax

Figure 5.14 GoogLeNet architecture [12].

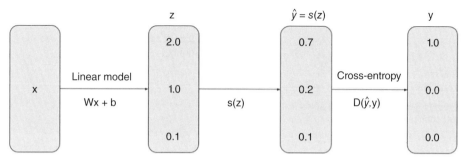

Figure 5.15 x: input, z: logit, $\hat{y} = s(z)$: softmax, y: 1-hot labels;

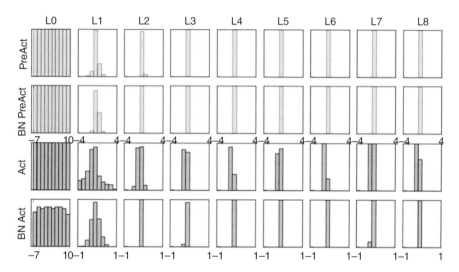

Figure 5.16 Example of interpretation of histogram distribution [Morvan].

5.3 Data Processing and Visualization

These are some of the difficulties of understanding the deep learning work. Activation visualization is currently using exciting methods to understand how the deep networks are performing well, shown in Figure 5.17.

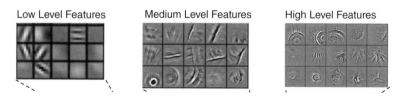

Figure 5.17 Illustrated the multiple layers features in representation [medium].

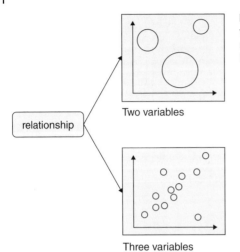

Figure 5.18 Relationship visualizations: two variables using the scatter diagram and three or more variables, it can visualize by using the bubble diagram.

Two variables

relationship

Three variables

There are three types of data visualization:

- Relationships are showing the correlation or connections comparing two or more variables. In Figure 5.18, it shows two separate cases for representing the relationships. While we use the scatter plot, which is to plot the Cartesian coordinates for data sets in two variables, more variables (greater or equal than three) are efficiently used to add to the bubble plot.
  However, there is one way to visualize the efficiency for many variables and we can use the correlation heap map, which we can show on one chart. We will show it later when sharing our experiences using deep learning and object segmentation in the Mask-RCNN framework.
- Comparison: this is a set of comparison that display the one or more variables, which is compare how the variables interact with each other (shown in Figure 5.19).
- Composition: it will collect the difference types of data and making (illustrated in Figure 5.20)

MNIST data sets is one of the most popular data sets based on using machine learning and deep learning and it is visualized as the Figures 5.21, 5.22, and 5.23.

The embedding of image or numbers is used in MNIST in three dimensions.

5.3.1 Regularization for Deep Learning and Visualization

5.3.1.1 Regularization for Linear Regression
To prevent the overfitting, the regularization is used. For example, there are single modules used listed as L1, L2, and elastic network regularization, while drop out [6], batch normalization, stochastic depth, and factional fooling as they are implemented with more complex levels. The question here is how to automatically choose the regularization parameter to have good performance. Let's look at model:

$$h_0(x) = \theta_x + \theta_1 x + \theta_2 x^2 + \ldots + \theta_n x^n \tag{5.19}$$

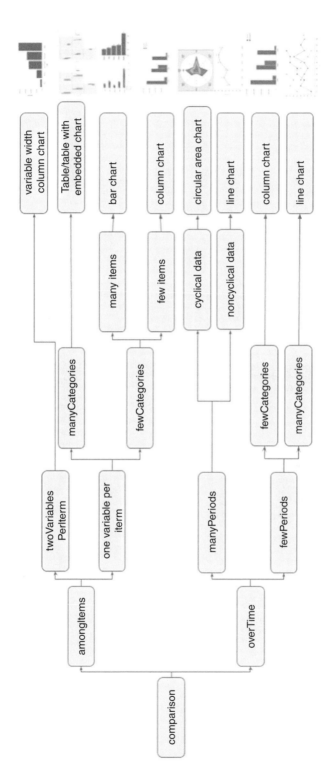

Figure 5.19 Comparison method: overview of charts is represented the most commons on data visualization.

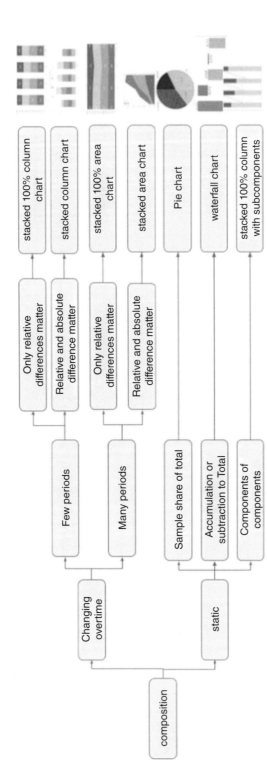

Figure 5.20 Composition methodology: overview of charts is represented most commons on data visualization.

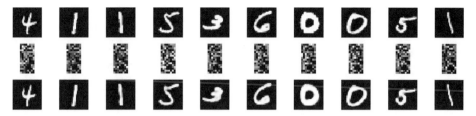

Figure 5.21 Example of visualization applied MNIST data set by using deep learning that is unsupervised learning algorithm: auto encoder.

Figure 5.22 MNIST visualization.

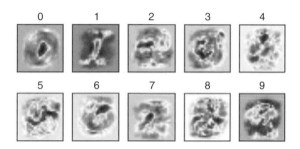

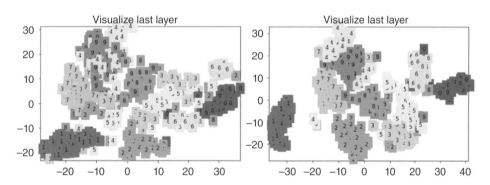

Figure 5.23 Example of visualization using MNIST in 3D.

and

$$J(\theta) = \frac{1}{2m} \sum_{i=1}^{m} (h_\theta(x^{(i)}) - y^{(i)})^2 + \frac{\lambda}{2m} \sum_{i=1}^{m} \theta_j^2 \tag{5.20}$$

Usually, for training,

$$J_{train}(\theta) = \frac{1}{2m} (h_\theta(x^{(i)}) - y^{(i)})^2. \tag{5.21}$$

For cross-validation,

$$J_{cross-validation}(\theta) = \frac{1}{2m_{cross-validation}} (h_\theta(x^{(i)}_{cross-validation}) - y^{(i)}_{cross-validation})^2 \tag{5.22}$$

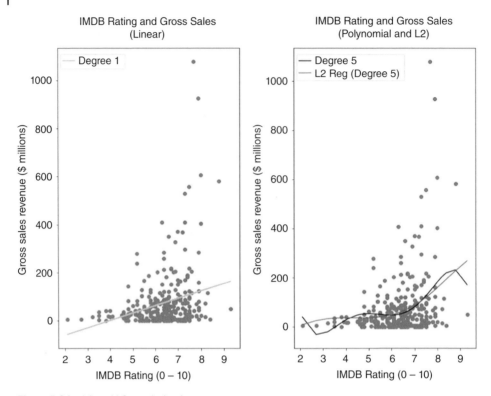

Figure 5.24 L1 and L2 regularization.

Testing

$$J_{test}(\theta) = \frac{1}{2m_{test}}\left(h_\theta\left(x_{test}^{(i)}\right) - y_{test}^{(i)}\right)^2. \tag{5.23}$$

It is easy to understand the simplified model based on using visualization techniques pictured in Figures 5.24 and 5.25 as the dropout. There is a complexity method of batch normalization in deep network.

5.4 Experiments and Results

5.4.1 Mask RCNN Based on Object Detection and Segmentation

In this section, we focus on describing Mask R-CNN [10] architecture and how it performs object segmentation. Prior to that, we also discuss Faster R-CNN, a neural network object detector from which Mask R-CNN stems. Objects in general can be things in various shapes, depending on what type of data set is fed to train the neural network model.

Firstly, Faster R-CNN is a deep neural network designed for multi-class object detection and introduced by Chollet, F. [11]. It consists of two main modules: a region proposal network (RPN) followed by the Fast R-CNN [21]. The region proposal module is a convolutional network fed with an image from which it extracts features and return locations

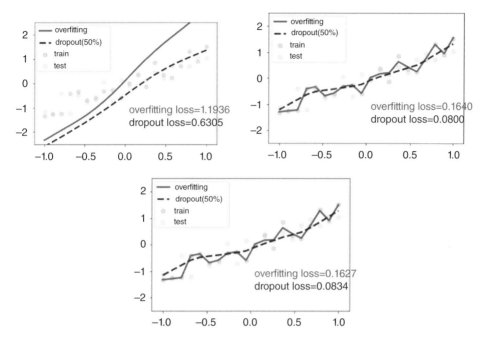

Figure 5.25 Dropout processing and visualization: sampling dropout loss base on overfitting problems.

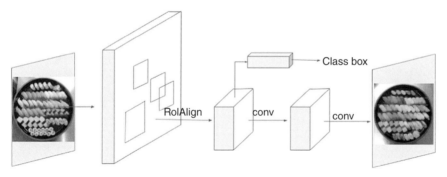

Figure 5.26 Mask-RCNN for object detection and segmentation [21].

where the object lies. These areas will be further analyzed by the Fast R-CNN detector to determine object type (classification) and to adjust rectangular bounding boxes (regression) to better fit the object shape. The system loss function L is a combined loss of classification $\mathcal{L}_{class}$ and regression $\mathcal{L}_{bbox}$:

$$\mathcal{L} = \mathcal{L}_{class} + \mathcal{L}_{bbox}. \tag{5.24}$$

Thanks to the share of convolutional feature map at classification, regression, and RPN stage, the Faster R-CNN [4] is faster than Fast R-CNN and therefore, it requires less computational effort.

In this section we consider the Mask-RCNN [10] framework for object detection and segmentation (Figure 5.26).

Mask R-CNN [10] is extended from Faster R-CNN. Besides the class label and the bounding box offset, the Mask RCNN is able to detect the shape of objects, called object masks.

This information is useful for designing high-precision robotic systems, especially autonomous robotics grasping and manipulation applications. The general loss function L considers the mask loss L mask $\mathcal{L}_{mask}$:

$$\mathcal{L} = \mathcal{L}_{class} + \mathcal{L}_{bbox} + \mathcal{L}_{mask}. \tag{5.25}$$

Additionally, the Mask R-CNN can achieve a high pixel level accuracy by replacing RoIPool [14–16] with RoIAlign. The RoIAlign is an operation for extracting a small feature map while aligning the extracted features with the input by using bilinear interpolation. Readers may refer to the paper [6] for further details. To train the detector, we reuse a Mask R-CNN implementation available at [14–16].

Let's look at the chart in Figure 5.27 below carefully. It can show what a chart says explicitly. Basically, there are three concepts of the data visualizations based on the large motivations like art science, including relationship, comparison, and composition.

Relationship: it shows the connections or correlation between two or more variables through data presented.

One of the useful relationships to understand and visualize our efficient networks during training is using graph modeling. Figure 5.27 shows the training process it will be useful for understanding which one be the best, and when we need to stop training.

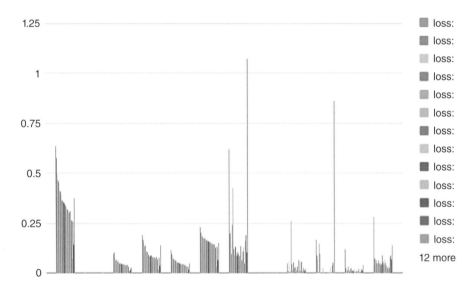

Figure 5.27 Mask-RCCN result progress: training with Mask-RCNN according to variety loss values.

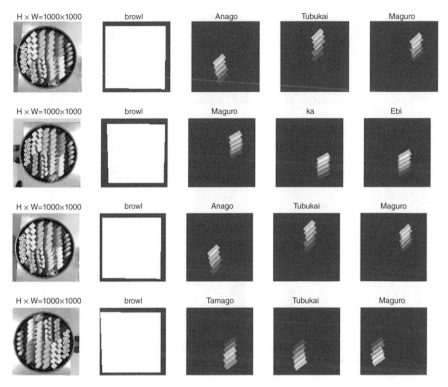

Figure 5.28 Deep learning and object visualization based on sampling during training set using Mask-RCNN.

According to Figure 5.27, it represents the many parameters that is easy to understand what is happening our model. In this case, each color is shown as different loss values based on decreasing or increasing in progress and it is very useful in particles such as a list of order here: total loss, rpn_class_loss, rpn_bbox_loss, mrcnn_class_loss, value_loss, val_rpn_class_loss, val_mrcnn_class_loss, val_mrcnn_bbox_loss, and val_mrcnn_mask_loss.

One of the experiences we are using to manage the convergence based on using a graph shown in Figures 5.28–5.30:

Sampling for human detection based on using two classes, involved faces, and human bodies.

We now discuss the visualization of activation function. In this case, we relay our experiences with our own data sets. Specifically, food detection and segmentations using more than 30 000 objects and 7000 images based on creating the 20 classes. From Figure 5.31–5.36, it is used to visualize the Mask-RCNN segmentation and detection in image processing.

This is the correlation heat map that show relationships of many variables in an all-in-one chart.

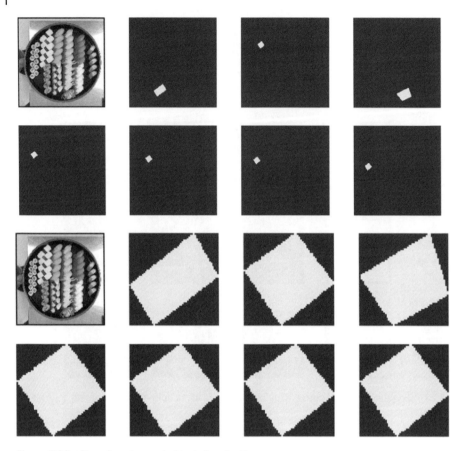

Figure 5.29   Deep learning and object visualization.

Figure 5.30   Human detection using Mask RCNN: noised data during the human detection and segmentation.

Figure 5.31 Showing the activation function of layers based on food recognition and segmentation using the Mask-RCNN.

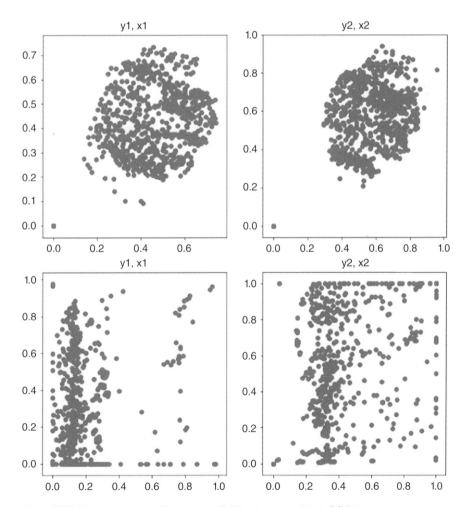

Figure 5.32 Interpretation of histogram distribution using Mask-RCNN.

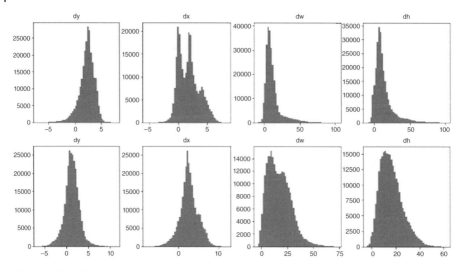

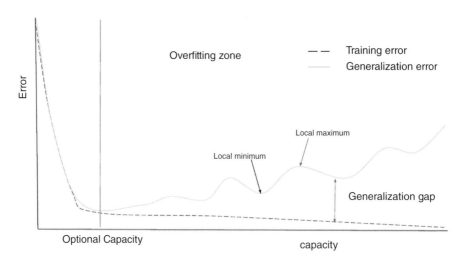

Figure 5.32 (*Continued*)

Figure 5.33 Overfitting representation based on experience from Mask-RCNN [21].

5.4.2 Deep Matrix Factorization

5.4.2.1 Network Visualization

In our experiments, we show our deep matrix factorization for recommendation systems. It is used to construct the relationship between users and items (Figure 5.37 for matrix factorization using deep learning).

Data set visualization is loading based on the table below:

	N. of users	N. of item	N. of rating	N. of density
MovieLens 100 K	944	1683	100 000	0.062 94

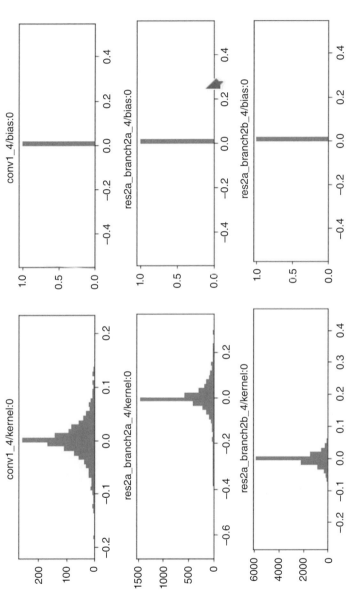

Figure 5.34 Weights histogram based on distributed parameters of training sets.

Figure 5.35   Correlations.

Figure 5.36   Visualization of food recognition.

For comparison between the loss function and validation function, we can also easily understand our model based on visualizing the plot function.

For deeper visualization, there are figures listed here (Figures 5.38 and 5.39) that illustrate the loss model and rating density of the data set [17].

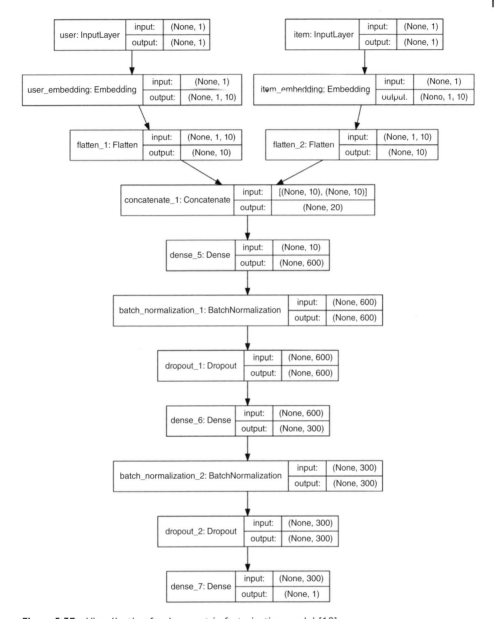

Figure 5.37 Visualization for deep matrix factorization model [18].

5.4.3 Deep Learning and Reinforcement Learning

The line in charts is really easy to visualize with many categories that change over time. It is all at once and precisely colored accordingly to what they tell us a story of effort.

It is an easy way to compare all at once as shown in Figure 5.40 to understanding the reinforcement learning.

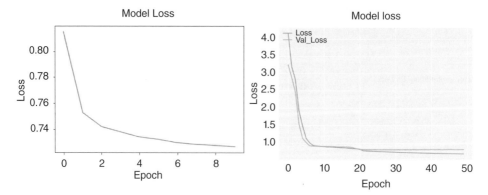

Figure 5.38 Visualization and loss function in deep learning for recommendation system.

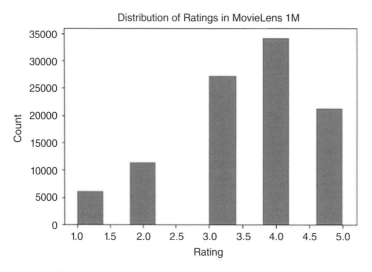

Figure 5.39 Data visualization in MovieLens 1 M of recommendation system based on deep matrix factorization module.

5.5 Conclusion

In summary, we understand the principles of visualization problems in probabilistic approaches based on using deep learning technologies. We also show the modeling by representing the state of the art not only in object detection and segmentation but also we study the methods of data visualization in recommendation systems. There are also three features used in data visualization including correlation, relationship, and composition. In the future, we integrated the methodologies to design for any users, such as a software engineer, who will be able to design the simple or complex networks.

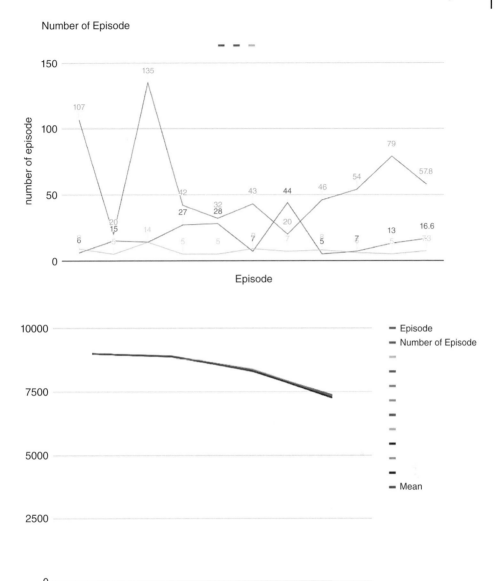

Figure 5.40 Line in charts, and modeling and visualization for reinforcement learning [19].

References

1 Goodfellow, I., Bengio, Y., and Courville, A. (2016). *Deep Learning*. MIT Press.

2 Mnih, V., Kavukcuoglu, K., Silver, D. et al. (2015). *Human-level control through deep reinforcement learning. Nature* https://doi.org/10.1038/nature14236.

3 Ross Girshick, 'Fast R-CNN: Towards Real-Time Object Detection with Region Proposal Networks', International Conference on Computer Vision (ICCV), IEEE, 2015.

4 Waleed Abdulla, 'Mask R-CNN for object detection and instance segmentation on Keras and TensorFlow', Github, GitHub repository, 2017.

5 Le, A.T., Bui, M.Q., Le, T.D., and Peter, N. (2017). D*life with reset: improved version of * Lite for complex environment. In: *First IEEE International Conference Robotic Computing(IRC); Taichung, Taiwan*, 160–163. IEEE https://doi.org/10.1109/IRC.2017.52.

6 Srivastava, N., Hinton, G., Krizhevsky, A. et al. (2014). *Dropout: a simple way to prevent neural networks from overfitting. The Journal of Machine Learning Research*: 1929–1958.

7 Vincent Dumoulin and Francesco Visin. A guide to convolution arithmetic for deep learning. 2016. arXiv:1603.07285.

8 Olga Russakovsky, Jia Deng, Hao Su, Jonathan Krause, Sanjeev Satheesh, Sean Ma, Zhiheng Huang, Andrej Karpathy, Aditya Khosla, Michael Bernstein, Alexander C. Berg and Li Fei-Fei. ImageNet Large Scale Visual Recognition Challenge. IJCV, 2015.

9 Ashia C. Wilson, Rebecca Roelofs, Mitchell Stern, Nathan Srebro, and Benjamin Recht, The Marginal Value of Adaptive Gradient Methods in Machine Learning, arXiv, 2018.

10 Le, T.D., Huynh, D.T., and Pham, H.V. (2018). Efficient human-robot interaction using deep learning with mask r-cnn: detection, recognition, tracking and segmentation. In: *2018 15th International Conference on Control, Automation, Robotics and Vision (ICARCV)*, 162–167.

11 Chollet, F. (2017). Xception: deep learning with depthwise separable convolutions. In: *2017 IEEE Conference on Computer Vision and Pattern Recognition (CVPR)*, 1800–1807.

12 Doan, K.N., Le, A.T., Le, T.D., and Peter, N. (2016). Swarm robots communication and cooperation in motion planning. In: *Mechatronics and Robotics Engineering for Advanced and Intelligent Manufacturing* (eds. D. Zhang and B. Wei), 191–205. Cham: Springer https://doi.org/10.1007/978-3-319-33581-0_15.

13 Than D. Le, Duy T. Bui Pham, VanHuy Pham, Encoded Communication Based on Sonar and Ultrasonic Sensor in Motion Planning, IEEE Sensors, 2018. DOI: https://doi.org/10.1109/ICSENS.2018.8589706

14 Alex Krizhevsky, Ilya Sutskever, Geoffrey E Hinton, Imagenet classification with deep convolutional neural networks 2012.

15 Alex Krizhevsky, 'One weird trick for parallelizing convolutional neural networks', arxiv, 2014.

16 Diederik P. Kingma and Jimmy Lei Ba. Adam: A method for stochastic optimization. 2014. arXiv: 1412.6980v9

17 Shaoqing Ren, Kaiming He, Ross Girshick, Jian Sun, 'Faster R-CNN: Towards Real-Time Object Detection with Region Proposal Networks,' Advances in Neural Information Processing Systems, 2015.

18 Duc Minh Nguyen, Evaggelia Tsiligianni, Nikos Deligiannis, Matrix Factorization via Deep Learning, 2018. arXiv:1812.01478.

19 Than D. Le ; An T. Le ; Duy T. Nguyen, Model-based Q-learning for humanoid robots, 18th International Conference on Advanced Robotics (ICAR), IEEE Xplore, 2017. DOI: https://doi.org/10.1109/ICAR.2017.8023674

20 Pytorch, www.pytorch.org

21 He, K., Gkioxari, G., Dollr, P., and Girshick, R. (2017). Mask-RCNN. In: *2017 IEEE International Conference on Computer Vision (ICCV)*, 2980–2988.

6

A Systematic Review on the Evolution of Dental Caries Detection Methods and Its Significance in Data Analysis Perspective

Soma Datta[1], Nabendu Chaki[1], and Biswajit Modak[2]

[1]*Department of Computer Science and Engineering, University of Calcutta, Kolkata, India*
[2]*Department of Dental Science, Nabadwip State General Hospital, Nabadwip, West Bengal, India*

6.1 Introduction

Dental caries is the most familiar oral disease. It is a painful bacterial disease caused mainly by Streptococcus mutants, acid, and carbohydrates. If caries remains untreated then it affects the root of the teeth. The World Health Organization (WHO) report [1] reveals that 98% adults and 60–90% [2] of school children are suffering from dental caries. It is an infectious and chronic disease that can affect us at any age. In the near future it will be an epidemic.

6.1.1 Analysis of Dental Caries

The problems behind dental caries include unhealthy oral conditions and negligence of teeth care, like brushing at least twice in a day, wash full mouth after smoking, etc. There are some general symptoms of dental caries like tooth pain, sensitivity due to hot and cold, bleeding from gums, pain during chewing the food, etc. Multiple factors are responsible to increase the growth rate of dental caries. These are teeth condition, saliva, plaque, time, and food habits [3–8].

Teeth condition. Teeth structure differs from man to man. If there is gap in between two teeth, this increases the probability of caries. The reason is that small amount of food gets stuck in this gap and due to acidic reaction, this decays our teeth.

Saliva. Saliva helps to wash the plaque from the teeth. In case of dry mouth, plaque quickly spreads.

Food habits. This plays a major role in creating caries. After eating or drinking, small food material gets stuck in our teeth until we brush. Sometimes, it is also not possible to remove all food material, especially carbohydrates, after brushing. Foods that stick in our teeth increase the risk of dental caries.

Care of teeth. The proper care of teeth could reduce the risk of caries. Everybody has bacteria in their saliva. After eating some food with sugar, this bacterium makes the sugar turn into acid. This acid starts a reaction with the calcium present in our teeth, and as a result

Intelligent Data Analysis: From Data Gathering to Data Comprehension,
First Edition. Edited by Deepak Gupta, Siddhartha Bhattacharyya, Ashish Khanna, and Kalpna Sagar.

our teeth are damaged. Proper hygiene of teeth reduces the risk of caries. Early caries creates a small hole at the enamel layer.

Most of the people in the world suffer from painful dental caries due to the above-mentioned reasons. Dental caries are of three types: enamel caries, dentine caries, and root caries. Enamel is the first layer of the teeth and it is made with calcium. In its early stage of caries, the enamel layer starts to decay. This is called enamel caries. Fillings and good oral hygiene prevent this type of caries spreading. This loss can progress through the dentin layer to the pulp region and ultimately, people will lose their tooth [4, 9]. This condition can be avoided if it is detected in its early stages. In that case, surgical procedures could be skipped. If dental caries is detected at its early stage, the dentist can start the initial treatments to control the progress of caries lesions. In the next phase, infection goes to the dentine layer where the layer starts to decay. This is called dentine caries. If it remains untreated then the root of the teeth becomes affected. The cemento enamel junction is affected mainly by root caries [10]. As a result, the patient feels lots of pain. The solution for this is a root canal or an uprooted tooth. Figure 6.1 shows dental caries at different phases. Figure 6.1a shows the enamel caries, Figure 6.1b shows dentine caries, and Figure 6.1c shows the root caries.

Figure 6.2 shows the worldwide caries affect rate along with its severity in the middle age group of people [11]. This graph is prepared according to the WHO report [2]. According to this report, 11% of the total populations are highly affected (which means all the risk factors) in dental caries, 12% people of the total population are suffering moderately (which means they have pain as they were going through the filling procedure), and 9% of the people in the total population have caries at its early stages, whereas another 7% of people have very low risk dental caries and for the remaining 61%, data are not found.

Figure 6.3 shows the percentage of affected rate for smokers. Tobacco smoking increases the risk factor of dental caries. Tobacco increases the growth of the Streptococcus mutant

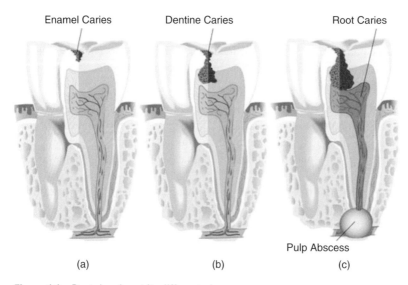

Enamel Caries　　　　Dentine Caries　　　　Root Caries

Pulp Abscess

(a)　　　　　　　　　(b)　　　　　　　　　(c)

Figure 6.1 Dental caries at its different phases.

Figure 6.2 Worldwide dental caries severity regions.

Dental caries Affected Rate in the World as per WHO Report

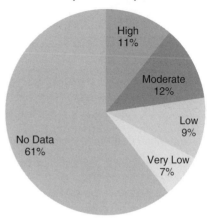

Percentage of affection

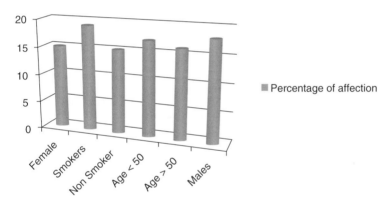

Figure 6.3 The affected risk of dental caries on smoking.

to the dental surface and increases the severity of dental caries. Nicotine is one of the toxins found in tobacco, which forms acid when it dissolves in the water base of saliva. This acidic object affects the calcium layer of teeth. The report in Figure 6.3 shows that the percentage of males affected is quite higher than the female population as the number of female smokers are less. The rate of those newly affected with dental caries is less than those of the age above fifty. Caries detection at its early stage prevents a lot of decay of our teeth. But if it remains untreated, it would become terrible. In the year of 2010, 2.4 billion people and 621 million children were attacked with caries in their permanent teeth due to lack of or ignorance of proper treatment. Figure 6.4 shows the recent data in CAPP (computer aided process planning) in 2014. This represents the global caries map for the children whose age is younger than 12 by average numbers of teeth affected using the decayed, missing teeth and filled teeth (DMFT) index of severity, soon to be called the DMFT metric. This world map clearly shows the severity rate of all the countries. For example, in North America, Australia, China, and some countries in Africa, children have very low affected rate,

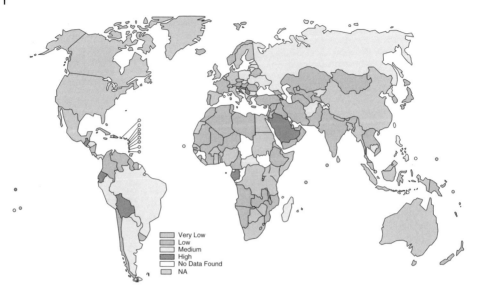

Figure 6.4 Worldwide dental caries affected Level that according to DMFT among 12-year-old children.

Table 6.1 Global DMFT trends for 12-year-old children [81–83].

Year	Global weighted mean DMFT	Number of countries	Countries having DMFT 3 or less	Percentage of countries having DMFT
1980	2.43	—	—	—
1985	2.78	—	—	—
2001	1.74	183	128	70%
2004	1.61	188	139	74%
2011	1.67	189	148	78%
2015	1.86	209	153	73%

whereas Brazil and Russia have more children with an affected rate, and Saudi Arabia and Peru have children with a very high affected rate. Table 6.1 shows the global DMFT trends for 12-year-old children. The data are taken from the year 1980 to 2015. Column 2 refers to the weighted mean DMFT, which equals 2.015. The third column refers to the number of countries worldwide. The fourth column refers to the number of countries having DMFT less than or equal to three. The fifth column represents the percentage of countries having DMFT less than or equal to three.

The major risk behind dental caries is that it increases the probability of chronic disease [12]. For example, dental infections increase the risk of pneumonia, premature birth, complications in diabetes, etc. Early caries detection and diagnosis reduces the risk factors of dental caries. This also reduces the time of patients and doctors, along with the treatment cost. Most of the time, dental caries starts at the occlusal surface of the teeth. The current caries detection techniques like radiographs, as well as tactile and visual inspection, cannot help in detecting caries at its early stage. Sometimes discoloration of the pit region could be misinterpreted as carious lesion [13]. The purpose of new detection methods is to remove the deficiencies of detection. The methods should satisfy the following criteria. These are as follows.

1. It should precisely and accurately detect the caries lesion.
2. It should icude ideal methods thatmonitor the whole progress rate of the caries lesion.
3. It should be easy to access as well as cheap.
4. It should be able to detect the caries lesion at every layer, including caries restoration.

6.2 Different Caries Lesion Detection Methods and Data Characterization

Caries detection methods have been broadly classified [4, 9, 14–18] as point methods, based on visible property of light methods, light emitting methods, OCT, software tools, and radiograph methods. Figure 6.5 shows the subcategorization of these methods. The next subsections explain the details of these methods. This classification involves light and transducer. It is classified into different groups on the basis of implementation technique and data pattern. In the point method, light absorption and secondary emission technique is used to determine the mineral concentration in the tooth. This mineral concentration is different for both caries and the healthy teeth region. This method is suitable for early caries lesion detection. Based on visible property of light is a kind of imaging technique that determines the caries lesion according to the visible light scattered or absorption quantity. It is capable of distinguishing different phases of dental caries evolution.

Radiographs are also used as an imaging technique to detect caries lesion. In this technique, very high frequency light is used for imaging. This high frequency light penetrates throughout the entire teeth. Caries lesion allows for more amounts of penetration of the high frequency light. OCT is a high-resolution imaging technique that uses white light for imaging. White light contains most of the different frequencies of light, so it is capable of capturing the true color of this region. This white light is noninvasive in nature, not like the x-ray image. These images are helpful to reconstruction 3D images. Light-emitting devices

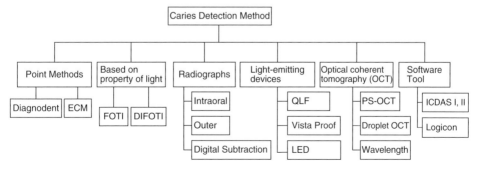

Figure 6.5 Classification of caries detection method.

are used to measure the depth and area of the caries lesion. This emitting technique can be used in periodic intervals to determine the nature of caries lesions as active or inactive. Software tools are kind of a hybrid approach, which include sound data, visual change data, localized enamel breakdown change data, etc., in order to decide the level of the caries lesion. It is a kind of automation of the techniques that are used by experienced dentists to detect the caries lesion.

6.2.1 Point Detection Method

Point detection method is very complex. Here, measurements are taken at different time periods during investigation. This method is of two types: diagnodent and electronic caries monitor (ECM) [9, 17, 19–22]. The diagnodent method is camera dependent and was invented by Kavo Diagnodant. This method recognizes the tooth demineralization area. Here the images are captured from the illuminated teeth using transmitted visible light. It is used to detect and quantify dental caries on smooth and occlusal dental surfaces [10, 23–27]. Both the molecules of the teeth area are able to absorb the light. In the next phase the emitted fluorescence is collected through the tip region. Then it is passed from ascending filter to photo diode section. Here a filtering technique is used to absorb ambient light [28, 29]. A laser diode is used to separate the fluorescent light from the ambient light. The diagnodent [30, 31] provides optimum performance for detecting carious lesion and monitoring its progression. However, it is much more sensitive than other traditional methods. It uses a small laser system that produces a red light of wavelength 655 nm [28, 32] (Figure 6.6).

The tooth fluoresces in the infrared range and the carious lesion properly appears under fluorescence. This instrument gives good performance during the detection of lesions at the dentin region. As it is a single point area measurement, the measurements taken at different time periods is very complex and tough. The main constrains of this method is that the tooth surface needs to clean and dry, otherwise it may produce a false-positive result. Besides that, this method fails to identify the deep dentinal caries [33–35]. It is mainly used as a second opinion where a false-positive result is higher than the visual inspection. This technique takes a few minutes to scan the entire mouth. An electronic caries monitoring method consists of probes and substrates. Probes mean from which the current will flow and substrate means contra electrode region. Here measurements are taken in two ways. It is taken either from enamel or exposed dentine surfaces. In this method, the result comes

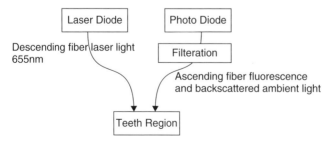

Figure 6.6 Internal diagram of point detection method.

as a number between 1 and 13; higher numbers refer to deeper lesion. The ECM [27, 36] method monitors the suspected carious lesion due to its electrical resistance behavior. It also computes the bulk resistance of tooth tissue. Sound enamel is a poor conductor of electricity and its pore size is not large. Demineralization happens because a carious lesion implies a large size of pore. Measurements can be taken from both enamel and exposed dentin surfaces. Based on the pore size, dentists make the decisions. It is working either in site mode or in surface specific mode. It finds the depth of root caries. ECM shows quantitative data that helps to monitor the lesion growth rate [24, 37, 38].

6.2.2 Visible Light Property Method

The visible light property method works on the basic principle of light spreading. It is divided into fiber optic trans-illumination (FOTI) [14, 39, 40] and digital imaging fiber optic trans-illumination (DIFOTI) [23]. FOTI is a fast, simple, noninvasive, and comparatively inexpensive method. Visual examination could be improved for the entire tooth surface in this method. FOTI works on the principle of light scattered or light absorption quantity. This may result in black plots that can be visually noticed and spotted as a presence of caries. Fiber optics transferred the illumination from a lighting source to the tooth surface. The light propagates from the fiber illuminator across tooth tissue to nonilluminated surfaces. The light distribution images are taken and used for diagnosis. A narrow beam of bright white light is directed across areas of contact between a proximal surfaces and disruption of crystal structure that occurs demineralization, defects the light beam, and produces shadows. Here enamel lesions appear as gray shadows and dentin lesions appear as orange brown. It is used to detect marginal dental caries at different phase of its evolution. This method is not adopted widely in dental clinic due to unavailability of equipment. DIFOTI is relatively new technology [38, 41]. It is the combination of FOTI and a digital CCD camera. The other parts of this method are handpiece, electronic control box, software, and a customized image capture card, etc. It is a method that employs digital image processing quantitative diagnosis and prognosis in dentistry. It is based on light propagation just below the tooth surface and is used to determine the lesion depth. It uses fiber optics trans-illumination of safe visible light to image the tooth. In this method, the light delivered by fiber optics is collected [42]. These images can be acquired in a repeatable fashion. Here the tooth is illuminated by a fiber-optics handpiece and tooth images are gathered by a CCD camera for further analysis. A computer generates the report based on the algorithm. One big advantage of using CCD camera is that a huge amount of images can be taken of the same tooth during the dental examination. DIFOTI [43] helps to view the dental carious images at a later date by storing it. This method can detect the location of the carious lesion but fails to find its depth. However, it is unable to find caries at subgingival area (Figures 6.7 and 6.8).

6.2.3 Radiographs

Dentists prefer radiographs for caries detection. Accurate radiographs are very important diagnostic tools to detect dental disease and other disease. It is useful to detect caries because it causes tooth demineralization that can be captured by radiographs very

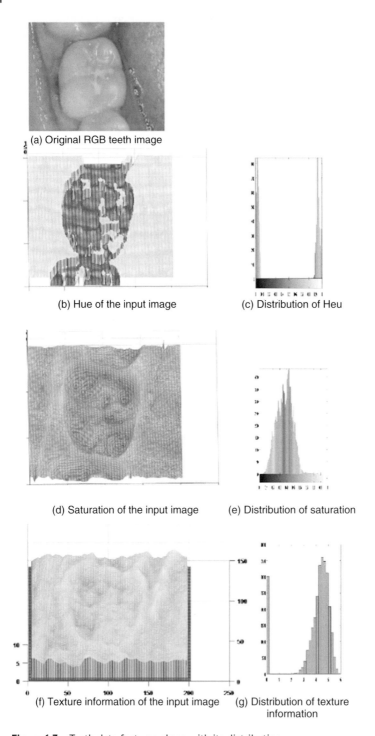

(a) Original RGB teeth image

(b) Hue of the input image

(c) Distribution of Heu

(d) Saturation of the input image

(e) Distribution of saturation

(f) Texture information of the input image

(g) Distribution of texture information

Figure 6.7 Teeth data features along with its distribution.

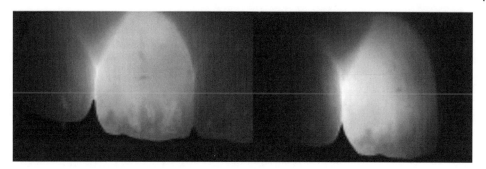

Figure 6.8 Discoloration of enamel under FOTI machine.

easily [44–46]. In radiographs, the affected carious lesion appears as darker than the unaffected portion of the tooth. However, early detection of carious lesion using radiographs is very difficult, especially when they are small in size. Three main types of dental x-rays [47, 48] are used to detect the dental caries. These are panoramic dental x-ray that shows a broad view of the jaws, teeth, sinuses, nasal area, infectious lesion, fractures, and dental caries. The bitewing x-rays show the teeth of upper and lower back where they touch each other, bone loss if any, and dental infection. Periapical x-rays show the entire tooth. These x-rays [49, 50] are very helpful to find impacted teeth, tumors, cysts, and bone changes. Digital subtraction radiographs [51] are a more advanced imaging tool. This method has been used in the assessment of the progression, arrest, or regression of caries lesion [52, 53]. This method provides vital information on any changes happening over time. Therefore, it is suitable for monitoring lesion behavior. However, dental radiographs fail to measure the carious lesion depth in a tooth

6.2.4 Light-Emitting Devices

The quantitative light-induced fluorescence (QLF) method comes under light-emitting method. The operating principle of QLF depends on the shape of the tooth enamel that becomes auto-fluorescenced under the lighting condition [35, 41]. The QLF method could readily detect lesions depth up to 500 m on smooth occlusal surface. The illumination systems contain a 50 W micro discharge arc lamp. A high-pass filter resides in front of the camera blocks. Here only the ambient light at 520 m is transmitted to the detector. In this method an object is wound up by a 370 nm wavelength of red or green light from a handpiece camera along with a high-pass filter. Here the images are captured, processed, and saved to provide the information about lesion area, lesion volume, and lesion depth [54, 55]. The preferred image is captured and saved by the operator, as well as detail about the teeth and the surface examination set in the program while the position and the orientation of the processed image is thereafter automatically stored in a preset pattern. The reason behind this is that when the patient comes back on recall, a contour guides the operator to the right position again. Also it can be used to measure an eroded lesion. This method is able to detect carious lesion at its early stage and can determine the state of the carious lesion that is either active or inactive. QLF is a very flexible and well-suited system and delivers a good result for follow-up purpose. This method is very user-friendly and

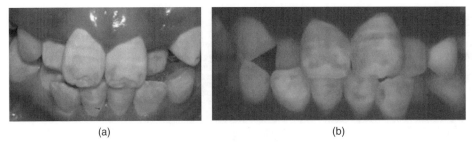

(a)   (b)

Figure 6.9 (a) (35–40) mm teeth image, (b) QLF teeth image.

well-suited for kids [56]. QLF can easily detect early demineralization and can quantify the total loss. However, it takes a long time to capture images and analyze it. QLF cannot differentiate between tooth decay and hypoplasia [57, 58]. It is a sensitive, reproducible method for quantification of 400 m depth lesion. As this method uses secondary emission that is fluoresced to detect the health of the teeth, thus the CCD module is a part of this method. The data pattern of this method is similar to the visible light property method. Figure 6.9 shows the teeth enamel layer under QLF light. Vista Proof and LED technology [59–61] are also categorized as light-emitting devices. The Vista Proof device is mainly used to detect the occlusal lesions, and us sometimes used for enamel caries. The sensitivity of Vista Proof devices is 86%. It has the better sensitivity than the diagnodent pen device. The design of the Vista Proof and QLF machine is almost the same and exits at a wavelength peak of 405 nm [62, 63]. The changes in the enamel layer can be detected when the tooth is illuminated by violet blue light. Vista proof is a device based on 6 blue GaN LED emitting a 405 nm light [64]. With this camera, it is possible to digitize the video signal from the dental surface during fluorescence emission using CCD sensor. These images show different areas of dental surface that fluorescent in green and in red for carious lesion. The software highlights the carious lesions and classifies them in a scale range from 0 to 5 giving a treatment orientation in the first evaluation, monitoring, and invasive treatment [25, 26, 65]. LED technology is the newest technology to detect the carious lesion. It consists of one camera with LED technology. This system illuminates the tooth enamel firstly, and then it records the fluorescence of the dental tissue and enhances the image. Clinical studies are ongoing to verify its values in application domain [66]. Vista Proof and LED technology are not much used in dental clinics. Different types of caries detection technologies are shown in Figure 6.10.

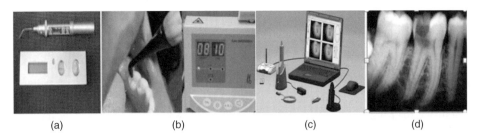

(a)   (b)   (c)   (d)

Figure 6.10 (a) FOTI device, (b) diagnodent device, (c) QLF machine, (d) caries detection using Radiograph.

6.2.5 Optical Coherent Tomography (OCT)

The OCT method is very popularly used in dentistry. OCT is a noninvasive, nonradioactive optical diagnostic tool based on interferometers and helps to get high-resolution images. It operates on the basic principle of white light. It utilizes the noninvasive light and biomedical optics to provide cross-sectional optical biopsy images. OCT system consists of a computer, compact diode light source, photo detector, and a handpiece scanner [67–69]. OCT dictionary address the image quality issue. It is typically designed as a filter system. Therefore, a handheld fiber–guided probe with light weight can be fabricated for clinical purpose. The probe will facilitate a small scanner head for the doctor to directly aim at regions of interest on a patient's teeth. OCT is based on low coherence interference and achieves microscale cross-sectional images. It increases the resolution quality much better than x-ray techniques. It is very difficult to detect early carious lesion, periodontal disease, and oral cancer with clinical examination and radiographs. This method may provide a solution to these problems due to its excellent spatial resolution. Three-dimensional imaging reconstruction is another advantage of dental OCT. It assists dentists to locate problems in soft and hard tissues more accurately and rapidly. However, the main problem with OCT is the partial penetration depth, a low scanning range, and the expense [19, 70].

6.2.6 Software Tools

Nowadays software tools are very popular to detect dental caries. Popular software is ICDAS, Logicon, etc. [27]. ICDAS means International Caries Detection and Assessment System. One important goal in developing ICDAS is that there exists for clinicians and researchers the ability to choose the stage of caries process and other features as well. ICDAS makes the decision based on some code generated by itself. It is a simple facts-based system to detect and classify the carious lesion. There are minor variations between the visual signs associated with each code. These depend on a number of factors including the surface characteristics, whether there are adjacent teeth present. The ICDAS detection codes for caries range from 0 to 6 depending on the severity of the lesion [17, 27, 71, 72]. Table 6.2 shows the details.

Table 6.2 Code meaning.

Code	Description
0	Sound
1	First visual change in enamel
2	Distinct visual change in enamel
3	Localized enamel breakdown
4	Underlying dark shadow from dentin
5	Distinct cavity with visible dentin
6	Extensive distinct cavity with visible dentin

6.3 Technical Challenges with the Existing Methods

Most of the caries detection methods are primarily based on visual examination, subject surface textures, position, touching consciousness, etc. [4]. These methods show the carious lesion area on the tooth. In real life, it is more critical to determine the carious affected area in between two teeth, i.e., interproximal caries. The methods described in the previous section confirm good resolution toward the detection of dental caries. But there are some known drawbacks as mentioned in Table 6.2. The infrastructure investments toward ECM and QLF [16] implementations are very expensive; the cost of treatment goes high. Besides that, the ECM method is only able to detect caries on a single extracted tooth, which is logically not feasible. On the other hand, the reliability of diagnodent is very high and its diagnostic validity, i.e., the sum of sensitivity and specificity is also high. In this case, good experts are needed to make the decision. QLF has high level specificity, but low sensitivity and reproducibility. Radiographs, QLF, DIFOTI, etc. are methods that are good for detecting dental caries. The main shortcomings of the conventional methods are that it is difficult to evaluate a lesion's growth rate. Now, x-ray is used to measure the depth of dental caries. It is useful for monitoring the progression and arrest of the occlusal caries in both primary and permanent teeth. Moreover, radiography, which is the most common method used to diagnose proximal caries, also exposes the patient to a relatively high dose of ionizing radiation [15, 73]. However, frequently taken x-ray is very harmful and it should be avoided for pregnant women and kids. There is not a single method that can detect caries at its early stage, which could help prevent caries from childhood. It has been observed that it is very difficult to measure the actual carious lesion depth to form an x-ray image. IN that case 3D measurement, like cone beam computer tomography (CBCT), is helpful. CBCT provides much more accurate results than conventional x-rays. This provides much more detailed information, like diseases of the jaw, caries lesion depth, bone structures of the face, nasal cavity, and sinuses, as well as forming a 360° view of a tooth [74, 75]. CBCT not only provides good results to measure carious lesion depth but also use to diagnose and treat tooth impactions, identification of maxillary sinus, trauma evaluation, and treatment. There is no question that cone beam provides a better potential to improve prevention and treatment of oral diseases and conditions [76, 77]. However, it is not so popular and is not used in dental clinics because of the following reasons:

1. It is very expensive.
2. It is not suitable for soft-tissue evaluation.
3. This method can only demonstrate limited-contrast resolution. [78]
4. CBCT is not telemedicine compatible hence it cannot be installed in rural health centers.

Sometimes, dentists go for an invasive intervention, which is actually necessary for that disease. This happens due to lack of data availability and poor image quality. Hence automation on dental image analysis based on image processing technique is more advantageous to detect caries at its early stages so that we can avoid surgical complications [20, 79]. This automation will also help to monitor caries lesion and demineralization lesions for better maintenance, as well as follow-ups, on dental health. In short we can say that although these methods offer relatively good diagnostic performances, they provide varying sensitivities for detecting same carious lesions, partially because these methods are invasive in

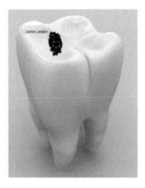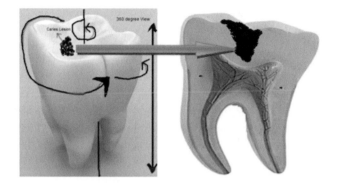

Figure 6.11 Caries affected lesion, 3D view of the same lesion and it spreads in dentine layer.

nature, and sometimes injurious to our health. There is a lack of a single method that can detect early caries as well as acute caries at any stage [80]. There is a huge demand for a low-cost device that can provide a 360° tooth-wise view that can help dental practitioners measure the caries depth in a tooth. Figure 6.11 shows the overview of the required system. This also minimizes unwanted oral intervention.

Kodak provides a customized dental radiography setup. In this setup technicians use RGV format to store the dental radiographic images. As per Versteeg Ch et al., the retaking rate of CCD-based dental radiography like RGV costs much more than the traditional film-based dental radiograph. Hence we can conclude that the considerable amount of data error is present in RGV system. Another article by S. Tavropoul et al., also concludes a similar type of opinion. Accuracy of cone beam dental CT, intraoral digital, and conventional film radiography is useful for for the detection of periapical lesion. Radiography uses x-ray that penetrates the dental tissues. So, it is very difficult to detect interproximal caries specifically between overlapped teeth. Dental radiography imaging is a kind of shadow imaging. Hence proper shape of the region may differ from the actual one when the film or CCD transducer is not aligned properly. Hence, a retake of the image is very common to optimize the data error in dental radiography. QLF is also an imaging-based diagnostic method. So image-related data error challenges are also present in it. In QLF the lesion size is monitored at regular intervals and the decision is made on the basis of a lesion contour boundary change. Hence image data registration is one of the challenges to proper diagnosis.

6.3.1 Challenges in Data Analysis Perspective

Now we focus on the data pattern. In KAVO diagnodent, no camera module is attached. The dentist has to touch the system to the suspected regions of caries and this system will beep if the caries lesion is found. Calibration is essential on regular intervals; otherwise, this system will not perform satisfactorily. As there is no camera module attached with this system, thus, systematic caries lesion monitoring for individual patients is very difficult. If caries depth is less, then generally these are not prominent in x-ray images, but this method can detect this caries lesion. A DSP filter is used to prevent the signals generated through the ambient light. ECM has a similar type of problem like calibration and monitoring in periodic intervals. This method works on the variation of impedance, so leakage of current may affect the result. A band-pass filter is used to clean the data obtained from the trans-user. Next-generation ECM and diagnodent will use a camera along with the method to reduce the overload of monitoring the caries lesion in periodic intervals. Image scaling is one of the most important parts at the time of monitoring a region. Hence image registration will be introduced when a camera module will be used with the point detection methods. This method uses CCD for imaging so that the data volume is huge. An image compression mechanism may be used, but it may degrade the quality of the region of interest in the image. It will be advantageous if the ROI is extracted from background and saved in high resolution. The segmentation of ROI by means of a manual method is laborious. Hence an automatic method is preferred. Automatic techniques will analyze the data of the image in terms of features like texture, color, intensity, etc., and then segment accordingly. Earlier ROI segmentation method segments the image on the basis of some thresholds that are fixed and given manually. But modern systems will set these thresholds (parameters) automatically on the basis of training data sets. Figure 6.7 shows the data features along with its distribution of a teeth region obtained from visible light property method. As already discussed, this technique is purely based on image. The data pattern of radiograph is of two types; the older one is an analog type and the newer one is a digital type. The analog type radiograph contains more detailed information than the digital type. However, the analog type radiograph is very prone to error and recovery and error detection is difficult. Due to this reason, digital radiography is preferred. Digital radiography supports operations used in digital image processing and the operations are similar to the information the previous sections explained. For monitoring the caries lesion using radiography, radiographic images are needed to be compared in regular intervals. Hence image scaling and registration is essential. This method uses different type of transducers. Some of them provide single dimensional data like sound, two-dimensional data like images, and some of them use multidimensional data. It is difficult to synchronize them by hand, thus, an automatic system is preferred for synchronization. Advanced soft computing techniques are also used to synchronize the data and assist to make a decision. Diagnodent senses the intensity level of secondary emissions from the teeth. This secondary emission varies from person to person so calibration is essential for each patient. It is a time-consuming process. This method senses the intensity of light, so that ambient light from the environment may change the value of the data. This is another challenge in preserving the property of data. The ECM method measures the cumulative impedance change in the teeth region and from the variation of cumulative impedance it determines the health of the affected teeth. This method is just like the diagnodent, where the external ambient light will not affect the system, but the moisture within the mouth may affect the reading. The saliva layer on the teeth may change the cumulative impedance, which causes an error in data. At the time of this test, maintaining standard dryness within the mouth is essential to reduce the data error. FOTI uses ambient light to determine the amount of light scattered or absorption rate from the teeth surface. After that the data is analyzed to determine the teeth health. The saliva layer on the teeth as well as the structure of the dental tissue may vary from person to person. Hence calibration is also essential for each subject in order to set the reference value. At the time of calibration, a healthy tooth is required. The selection of a healthy tooth depends on the experience of the operators. Wrong selection leads to wrong data acquisition and an erroneous data analysis report.

6.4 Result Analysis

Effectiveness of such existing methods depends on some metrics like specificity and sensitivity. Specificity is the true negatives rate whereas sensitivity is the true positive rate. Figure 6.12 is showing the lesion prevalence, sensitivity, and specificity for occlusal surfaces. These data are based on an excellent systematic review by Bader et al. [3] where the diagnostic performance of Bader studies are restricted in terms of diagnostic methods, especially for primary teeth, anterior teeth, root surfaces, visual/tactile, and FOTI methods. They limited the measurement study on histological validation. Figure 6.13 shows the lesion prevalence, sensitivity, and specificity [53] for proximal surfaces. These graphs show that while specificity is sufficient, the sensitivity of the traditional methods is much less.

6.5 Conclusion

Dental caries is a very common infectious disease among adults, as well as children. Hence caries detection at its early stage is very crucial. Several methods are available to detect caries. Among them, as discussed in this review, some are highly expensive, injurious to health, fail to detect caries in its early stage, and are partially invasive in nature. Detection and treatment of caries affected lesions are not confined onto only the boundary of normal disease detection and treatment methods; it also includes the essence of beauty or overall satisfaction with one's appearance and the capability to spend the same amount of money with a less painful treatment. Nowadays, carious lesion detection and treatments have become cosmetic services with different pricing packages. Hence selection of diagnosis and treatment packages is different, thus, lots of factors and constraints need to be considered here. To address these issues, a mathematical model has been proposed below. The

Figure 6.12 Performance of traditional caries detection methods after Bader et al.

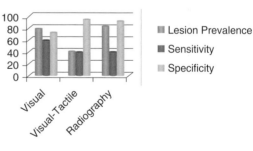

Figure 6.13 Performance of traditional method for Proximal Surfaces after Bader et al.

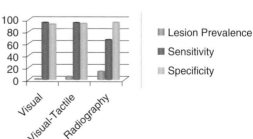

budget is the maximum amount of money that can be spent for the carious lesion treatment. A mathematical model on the selection procedure of caries treatment has been constructed based on the experiences of dentists and by use of a survey in urban and semi-urban areas near Kolkata. The variable names are self-explanatory.

$$DiagnosisCost + TreatmentCost + PostTreatmentCost + \cdots \leq Budget \tag{6.1}$$

For simplicity consider only the first two parameters:

$$DiagnosisCost + TreatmentCost \leq Budget \tag{6.2}$$

$$DiagnosticCost + TreatmentCost \geq \frac{Budge}{QualityCompromizeFactor} \tag{6.3}$$

value of Quality compromise factor ≥ 1

$$DiagnosisCost = A_1 * \text{X-Ray cost} + A_2 * QLF \text{ cost} + \cdots$$

where A_1, A_2, A_3 are Boolean variable and

$$\sum_{i=1}^{n} A_i = 1 \tag{6.4}$$

n = total number of diagnosis methods

This can be rewritten as

$$Diagnostic\_Cost = \sum_{i=1}^{n} A_i * (\text{cost of ith Diagnostic method}) \tag{6.5}$$

$$\frac{A_1 * X\_RayCost}{Risk\_Factor\_forX\_ray} + \frac{A_1 * QLFCost}{Risk\_Factor\_forQLF} + \cdots \geq Satisfaction\_Factor \tag{6.6}$$

This can be formulated as follows

$$\sum_{i=1}^{n} A_i * \frac{Cost\_ith\_Diagnosis\_method}{Risk\_Factor\_of\_ith\_Diagnosis\_Method} \geq Satisfaction\_Factore \tag{6.7}$$

Similarly, the treatment cost-related equations are

$$\text{Treatment Cost} = B_1 * RTC \text{ cost} + A_2 * \text{Teeth plucked out cost} + \cdots$$

Where B_1, B_2, and B_3 are Boolean variable and $\sum_{i=1}^{m} B_i = 1$

Similarly,

$$Diagnostic\_Cost = \sum_{i=1}^{n} B_i * Cost\_ith\_Diagnostic\_Method \tag{6.8}$$

$$\frac{B_1 * RTC\_Cost}{Risk\_Factor\_RTC} + \frac{B_1 * Teeth\_Pluckedout\_Cost}{Risk\_Factor\_Teeth\_Plucked\_Cost} + \cdots \geq Safe\_Factor \tag{6.9}$$

Similarly,

$$\sum_{i=1}^{m} B_i * \frac{Cost\_ith\_Treatement\_Method}{Risk\_for\_ith\_treatement\_Method} \geq Safe\_Factor \tag{6.10}$$

This mathematical model helps dentists and patients to choose better treatment with lower cost. To prevent caries in its early stage, not only is primary awareness sufficient, but also it is more important to detect actual caries-affected lesions. It is also important to establish a system that can identify and measure various advanced and early stages of caries lesion detection. Though there are many new and good caries detection methods available, there is a huge gap. Hence there is need for a complete system that can identify the caries lesion in its early and advanced stages. In the next phase of our research work we are planning to develop such a system that can detect the affected carious lesion and extract the features of it to assist dentists. The system will be primarily based on image processing techniques. It will be safe for humans and will be a low-cost device. The system can also be used in remote areas by the use of telemedicine technology.

Acknowledgment

Funding was provided by the Visvesvaraya fellowship for Ph.D. scheme under Ministry of Electronics and Information Technology of the Government of India.

References

1 Who: Fact sheet, World Health Organization Media Centre, http://www.who.int/mediacentre/factsheets/fs318/en.

2 Petersen, P.E. (2003). The world oral health report 2003: continuous improvement of oral health in the 21st century–the approach of the who global oral health programme. *Community Dentistry and Oral Epidemiology* 31: 3–24.

3 Bader, J.D., Shugars, D.A., and Bonito, A.J. (2001). Systematic reviews of selected dental caries diagnostic and management methods. *Journal of Dental Education* 65: 960–968.

4 Karlsson, L. (2010). Caries detection methods based on changes in optical properties between healthy and carious tissue. *International Journal of Dentistry* 2010: 1–9.

5 Kidd, E. and Fejerskov, O. (2004). What constitutes dental caries? Histopathology of carious enamel and dentin related to the action of cariogenic bio-films. *Journal of Dental Research* 83: 35–38.

6 Lussi, A., Imwinkelried, S., Pitts, N. et al. (1999). Performance and reproducibility of a laser fluorescence system for detection of occlusal caries in vitro. *Caries Research* 33: 261–266.

7 Osterloh, D. and Viriri, S. (2018). Unsupervised caries detection in nonstandardized periapical dental x-rays. In: *International Conference on Computer Vision and Graphics* (ed. M.E. McHenry), 329–340. New York: Springer.

8 Pretty, I.A. (2006). Caries detection and diagnosis: novel technologies. *Journal of Dentistry* 34: 727–739.

9 Diniz, M.B., de Almeida Rodrigues, J., and Lussi, A. (2012). Traditional and novel caries detection methods. In: *Contemporary Approach to Dental Caries*. InTech.

10 Gomez, J., Tellez, M., Pretty, I. et al. (2013). Non-cavitated carious lesions detection methods: a systematic review. *Community Dentistry and Oral Epidemiology* 41: 55–66.

11 Smejkalova, J., Jacob, V., Hodacova, L. et al. (2012). The inuence of smoking on dental and periodontal status. In: *Oral Health Care-Pediatric, Research, Epidemiology and Clinical Practices* (ed. M. Virdi). InTech.

12 Petersen, P.E., Bourgeois, D., Ogawa, H. et al. (2005). The global burden of oral diseases and risks to oral health. *Bulletin of the World Health Organization* 83: 661–669.

13 Prabhakar, N., Kiran, K., and Kala, M. (2011). A review of modern noninvasive methods for caries diagnosis. *Archives of Oral Sciences & Research* 1: 168–177.

14 Cortes, D., Ellwood, R., and Ekstrand, K. (2003). An in vitro comparison of a combined foti/visual examination of occlusal caries with other caries diagnostic methods and the effect of stain on their diagnostic performance. *Caries Research* 37: 8–16.

15 de Gonzalez, A.B. and Darby, S. (2004). Risk of cancer from diagnostic x-rays: estimates for the UK and 14 other countries. *The Lancet* 363: 345–351.

16 E. D. J. de Jong, & L. Stöber. (2003). Quantitative light-induced uorescence (QLF) – a potential method for the dental practitioner. Quintessence International, 34(3) (2003).

17 Ismail, A., Sohn, W., Tellez, M. et al. (2007). The international caries detection and assessment system (ICDAS): an integrated system for measuring dental caries. *Community Dentistry and Oral Epidemiology* 35: 170–178.

18 Selwitz, R.H., Ismail, A.I., and Pitts, N.B. (2007). Dental caries. *The Lancet* 369: 51–59.

19 Drexler, W. and Fujimoto, J.G. (2008). *Optical Coherence Tomography: Technology and Applications*. New York: Springer Science & Business Media.

20 Ely, E.W., Stump, T.E., Hudspeth, A.S., and Haponik, E.F. (1993). Thoracic complications of dental surgical procedures: hazards of the dental drill. *The American Journal of Medicine* 95: 456–465.

21 Haruyama, O., Kimura, H., Nishiyama, N. et al. (2001). The anomalous behavior of electrical resistance for some metallic glasses examined in several gas atmospheres or in a vacuum. In: *Amorphous and Nanocrystalline Materials* (ed. M.E. McHenry), 69–86. New York: Springer.

22 H. Nathel, J. H. Kinney, & L. L. Otis, (1996) Method for detection of dental caries and periodontal disease using optical imaging, Oct. 29 1996. US Patent 5,570,182.

23 Carounanidy, U. and Sathyanarayanan, R. (2009). Dental caries: a complete changeover (part ii)-changeover in the diagnosis and prognosis. *Journal of Conservative Dentistry: JCD* 12: 87.

24 Ghosh, P., Bhattacharjee, D., and Nasipuri, M. (2017). Automatic system for plasmodium species identification from microscopic images of blood-smear samples. *Journal of Healthcare Informatics Research* 1: 231–259.

25 Gimenez, T., Braga, M.M., Raggio, D.P. et al. (2013). Fluorescence-based methods for detecting caries lesions: systematic review, meta-analysis and sources of heterogeneity. *PLoS One* 8: e60421.

26 Guerrieri, A., Gaucher, C., Bonte, E., and Lasfargues, J. (2012). Minimal intervention dentistry: part 4. Detection and diagnosis of initial caries lesions. *British Dental Journal* 213: 551.

27 Zandonfia, A.F. and Zero, D.T. (2006). Diagnostic tools for early caries detection. *The Journal of the American Dental Association* 137: 1675–1684.

28 Lussi, A., Hibst, R., and Paulus, R. (2004). Diagnodent: an optical method for caries detection. *Journal of Dental Research* 83: 80–83.

29 Zeitouny, M., Feghali, M., Nasr, A. et al. (2014). Soprolife system: an accurate diagnostic enhancer. *The Scientific World Journal* 2014: 651–667.

30 Harris, R., Nicoll, A.D., Adair, P.M., and Pine, C.M. (2004). Risk factors for dental caries in young children: a systematic review of the literature. *Community Dental Health* 21: 71–85.

31 Whitters, C., Strang, R., Brown, D. et al. (1999). Dental materials: 1997 literature review. *Journal of Dentistry* 27: 401–435.

32 Anttonen, V., Seppä, L., and Hausen, H. (2003). Clinical study of the use of the laser fluorescence device diagnodent for detection of occlusal caries in children. *Caries Research* 37: 17–23.

33 Bamzahim, M., Shi, X.-Q., and Angmar-Mansson, B. (2002). Occlusal caries detection and quantification by diagnodent and electronic caries monitor: in vitro comparison. *Acta Odontologica Scandinavica* 60: 360–364.

34 Sheehy, E., Brailsford, S., Kidd, E. et al. (2001). Comparison between visual examination and a laser fluorescence system for in vivo diagnosis of occlusal caries. *Caries Research* 35: 421–426.

35 Shi, X., Tranaeus, S., and Angmar-Mansson, B. (2001). Comparison of qlf and diagnodent for quantification of smooth surface caries. *Caries Research* 35: 21.

36 Pretty, I. and Ellwood, R. (2013). The caries continuum: opportunities to detect, treat and monitor the re-mineralization of early caries lesions. *Journal of Dentistry* 41: S12–S21.

37 Ashley, P., Blinkhorn, A., and Davies, R. (1998). Occlusal caries diagnosis: an in vitro histological validation of the electronic caries monitor (ECM) and other methods. *Journal of Dentistry* 26: 83–88.

38 Stookey, G.K. and Gonzfialez-Cabezas, C. (2001). Emerging methods of caries diagnosis. *Journal of Dental Education* 65: 1001–1006.

39 Pitts, N.B. (2001). Clinical diagnosis of dental caries: a european perspective. *Journal of Dental Education* 65: 972–978.

40 Tranfius, S., Shi, X.-Q., and Angmar-Mfiansson, B. (2005). Caries risk assessment: methods available to clinicians for caries detection. *Community Dentistry and Oral Epidemiology* 33: 265–273.

41 Yu, J., Tang, R., Feng, L., and Dong, Y. (2017). Digital imaging fiber optic transillumination (DIFOTI) method for determining the depth of cavity. *Beijing da xue xue bao. Yi xue ban Journal of Peking University. Health sciences* 49: 81–85.

42 C. Gutierrez, DIFOTI (Digital Fiberoptic Transillumination): Validity at In Vitro, PhD thesis, Dissertation Zum Erwerb des Doktorgrades an der Zahnheilkunde an der Medizinischen Fakuly at der Ludwig- Maximilians-University zu Munchen, 2008.

43 Attrill, D. and Ashley, P. (2001). Diagnostics: Occlusal caries detection in primary teeth: a comparison of diagnodent with conventional methods. *British Dental Journal* 190: 440.

44 Pitts, N. (1996). The use of bitewing radiographs in the management of dental caries: scientific and practical considerations. *Dentomaxillo-facial Radiology* 25: 5–16.

45 Wenzel, A. (1998). Digital radiography and caries diagnosis. *Dentomaxillofacial Radiology* 27: 3–11.

46 Wenzel, A., Larsen, M., and Feierskov, O. (1991). Detection of occlusal caries without cavitation by visual inspection, film radiographs, xeroradiographs, and digitized radiographs. *Caries Research* 25: 365–371.

47 E. Hausmann, D. Wobschall, L. Ortman, E. Kutlubay, K. Allen, & D. Odrobina, (1997) Intraoral radiograph alignment device, May 13 1997. US Patent 5,629,972.28

48 Wenzel, A., Hintze, H., Mikkelsen, L., and Mouyen, F. (1991). Radiographic detection of occlusal caries in noncavitated teeth: a comparison of conventional film radiographs, digitized film radiographs, and radiovisiography. *Oral Surgery, Oral Medicine, Oral Pathology* 72: 621–626.

49 Patel, S., Dawood, A., Wilson, R. et al. (2009). The detection and management of root resorption lesions using intraoral radiography and cone beam computed tomography–an in vivo investigation. *International Endodontic Journal* 42: 831–838.

50 Preston-Martin, S., Thomas, D.C., White, S.C., and Cohen, D. (1988). Prior exposure to medical and dental x-rays related to tumors of the parotid gland1. *JNCI: Journal of the National Cancer Institute* 80: 943–949.

51 Gröndahl, H.-G. and Huumonen, S. (2004). Radiographic manifestations of periapical inammatory lesions: how new radiological techniques may improve endodontic diagnosis and treatment planning. *Endodontic Topics* 8: 55–67.

52 Datta, S., Chaki, N., and Modak, B. (2019). A novel technique to detect caries lesion using Isophote concepts. *IRBM* 40 (3): 174–182.

53 Wenzel, A. (2004). Bitewing and digital bitewing radiography for detection of caries lesions. *Journal of Dental Research* 83: 72–75.

54 Coulthwaite, L., Pretty, I.A., Smith, P.W. et al. (2006). The microbiological origin of uorescence observed in plaque on dentures during qlf analysis. *Caries Research* 40: 112–116.

55 Pretty, I.A., Edgar, W.M., and Higham, S.M. (2002). Detection of in vitro demineralization of primary teeth using quantitative light-induced uorescence (QLF). *International Journal of Paediatric Dentistry* 12: 158–167.

56 Pretty, I.A., Edgar, W.M., and Higham, S.M. (2002). The effect of ambient light on qlf analyses. *Journal of Oral Rehabilitation* 29: 369–373.

57 Pretty, I., Edgar, W., and Higham, S. (2001). Aesthetic dentistry: the use of qlf to quantify in vitro whitening in a product testing model. *British Dental Journal* 191: 566.

58 Yin, W., Hu, D., Li, X. et al. (2013). The anti-caries efficacy of a dentifrice containing 1.5% arginine and 1450 ppm fluoride as sodium monouorophosphate assessed using quantitative light-induced fluorescence (QLF). *Journal of Dentistry* 41: S22–S28.

59 Curzon, P. (1994). *The Verified Compilation of Vista Programs, in 1st ProCoS Working Group Meeting*. Denmark, Citeseer: Gentofte.

60 Raggio, D.P., Braga, M.M., Rodrigues, J.A. et al. (2010). Reliability and discriminatory power of methods for dental plaque quantification. *Journal of Applied Oral Science* 18: 186–193.

61 Ruckhofer, E. and Städtler, P. (2010). Ist vista proof eine hilfe bei der kariesdiagnostik? *Stomatologie* 107: 13–16.

62 Presoto, C.D., Trevisan, T.C., Andrade, M.C.D. et al. (2017). Clinical effectiveness of uorescence, digital images and icdas for detecting occlusal caries. *Revista de Odontologia da UNESP* 46: 109–115.

63 Rodrigues, J., Hug, I., Diniz, M., and Lussi, A. (2008). Performance of fluorescence methods, radiographic examination and ICDAS II on occlusal surfaces in vitro. *Caries Research* 42: 297–304.

64 Jablonski-Momeni, A., Heinzel-Gutenbrunner, M., and Klein, S.M.C. (2014). In vivo performance of the vistaproof fluorescence based camera for detection of occlusal lesions. *Clinical Oral Investigations* 18: 1757–1762.

65 Jablonski-Momeni, A., Heinzel-Gutenbrunner, M., and Vill, G. (2016). Use of a fluorescence-based camera for monitoring occlusal surfaces of primary and permanent teeth. *International Journal of Paediatric Dentistry* 26: 448–456.

66 Guerra, F., Corridore, D., Mazur, M. et al. (2016). Early caries detection: comparison of two procedures. A pilot study. *Senses and Sciences* 3: 317–322.

67 Hsieh, Y.-S., Ho, Y.-C., Lee, S.-Y. et al. (2013). Dental optical coherence tomography. *Sensors* 13: 8928–8949.

68 Otis, L.L., Everett, M.J., Sathyam, U.S., and Colston, B.W. Jr., (2000). Optical coherence tomography: a new imaging: technology for dentistry. *The Journal of the American Dental Association* 131: 511–514.

69 Sun, C.-W., Ho, Y.-C., and Lee, S.-Y. (2015). Sensing of tooth microleakage based on dental optical coherence tomography. *Journal of Sensors* 2015: 120–132.

70 Wang, X.-J., Milner, T.E., De Boer, J.F. et al. (1999). Characterization of dentin and enamel by use of optical coherence tomography. *Applied Optics* 38: 2092–2096.

71 Alammari, M., Smith, P., De Jong, E.D.J., and Higham, S. (2013). Quantitative light-induced fluorescence (QLF): a tool for early occlusal dental caries detection and supporting decision making in vivo. *Journal of Dentistry* 41: 127–132.

72 Pitts, N., Ekstrand, K., and I. Foundation (2013). International caries detection and assessment system (ICDAS) and its international caries classification and management system (ICCMS)–methods for staging of the caries process and enabling dentists to manage caries. *Community Dentistry and Oral Epidemiology* 41: e41–e52.

73 Nardini, M., Ridder, I.S., Rozeboom, H.J. et al. (1999). The x-ray structure of epoxide hydrolase from agrobacterium radiobacter ad1 an enzyme to detoxify harmful epoxides. *Journal of Biological Chemistry* 274: 14579–14586.

74 Mah, J.K., Huang, J.C., and Choo, H. (2010). Practical applications of cone-beam computed tomography in orthodontics. *The Journal of the American Dental Association* 141: 7S–13S.

75 Sherrard, J.F., Rossouw, P.E., Benson, B.W. et al. (2010). Accuracy and reliability of tooth and root lengths measured on cone-beam computed tomographs. *American Journal of Orthodontics and Dentofacial Orthopedics* 137: S100–S108.

76 Lascala, C., Panella, J., and Marques, M. (2004). Analysis of the accuracy of linear measurements obtained by cone beam computed tomography (CBCT-newtom). *Dentomaxillofacial Radiology* 33: 291–294.

77 Sukovic, P. (2003). Cone beam computed tomography in craniofacial imaging. *Orthodontics & Craniofacial Research* 6: 31–36.

78 Misch, K.A., Yi, E.S., and Sarment, D.P. (2006). Accuracy of cone beam computed tomography for periodontal defect measurements. *Journal of Periodontology* 77: 1261–1266.

79 Cogswell, W.W. (1942). Surgical problems involving the mandibular nerve. *The Journal of the American Dental Association* 29: 964–969.

80 Hollander, F. and Dunning, J.M. (1939). A study by age and sex of the incidence of dental caries in over 12,000 persons. *Journal of Dental Research* 18: 43–60.

81 Al-Ansari, A.A. et al. (2014). Prevalence, severity, and secular trends of dental caries among various saudi populations: a literature review. *Saudi Journal of Medicine and Medical Sciences* 2: 142.

82 da Silveira Moreira, R. (2012). Epidemiology of dental caries in the world. In: *Oral Health Care-Pediatric, Research, Epidemiology and Clinical Practices*. InTech.

83 Petersen, E.P. (2009). Oral health in the developing world, World Health Organization global oral health program chronic disease and health promotion. Geneva:. *Community Dentistry and Oral Epidemiology* 58 (3): 115–121.

7

Intelligent Data Analysis Using Hadoop Cluster – Inspired MapReduce Framework and Association Rule Mining on Educational Domain

Pratiyush Guleria and Manu Sood

Department of Computer Science, Himachal Pradesh University, Shimla, India

7.1 Introduction

Intelligent data analysis (IDA) is the field of artificial intelligence (AI). With the help of IDA, meaningful information is discovered from a large amount of data. This field includes pre-processing of data, data analysis, mining techniques, machine learning, neural networks, etc. Statistical techniques [1] are implemented for data analysis. The types of statistics are as follows: (i) descriptive statistics and (ii) inferential statistics. In descriptive statistics, assumptions are not used for summarizing data, whereas, in inferential statistics, conclusions are drawn based on certain assumptions.

IDA has emerged as a combination of many fields: (i) statistics, (ii) AI, (iii) pattern recognition, and (iv) machine learning [2]. IDA in the educational field can be achieved with the help of educational data mining and learning analytics [3], which will become a milestone in the field of education and a teacher-taught paradigm.

With the help of these fields, students data can be analyzed in an intelligent manner to achieve the following: (i) academic analytics, (ii) course redesigning, (iii) implementing learning management systems, (iv) interactive sessions between instructors and students, (v) implementing new patterns of students assessments, (vi) improving pedagogies, (vii) predicting students dropouts, and (viii) statistical reports of the learners.

The author in [4] has quoted the definition of *Learning Analytics* by George Siemens. According to George Siemens, "Learning analytics is the use of intelligent data, learner-produced data, and analysis models to discover information and social connections for predicting and advising people's learning."

Data mining is the most striking area coming to the fore in modern technologies for fetching meaningful information from a large volume of data that is unstructured with lots of uncertainty and inconsistencies. There is stupendous leverage to the educational sector in using data mining techniques to inspect data from students, assessments, latest scholastic patterns, etc. This proves to provide a quality education as well as decision-making advice for students in order to escalate their career prospects and select the right course for training to fulfill the skill gap that exists between primary education and the industry that will

Intelligent Data Analysis: From Data Gathering to Data Comprehension,
First Edition. Edited by Deepak Gupta, Siddhartha Bhattacharyya, Ashish Khanna, and Kalpna Sagar.
© 2020 John Wiley & Sons Ltd. Published 2020 by John Wiley & Sons Ltd.

hire them. Data mining has a great impact on scholastic systems where education is measured as the primary input for informative evolution. In such a scenario where data is being generated at an alarming rate from many sources like media files, data from social media websites, e-mails generated, Google search, instant messaging, mobile users, internet of things (IOT), etc., data grows enormously and it may not fit on a single node, which is why the the term "big data" was coined. It is a term for data sets that are so large in volume or complex that it becomes arduous to manage them with customary data processing application software.

Using big data techniques, unstructured data is gathered and analyzed to reveal informative data for operations, which includes the gathering of data for storage and analysis purposes that gains control over operations, such as experimental, fact-finding, allocation, data visualization, updating, and maintaining the confidentiality of information.

The problem pertaining to the enormously large size of a data set can be solved by Hadoop using a MapReduce model, which works on a Hadoop layer and allows parallel processing of the data stored in a Hadoop distributed file system that allows dumping any kind of data across the cluster. MapReduce tasks run over Hadoop clusters by splitting the big data, i.e., input file, into small pieces and processing the data on parallel distributed clusters. It is an open-source programming prototype that performs parallel processing of applications on clusters, and with the distribution of data, computation becomes faster. A MapReduce programming framework executes its operations into three stages, i.e., map phase, shuffle phase, and reduce phase.

In data mining, association rule learning is a method for discovering interesting relations among variables in large databases [5]. A big data approach using association rule mining can help colleges, institutions, and universities get a comprehensive perspective of their students. It provides compelling reasons to the institutions to put in concentrated efforts so as to enhance the quality of education by enhancing the skill sets of the potential students, keep them intensely focused, matching their skill sets with the demands of the market and/or society, and making them ready for future challenges. It helps in answering questions related to the (i) learning behaviors, (ii) understanding patterns of students, (iii) curriculum trends, and (iv) selection of courses to help create captivating learning experiences for students. In this chapter, the authors have, for experimental exploration, synthesized a data set based upon the preferences of students so that MapReduce and Apriori algorithms can be applied in order to derive the appropriate rules for helping students make well-informed decisions.

7.1.1 Research Areas of IDA

AI is the growth engine of present and future. The objective of AI is to enrich machines with intelligence. The four factors important to exhibiting intelligence are as follows: (i) thinking, (ii) decision-making, (iii) problem solving, and (iv) learning.

Expert systems are developed using AI techniques. The application roadmap for expert systems includes an inference engine, which is categorized as follows: (i) intelligence assistance, (ii) expert systems, and (iii) learning analytics.

The fields of AI using IDA techniques are shown in Figure 7.1.

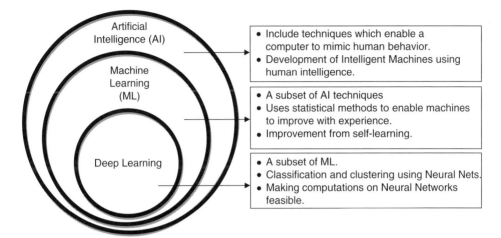

Figure 7.1 Artificial intelligence and its subsets using intelligent data analytics.

7.1.2 The Need for IDA in Education

IDA helps in the development of intelligent data analytical systems for improving learning in education. IDA results in a revolutionary change in the educational system using learning analytics and educational data mining. In educational data mining, data from an educational domain are analyzed to better understand the learning behaviors of students.

7.2 Learning Analytics in Education

Learning analytics are powerful tools and play an important role in the educational sector. An example is a learning management system such as Moodle. Learning analytics improve the learner abilities, as every learner has different levels in attaining knowledge for a course. With the help of learning analytics, an intelligent curriculum can be developed to cater to the needs of each learner.

Learning analytics is the measurement of data about learners that is used to optimize their learning experience. In classroom learning, sometimes it is not possible to focus on an individual student or to attend to the individual weaknesses, but with the help of learning analytics, work can be done on individual weaknesses and individual student queries can be monitored/identified.

With the help of learning analytical techniques, learners' performance can be monitored and predictions can be performed on individual learner performance, which helps students to fulfill their educational objectives.

The learning analytics role in improving student learning is shown in Figure 7.2. Every student has a different pace of learning and understanding; therefore, learning analytics provide learners an environment where he/she can learn from the program as per their requirement, which helps to achieve the following results:

- It can be cost-effective,
- It can be the best utilization of time and resources,

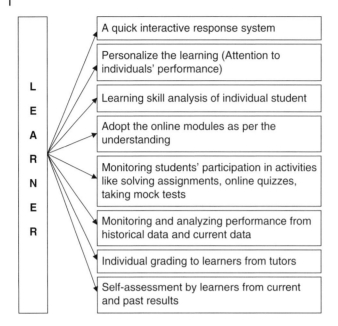

Figure 7.2 Learner support provided by learning analytics.

- It can have an increased retention rate,
- It can run the tutorial repeatedly to solve and better understand the problem,
- It can solve the assignment online,
- It can take mock tests,
- It can set the difficulty levels of tests,
- It can learn the topic module-wise,
- It can make predictions on the success/failure rate of learners,
- It can set the skill level,
- It can ensure learning becomes easy through audio and visual approaches,
- It can help in predicting new academic trends.

There are two types of analytics: learning analytics and academic analytics. Learning analytics focus on consistent improvements in learning efficiencies from primary to higher studies. Learning analytics facilitate academicians, learners, and management authorities for improvement in the classroom by using course-level activities from the intelligent analysis of data generated, whereas, in academic analytics, there is a comparative analysis of quality and standard norms followed while implementing educational policies in universities and colleges on national and international levels.

Apart from it, there is an analysis of learner abilities with respect to traditional and modern educational systems that result in effective decision-making by policymakers when funding organizations [6]. The sample techniques for the analytics engine are shown in Figure 7.3.

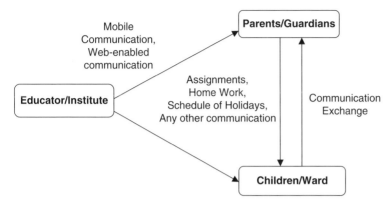

Figure 7.3 Learning through web and mobile computing.

7.2.1 Role of Web-Enabled and Mobile Computing in Education

Mobile computing has played an important role in effective learning among learners and educators. Through M-computing and web-enabled learning, educators communicate with parents and learners through online and SMS facility. The effective communication between parents and educators in a modern educational system through web-enabled tools is shown in Figure 7.4.

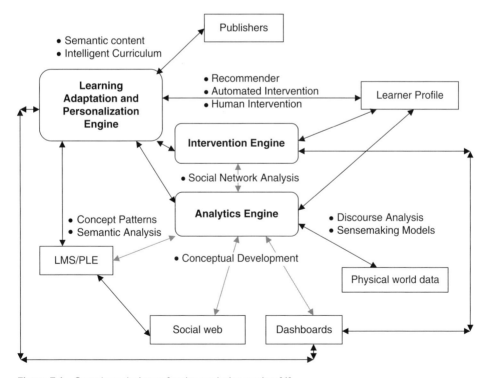

Figure 7.4 Sample techniques for the analytics engine [6].

7.2.2 Benefits of Learning Analytics

- It can enhance personalized efforts for at-risk students.
- The learner gets all the resources of attaining knowledge like reading materials, assignments, other information related to examinations, and registrations in time to study.
- It can encourage students with timely feedback of their performance.
- Teachers also get feedback from students about the teaching methodologies that have more impact on them.
- It can predict academic trends on the basis of data generated during the learning activities of learner and educators.
- It can support frequent revisions in the course syllabi as per the industry requirement.
- It can enhance interactive sessions between educators and learners in order to generate results to design the schedule of theory and practical concepts.

7.2.3 Future Research Directions of IDA

The future research areas of IDA are as follows:

- Deep artificial neural networks
- Industrial automation (manufacturing), information and communication technology and medical sciences
- Smart city projects
- Expert systems for medical sciences
- Prediction of climate changes using big data AI techniques
- Business intelligence
- Robotics
- Fraud detections
- Intelligent personal assistants using speech recognition
- E-mail filtering

7.3 Motivation

In the present scenario, our educational system is facing threats related to quality-driven education, lack of skills in students as per the industry demand, and problems in effective decision-making due to complexity in educational patterns.

The Hadoop MapReduce framework and association rule mining contribution for educational data mining has not been explored and is underutilized to mine the meaningful information from huge unstructured data accumulated in the form of complex data sets, educational entities, and patterns.

With big data solutions, comprehensive aspects about the student are obtained, which help colleges, institutes, and organizations to improve education quality and graduate skilled workers. There are many questions that are unanswered and are pertinent to the (i) learning observance, (ii) perceptive learning, (iii) curriculum trends, (iv) syllabi, and (v) future courses for students, which is where we will focus this chapter.

7.4   Literature Review

Distance and e-learning have given an effective alternative to traditional learning. In [7], the authors have discussed the MOOC research initiatives and the framework of MOOC research. MOOC future research framework includes: (i) success rate of student learning, (ii) curriculum design, (iii) social learning, (iv) social network analysis, (v) self-regulated learning, (vi) motivation, and (vii) attitude.

There is a need to develop learning analytical methods and techniques to improve learning [8]. The field [9] of learning analytics has opened new routes for educators/teachers to understand the need of learners with effective use of limited resources.

Chatti et al. [10] have discussed four measures for learning analytics. These measures are (i) data, environment, content, (ii) stakeholders, (iii) objectives, and (iv) methods. In order to improve learning, learning analytics need to be linked with the existing educational research [11].

According to Cooper [12], the techniques of learning analytics include the following: (i) statistics, (ii) business intelligence, (iii) web analytics (Google analytics), (iv) operational research (mathematical and statistical methods), (v) AI, (vi) data mining, (vii) social network analysis (geographical dimensions), and (viii) information visualization.

In [13], the authors discussed the significance of learning analytics and educational data mining. These fields transform data into meaningful information that benefits students, educators, and policymakers.

The authors in [14] emphasized predictions using existing models in educational data mining.

The applications of educational data mining helps to achieve the following: (i) improving learner models, (ii) improving domain models, (iii) pedagogical support, and (iv) scientific methods for conducting research on learning analytics of learners and educators [14, 15]. There are approaches and methods in educational data mining, which are as follows: (i) predictions, (ii) clustering, (iii) association rule mining, (iv) learning behavior of learners/educators, and (v) discovering educational patterns.

The author in [16] has stressed upon the four areas of EDM tasks, which include the following: (i) the development of applications for assessing students learning performance, designing the courses based on students individual behavior, (ii) development of methods to analyze and evaluate online material, and (iii) feedback analytics of learners/educators in e-learning courses.

E-learning and learning management systems are the main sources of data used in EDM. Apart from it, there are intelligent tutoring systems that help in analyzing student learning behavior, evaluating learning resources of students, and self-learning of learners [17].

7.4.1   Association Rule Mining and Big Data

Big data help in mining learning information that involves student performance and learning approaches. The author has compared the modern learning analytics with a traditional one where instructors rely on the sessional performance of students, whereas, in modern learning, instructors can analyze the individual performance of the student and the techniques that are most effective [18].

There [19] are web-enabled tools that help in evaluating the students' activities online. The activities are as follows: (i) time devoted by students in reading online, (ii) utilization of electronic resources, and (iii) how fast students understand online study material.

In [20], there are some pedagogical approaches that are effective with students. The approaches are as follows:

- Prediction of students dropout rate
- Students who require extra help
- Students who require assignments, mock tests, etc.
- Immediate feedback to teachers about the academic performance of students
- Tailor-made assignments for individual students
- Course design
- Grading of students
- Visualizations through dashboard for tracking student performance

Data mining and learning analytics fields have the potential for improved research, evaluation, and accountability through data mining, data analytics, and dashboards [21].

The authors have stressed the role of big data and analytics in the shaping of higher education, which involves ubiquitous computing, smart classrooms, and innovation in smart devices for providing education [22]. Big data and learning analytics pave the path for guiding reform activities in higher education and guide educators for continuous improvements in teaching. In [23], the authors have proposed an algorithm for efficient implementations of the Apriori algorithm in the MapReduce framework. MapReduce frameworks performs mining using map and reduce functions on large data sets of terabytes in size.

Association rule mining is a mining technique that finds frequent itemsets in a database with minimum support and confidence constraints [24].

Using the Apriori algorithm and MapReduce technique of Hadoop, frequently occurring itemsets in a data set can be identified [25].

Katrina Sin and Loganathan Muthu have stressed the need of educational data mining in the present setting of education where a large amount of data accumulated from massive open online courses (MOOC) [26].

There are some open-source tools of data mining like MongoDB and Apache Hadoop [27].

Patil and Praveen Kumar have performed the classification of data using MapReduce programming and have proposed a mining model for effective data analysis related to students pursuing higher education [28].

In a MapReduce framework, there is parallel processing of clusters for managing the huge volume of data [29].

MapReduce scales to a large array of machines to solve large computational problems [30].

Provost Foster and Fawcett Tom [31] have discussed the relationship of data science to big data and their significance in decision-making. A single node computer is inadequate with the storage of a large volume of data and processing demand increases beyond the capacity of the node. In such a situation, distributed data and parallel computations on the network of machines for the completion of the task is the preferable solution for faster computational processing and saving storage space [32].

Enrique Garcia et al. has discussed an educational data mining tool based on association rule mining. This tool helps improve e-learning courses and allows teachers to analyze and discover hidden information based on an interaction between the students and the e-learning courses [33]. Authors have presented association rule mining technique for assessing student data. The student's performance can be analyzed using association rule mining [34].

7.5 Intelligent Data Analytical Tools

There are intelligent data analytical tools, with the help of which, data mining is performed to extract the meaningful information from a large volume of data that may be structured, semi-structured, or completely unstructured.

In the present scenario, R programming is widely used for statistical computing and data analysis. KNIME (Konstanz Information Miner) is an open-source tool for data analysis.

The tools [35] like Microsoft Excel, Google Sheets, etc., are used on the data sets and are suited for the following: (i) manipulation, (ii) cleaning, and (iii) formatting. For constructing models and patterns, there are other tools like RapidMiner, WEKA (The Waikato Environment for Knowledge Analysis), MATLAB (Matrix Laboratory), SPSS (statistical analysis in social science), KEEL (knowledge extraction based on evolutionary learning), KNIME, and Orange, etc. Python has also emerged as a powerful language of machine learning for data analysis and modeling.

WEKA is the project of the University of Waikato, which gives an environment to researchers and data analysts in data mining and machine learning tasks. It consists of a collection of machine learning algorithms for the following techniques: (i) classification, (ii) clustering, (iii) association rules, (iv) regression, and (v) visualization. The process of data mining using WEKA is shown in Figure 7.5.

RapidMiner is a platform for data science. This software provides an environment for machine learning, predictive modeling, and data mining. It consists of machine and deep learning algorithms along with support for R and Python programming.

With the RapidMiner tool, data is explored to identify patterns followed with model validation. The business applications can also be integrated with models generated from this tool.

The results generated can be visualized in BI tools like Qlik and Tableau. Using this software on educational data sets, the software generates the phases as follows: (i) modeling, (ii) distribution model, (iii) distribution table, (iv) decision trees, and (v) plot views, etc. Figures 7.6 and 7.7 display the samples of decision trees and distribution table for the synthesized educational data set shown in Table 7.1.

MATLAB (Matrix Laboratory) is developed by Mathworks. With the help of MatLab, numerical analysis is performed. The data mining techniques like K-means and neural networks clustering techniques on educational data sets are performed in MATLAB. Figure 7.8a and b display the clusters obtained from MATLAB software for a sample of synthesized educational data sets. The attributes of sample educational data sets are categorized as (i) attendance, (ii) assignment, (iii) lab work, (iv) sessional performance, and (v) class performance. The students' class results are predicted after analyzing

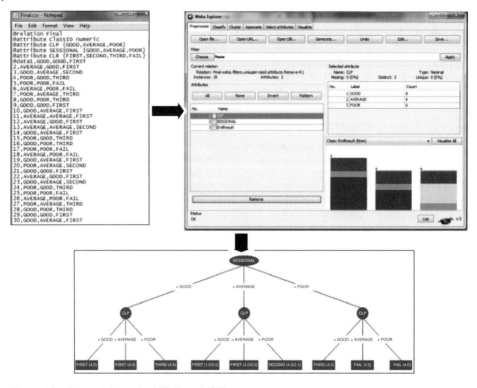

Figure 7.5   Data mining using WEKA tool [36].

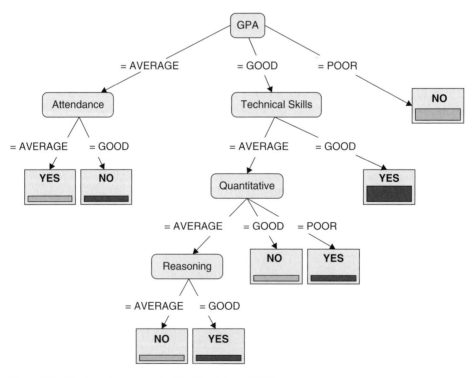

Figure 7.6   Decision tree generated for the data set [37].

Result Overview | SimpleDistribution (Naive Bayes)

Text View | Plot View | Distribution Table | Annotations

Attribute	Parameter	YES	NO
Attendance	value=GOOD	0.380	0.368
Attendance	value=AVERAGE	0.191	0.315
Attendance	value=POOR	0.428	0.315
Attendance	value=unknown	0.001	0.001
GPA	value=GOOD	0.854	0.315
GPA	value=POOR	0.001	0.525
GPA	value=AVERAGE	0.143	0.158
GPA	value=unknown	0.001	0.001
Reasoning	value=GOOD	0.665	0.263
Reasoning	value=AVERAGE	0.049	0.473
Reasoning	value=POOR	0.286	0.263
Reasoning	value=unknown	0.001	0.001
Quantitative	value=GOOD	0.238	0.420
Quantitative	value=POOR	0.380	0.263
Quantitative	value=AVERAGE	0.380	0.315
Quantitative	value=unknown	0.001	0.001
Communication Skills	value=GOOD	0.570	0.263
Communication Skills	value=AVERAGE	0.333	0.420
Communication Skills	value=POOR	0.096	0.315
Communication Skills	value=unknown	0.001	0.001

Figure 7.7 Distribution table for the data set [37].

Table 7.1 Educational data set [37].

Attendance	GPA	Reasoning	Quantitative	Communication Skills	Technical Skills	Placement
Good	Good	Good	Good	Good	Good	Yes
Average	Good	Average	Good	Average	Good	Yes
Poor	Good	Poor	Poor	Good	Good	Yes
Poor	Good	Good	Good	Average	Average	No
Poor	Good	Average	Average	Poor	Average	No
Poor	Good	Good	Average	Poor	Good	Yes
Good	Poor	Poor	Poor	Good	Average	No
...	...	...	...	...	...	...

performance in these attributes. The synthesized educational data set of students is shown in Table 7.2. Using the K-means clustering technique on this data set in MATLAB software has generated the following two clusters: (i) students who are short of attendance and (ii) students who have performed poorly in tests.

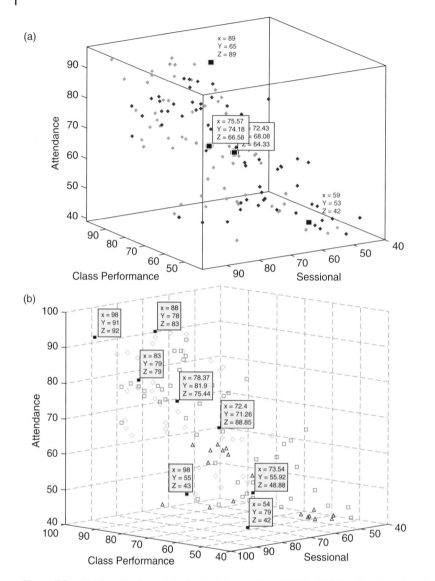

Figure 7.8 (a). Visualization of student attributes (K = 2) [38]. (b). Visualization of student attributes (K = 3) [38].

After applying the preprocessing and data mining models on the data set, Figure 7.8a shows the clustering of students, i.e., K = 2 and Figure 7.8b shows clustering of students, i.e., K = 3.

KEEL (http://www.keel.es/) means knowledge extraction based on evolutionary learning. It is Java-based software that supports the following:

a) Knowledge extraction algorithms
b) Preprocessing of data
c) Intelligence-based learning
d) Statistical methodologies

Table 7.2 Synthesized educational data set [38].

Class id	CLP	Sessional	ATTD	ASSGN	LW	CLR
1	78	79	89	9	23	79
2	79	80	89	10	22	81
3	71	70	76	10	23	71
4	52	79	55	4	10	80
5	52	50	45	3	12	51
⋮	⋮	⋮	⋮	⋮	⋮	⋮
100	78	80	78	9	20	79

SPSS means statistical package for social sciences and is developed by IBM. SPSS is mainly used for statistical analysis. In SPSS, we can import data from Excel spreadsheets for generating regression models, predictive analytics, and data mining on educational data sets.

7.6 Intelligent Data Analytics Using MapReduce Framework in an Educational Domain

7.6.1 Data Description

The data set shown in Table 7.3 consists of the courses preferred by students of computer science and information technology fields for industrial training purposes. This data set has been synthesized for the purpose of experimental exploration in the area of computer

Table 7.3 The data set for course selection.

Specialization	Preferable Technology
Mobile Computing<MC>	Android
Web Technology<WT>	PHP
Web Technology<WT>	Asp.Net
Mobile Computing<MC>	Android
Application Programming<AP>	Java
Application Programming<AP>	Java
Machine Learning<ML>	Python
Artificial Intelligence<AI>	Hadoop
Web Technology<WT>	PHP
Application Programming<AP>	Java
⋮	⋮
n specialized courses	n technologies

sciences only. In real life, the proposed system shall include all the areas of study at secondary and tertiary levels of education in formal as well as informal education sector. The output obtained using the proposed methodology can be tested on such practical applications, which may run into terabytes of data.

7.6.2 Objective

The objective here is to predict the interest of a student in training course(s) from various available combinations for training from the data set.

7.6.3 Proposed Methodology

7.6.3.1 Stage 1 Map Reduce Algorithm

Here, in the proposed methodology, work is performed on Hadoop 2.6.1 and JDK-8 on Ubuntu 12.04 machine. A Hadoop distributed file system stores the input data set shown in Table 7.3 for processing using the MapReduce framework. Using the framework, data will be parallelly processed on multiple clusters. Hadoop splits the task into two phases. The two phases used for parallel processing are as follows:

MAP phase: In MAP phase, the output is in the form of <key, value> pair.

REDUCE phase: In Reduce phase, the output from Map phase will be combined together and passed in Reduce phase. The Reduce phase will process each of the <key, value> pairs to generate the output. The final output obtained from Reducer will be stored in HDFS.

The working principle of the Map Reduce framework is shown in Figure 7.9, which depicts the methodology adopted, whereas the input data file obtained from Table 7.3 splits into different chunks on cluster machines and passes through different phases to get the summarized results. The unstructured data stored in HDFS after shuffling is refined to obtain an output and is called "Reduce phase."

The MapReduce technique uses two components when a request for information comes in, i.e., a Job Tracker and a Task Tracker. The Job Tracker works on the Hadoop master node, which is the Namenode, and the TaskTraker works on each node within the Hadoop network.

The advantage of Hadoop is the mechanism it follows for the distribution of data on multiple clusters, which results in the faster processing of data and the best utilization of storage space and the processing power of machines.

7.6.3.2 Stage 2 Apriori Algorithm

In stage 2 of the methodology, association rule mining is implemented on the synthesized educational data set shown in Table 7.3.

The association rules will be discovered through the Apriori algorithm where support and confidence are calculated for frequent itemsets. The support for each item is calculated in the data set and checked with a minimum threshold value set. The items not fulfilling the minimum threshold value will be eliminated. The rules will be selected only if minimum support and confidence threshold values will be satisfied.

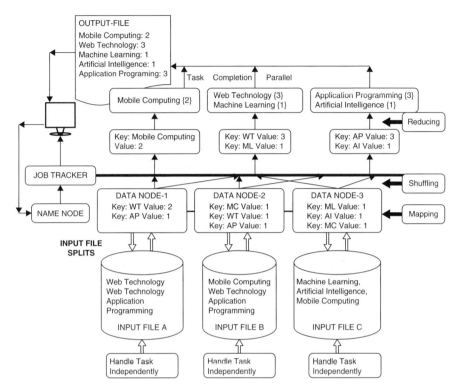

Figure 7.9 Working principle of MapReduce framework.

The equations for support and confidence of the rules are shown in Equations 7.1 and 7.2.

$$\text{Support } (i => j[s, c]) = p (i \cup j) = \text{support } (\{i, j\}) \tag{7.1}$$

$$\text{Confidence } (i => j[s, c]) = p (j \mid i) = p (i \cup j)/p(i)$$
$$= \text{support } (\{i \cup j\})/\text{support } (\{i\}) \tag{7.2}$$

Here s, c denotes the support and confidence, whereas p denotes the probability of occurrence of items in the database.

7.7 Results

The results shown in Figure 7.10 depict that a maximum of students have shown key interest toward machine learning and AI, in comparison to other program, i.e., application programming, web technology, and mobile computing, which helps management as well as faculty members to induct such specialized courses in course curriculum and conduct in-house training programs and workshops for the benefit of students.

Table 7.4 shows that maximum students have opted for specialization in AI, therefore students may opt for {Python, C++}-like technologies, whereas in application

Figure 7.10 Output obtained from MapReduce programming framework.

Table 7.4 Output of Map reduce task.

Specialized Courses<Key, Value>Pair	Students Count
Artificial Intelligence<AI,39>	39
Application Programming<AP,28>	28
Mobile Computing<MC,26>	26
Machine Learning<ML,33>	33
Web Technology<WT,28>	28
⋮	⋮
<N,N>specialized courses	<N>count

programming, students may opt for Java, DotNet, and PHP-like languages. In mobile computing, 26 students have shown interest, therefore, Android is the right program for training.

From Figure. 7.9, it can also be derived that after mapping, the shuffling task consolidated the relevant records obtained from the mapping phase. The summarized output is obtained through the reducing phase. After applying the Apriori algorithm to the output obtained in Figure 7.10, the best rules are displayed in Table 7.5. This table shows the preferable technologies for students to opt into and thus enhance their skills.

Table 7.5 Best rules found by Apriori.

Course Specialization = Artificial Intelligence(39)⇒**Preferable Technology** = {Python}
Course Specialization = Application Programming(28)⇒**Preferable Technology** = {Java}
Course Specialization = Mobile Computing(26)⇒**Preferable Technology** = {Android}
Course Specialization = Machine Learning(33)⇒**Preferable Technology** = {Python}
Course Specialization = Web Technology(28)⇒**Preferable Technology** = {Php,Asp.Net}

7.8 Conclusion and Future Scope

In this chapter, the authors have discussed in detail IDA and its role in the educational environment. IDA tools have also been discussed in the chapter.

The authors have synthesized an experimental data set that consists of courses related to the field of ICT and their attributes. The data set is processed through the proposed methodology of the MapReduce algorithm in the first stage and the Apriori algorithm in the second stage. The results and their analysis show that MapReduce and association rule mining can provide students the career counseling support that strengthens their decision-making to opt for the right course(s) for training activities as per industry requirements. Here, the experimentation has been limited only to the internship/training requirements of computer science engineering and information technology fields. In future, the authors intend to involve all the branches of engineering and technology in the first phase, other professional courses in the second phase, and lastly, a generic career counseling system with necessary appropriate enhancements in the third phase. Using not only Hadoop and Apriori algorithms, but with the inclusion of some machine learning/AI techniques, more meaningful information and academic patterns can be retrieved from relevant data sets on a larger scale in the future. The results so obtained are surely going to help educational institutions to find answers to some of the yet unanswered questions.

All these propositions have at the center of their focus, the improvement of the quality of education and employment prospects of the students. The proposed methodology is going to be the pivotal point in designing and implementing such support systems that will facilitate intelligent decision-making by parents, teachers, and mentors related to the careers of their children/wards/students and the strengthening of in-house training programs.

References

1 Berthold, M.R., Borgelt, C., Höppner, F., and Klawonn, F. (2010). Guide to intelligent data analysis: how to intelligently make sense of real data. *Texts in Computer Science* 42 https://doi.org/10.1007/978-1-84882-260-3_1, Berlin, Heidelberg: Springer-Verlag.

2 Hand, D.J. (1997). Intelligent data analysis: issues and opportunities. In: *Advances in Intelligent Data Analysis Reasoning about Data*, IDA 1997. Lecture Notes in Computer Science, vol. 1280. Berlin, Heidelberg: Springer.

3 Baepler, P. and Murdoch, C.J. (2010). Academic analytics and data mining in higher education. *International Journal for the Scholarship of Teaching and Learning* 4 (2) Article 17.

4 Scapin, R. (2015). "Learning Analytics in Education: Using Student's Big Data to Improve Teaching", *IT Rep Meeting – April 23rd*.

5 Liu, B. and Wong, C.K. (2000). Improving an association rule based classifier. *Journal In Principles of Data Mining and Knowledge Discovery*: 504–509.

6 Siemens, G., Gasevic, D., Haythornthwaite, C., et al. Open Learning Analytics: an integrated & modularized platform. Doctoral dissertation, Open University Press. https://solaresearch.org/wp-content/uploads/2011/12/OpenLearningAnalytics.pdf

7 Gašević, D., Kovanović, V., Joksimović, S., and Siemens, G. (2014). Where is research on massive open online courses headed? A data analysis of the MOOC research initiative. *The International Review of Research in Open and Distributed Learning* 15 (5).

8 Gasevic, D., Rose, C., Siemens, G. et al. (2014). Learning analytics and machine learning. In: *Proceedings of the Fourth International Conference on Learning Analytics and Knowledge*, 287–288. New York: ACM https://doi.org/10.1145/2567574.2567633.

9 Clow, D. (2013). An overview of learning analytics. *Teaching in Higher Education* 18 (6).

10 Chatti, M.A., Lukarov, V., Thus, H. et al. (2014). Learning analytics: challenges and future research directions. *eleed* (10).

11 Gasevic, D., Dawson, S., and Siemens, G. (2015). Let's not forget: learning analytics are about learning. *Tech Trends* 59 (1): 64–71. https://doi.org/10.1007/s11528-014-0822-x.

12 Cooper, A. (2012). *A Brief History of Analytics: A Briefing Paper*, CETIS Analytics Series. JISC CETIS http://publications.cetis.org.uk/wp-content/uploads/2012/12/Analytics-Brief-History-Vol-1-No9.pdf.

13 G. Siemens. (2011). What are learning analytics. Retrieved March 10.

14 Baker, R.S.J.D. and Yacef, K. (2009). The state of educational data mining in 2009: a review and future visions. *Journal of EDM* 1 (1): 3–17.

15 Baker, R.S.J.D. (2010). Data mining for education. *International Encyclopedia of Education* 7: 112–118.

16 Castro, F., Vellido, A., Nebot, A., and Mugica, F. (2007). Applying data mining techniques to e-learning problems. In: *Evolution of Teaching and Learning Paradigms in an Intelligent Environment*, 183–221. Berlin, Heidelberg: Springer.

17 Algarni, A. (2016). Data mining in education. *International Journal of Advanced Computer Science and Applications* 7 (6).

18 Manyika, J., Chui, M., Brown, B. et al. (2011). *Big Data: The Next Frontier for Innovation, Competition, and Productivity*. New York: Mckinsey Global Institute.

19 Castro, F., Vellido, A., Nebot, A., and Mugica, F. (2007). Applying data mining techniques to e-learning problems. *Studies in Computational Intelligence* 62: 183–221.

20 U.S. Department of Education Office of Educational Technology (2012). Enhancing Teaching and Learning through Educational Data Mining and Learning Analytics: An issue brief. In *Proceedings of conference on advanced technology for education*. https://tech.ed.gov/wp-content/uploads/2014/03/edm-la-brief.pdf

21 West, D.M. (2012). *Big Data for Education: Data Mining, Data Analytics, and Web Dashboards*. Washington, DC: Governance Studies at Brookings Institute https://pdfs .semanticscholar.org/5a63/35fa6a09f3651280effc93459f1278639cc4.pdf?_ga=2.36321056 .1417896260.1577346636-557630246.1577346636.

22 Siemens, G. and Long, P. (2011). Penetrating the fog: Analytics in learning and education. *EDUCAUSE Review* 46 (5): 30.

23 Lin, M.-Y., Lee, P.-Y., and Hsueh, S.-C. (2012). Apriori-based frequent itemset mining algorithms on MapReduce. In: *Proceedings of the 6th International Conference on Ubiquitous Information Management and Communication*, 76. New York: ACM.

24 Ma, B.L.W.H.Y. and Liu, B. (1998). Integrating classification and association rule mining. In: *Proceedings of the 4th International conference on knowledge discovery and data mining*.

25 Woo, J. (2012). Apriori-map/reduce algorithm. In: *Proceedings of the International Conference on Parallel and Distributed Processing Techniques and Applications (PDPTA)*. The Steering Committee of The World Congress in Computer Science. Computer Engineering and Applied Computing (WorldComp).

26 Sin, K. and Muthu, L. (2015). Application of big data in education data mining and learning analytics – aliterature review. *ICTACT Journal on Soft Computing*, ISSN: 2229-6956 (online) 5 (4).

27 Manjulatha, B., Venna, A., and Soumya, K. (2016). Implementation of Hadoop operations for big data processing in educational institutions. *International Journal of Innovative Research in Computer and Communication Engineering*, ISSN (Online) 4 (4): 2320–9801.

28 Patil, S.M. and Kumar, P. (2017). Data mining model for effective data analysis of higher education students using MapReduce. *International Journal of Engineering Research & Management Technology*, ISSN: 2278-9359 6 (4).

29 Vaidya, M. (2012). Parallel processing of cluster by MapReduce. *International Journal of Distributed and Parallel Systems (IJDPS)* 3 (1).

30 Dean, J. and Ghemawat, S. (2010). MapReduce: Simplified Data Processing on Large Clusters, Google, Inc,. In: *Proceedings of the 6th conference on Symposium on Opearting Systems Design & Implementation*.

31 Foster, P. and Tom, F. (2013). Data science and its relationship to big data and data-driven decision making. *Big Data*, Mary Ann Liebert, Inc. 1: 1.

32 Steele, B., Chandler, J., and Reddy, S. (2016). Hadoop and MapReduce. In: *Algorithms for Data Science*. Cham: Springer.

33 García, E., Romero, C., Ventura, S., and de Castro, C. (2011). A collaborative educational association rule mining tool. *Internet and Higher Education* 14 (2011): 77–88. https://doi.org/10.1016/j.iheduc.2010.07.006.

34 Kumar, V. and Chadha, A. (2012). Mining association rules in Student's assessment data. *IJCSI International Journal of Computer Science Issues*, ISSN (Online): 1694-0814 9 (5) No. 3.

35 Slater, S., Joksimovic, S., Kovanovic, V. et al. (2017). Tools for educational data mining: areview. *Journal of Educational and Behavioral Statistics* 42 (1): 85–106. https://doi.org/ 10.3102/1076998616666808.

36 Guleria, P., Thakur, N., and Sood, M. (2014, 2014). Predicting student performance using decision tree classifiers and information gain. In: *International Conference on Parallel, Distributed and Grid Computing, Solan, Himachal Pradesh, India*, 126–129. IEEE https://doi.org/10.1109/PDGC.2014.7030728.

37 Guleria, P. and Sood, M. (2015). Predicting student placements using Bayesian classification. In: *2015 Third International Conference on Image Information Processing (ICIIP)*, 109–112. Solan, Himachal Pradesh, India: IEEE https://doi.org/10.1109/ICIIP.2015 .7414749.

38 Guleria, P. and Sood, M. (2014). Mining educational data using K-means clustering. *International Journal of Innovations & Advancement in Computer Science* 3 (8): 2347–8616.

8

Influence of Green Space on Global Air Quality Monitoring: Data Analysis Using K-Means Clustering Algorithm

Gihan S. Pathirana[1] and Malka N. Halgamuge[2]

[1] School of Computing and Mathematics, Charles Sturt University, Melbourne, Australia
[2] Department of Electrical and Electronic Engineering, The University of Melbourne, Parkville, Australia

8.1 Introduction

In this century, industrialization has contributed to climate variation, and has adversely impacted the environmental, which has raised severe health problems. Even though the world is trying to find ways to heal the planet and preserve, protect, and enhance the global nature, deforestation, air pollution, and greenhouse gas emissions from human activities are now a greater threat than ever before.

Air pollution can be defined as the pollutants carbon monoxide (CO), sulfur dioxide (SO_2), particular matter (PM), nitrogen oxide (NO_2), and ozone (O_3) in such levels that they have a negative effect on environment and on human health [1]. Currently, air pollution kills around 7 million people worldwide each year, which is a horrendous figure and needs immediate resolution [2]. Health risk effects on pulmonary, neurological structure, cardiac, and vascular systems are some of the diseases that have emerged as a result of polluted air. In 2005, 4700 O_3-related deaths were attributed to air pollution [3]. Approximately 130 000 $PM_{2.5}$ deaths were reported in the United States due to the increase of ecological footprints. Unlike any other pollutant, PM creates health issues such as cardiovascular and respiratory diseases and specific types of serious lung cancers [4, 5]. Basically, PM consists of organic and inorganic elements such as sulfate (SO_4^{2-}), nitrates (NO_3), ammonia (NH_3), sodium chloride (NaCl), black carbon, mineral dust, liquid particles, and physical solids. Smaller particles less than 10 μm ($\leq PM_{10}$) can be identified as some of the most health-damaging particles; the human breathing system is unable to filter them [4, 5]. Uncontrolled population growth, urbanization, and the extermination of green spaces, fossil fuels that burn and emit exhaust gases – all of which are major causes of pollution and significantly impact global air quality.

Due to urbanization growth and city population, global green surfaces are substituted rapidly by continuous construction of massive concrete surfaces. Forests and human-made gardens (without cultivations) cover 31% of the land area, just over 4 billion hectares. Nonetheless, in preindustrial areas, it is going down to 5.9 billion hectares. According to the research data of the United Nations Food and Agriculture Organization, the results

Intelligent Data Analysis: From Data Gathering to Data Comprehension,
First Edition. Edited by Deepak Gupta, Siddhartha Bhattacharyya, Ashish Khanna, and Kalpna Sagar.
© 2020 John Wiley & Sons Ltd. Published 2020 by John Wiley & Sons Ltd.

of deforestation are at their highest rate since the 1990s. Annually, the biosphere has lost an average of 16 million hectares of green space [6]. Green spaces affect the air quality by direct elimination and by controlling air quality, reducing the spread of air pollutants, removing industrial pollutant use from local microclimates, and limiting the emission of volatile organic compounds (VOCs), which can add to O_3 and $PM_{2.5}$ formations [7]. Studies have exposed that trees, especially low VOC–releasing species, can be used as a viable approach to reduce urban O_3 levels. According to previous studies, forests absorb impurities in the air and this is a process that is mutual to all vegetation. In an exchange of gases, plants draw in carbon dioxide, alter it to make food, and discharge oxygen.

This shows that forests have a duty to enhance the environment air quality. Green spaces fight against pollution by sticking particles and aerosols to leaf surfaces, and by reducing the air movements that make them fall into the groundwater [1, 3]. Furthermore, the forests play a major role in decreasing greenhouse gas effects by eliminating CO_2 gas. Previous explorations have focused on urbanization and the capability of minimizing airborne PM and NO_2 by urban forestry. Several tree arrangements are able to adjust wind profiles. Also, the ability to generate wind inversions through forest structures assists in pulling pollutants from the air or can be used as physical resistance factors to prevent the diffusion of pollutants into the atmosphere [1, 3].

As a result of deforestation, air pollution is becoming a major environmental problem that affects human health and the global climate in significant ways. Additionally, some studies do not really focus on experimental samples of analyzed data related to the air pollutant elimination capacity of green space, but typically instead evaluate the Urban Forest Effects Model (UFORE) by Nowak [8]. Additionally, few studies show experiential evidence of the combination among overall ambient PM densities and urban forestry densities [9]. Studies put a reasonable amount of effort to prove or confirm the typical evaluations that have either been established to work considerably to reduce developments related to assessments from models [9] or those evaulations that have shown no positive effects [10].

According to the studies of Sanderson et al., if we ignore the current deforestation, it will result in overrages of 6% in the probable growth of overall isoprene releases and of 5–30 ppb surface ozone (O_3) levels because of climate modification over the years 1990–2090 [11]. Ganzeveld and Lelieveld initiated a study that had an important outcome on atmospheric chemistry from Amazonian deforestation; they did this by including solid reductions in ozone (O_3) dry sedimentation and isoprene emissions [12]. Lathiere et al. showed that tropical deforestation might gain a 29% decrease in universal isoprene emissions [11]. Ganzeveld et al. showed decreases in universal isoprene emissions and growths in boundary layer ozone (O_3) mixing ratios by up to 9 ppb in response to 2000–2050 variations in land usage and land cover [11].

Most of the previous studies discussed the consequences of anthropogenic land usage alteration on the global climate and ignored the potential atmospheric chemistry and air quality index. There are some studies that address why there needs to be a focus on the green space effects on global air pollution by only considering specific countries and cities. Nonetheless, there needs to be more research done in the area showing the connection between green space variation and a country's air pollution, as this is considered as critical and essential for this constantly growing problem. This study is expected to evaluate whether or not there is a visible relationship that exists or not between global air quality by considering PM and global forest density.

Table 8.1 Air quality categories (annual mean ambient defined by WHO).

Condition	PM value ($\mu g\, m^{-3}$)	Category
Not Polluted ($PM_{2.5}$)	<10	1
Starting Level	11–15	2
Medium Level	16–25	3
Risk Level	26–35	4
High Risk Level	36–69	5
Extremely High-Risk Level	70 more	6

8.2 Material and Methods

8.2.1 Data Collection

For this study a combination of the global green space and global air quality has been used that considers the big data set published by the World Health Organization (WHO) and World Food and Agriculture Organization (FAO) [13], which contains more than 1000 K (1 048 000) points of data collected during 1990–2015 from 160 countries.

8.2.2 Data Inclusion Criteria

In this research, the expected outcome focused on whether or not there is a relation between air quality and green space area. Thus, there are several approaches to measure the air quality. But this study has used data that has been reported in academic articles published 2014, which looked at air quality and measures data by considering PM. In the pursuit to collect data points for green space data, we used the database that was published by the United Nations FAO in 2015.

Air pollutants, such as carbon monoxide (CO), sulfur dioxide (SO_2), nitrogen oxide (NO_2), and ozone (O_3), are not considered, since those pollutants are not able to be absorbed by the tree surface. The records with missing values were removed and some attributes, such as country reference numbers and green space data records before 1990, were not considered as well, since those were not related to the aimed research. The flow of the process is described in Figure 8.1.

8.2.3 Data Preprocessing

The data set also contains hourly calculated air quality. In order to increase the accuracy of the measurements, this study has used the annual average values of air quality rates. The air quality data collected by using air quality measurement sensors located in the ground and

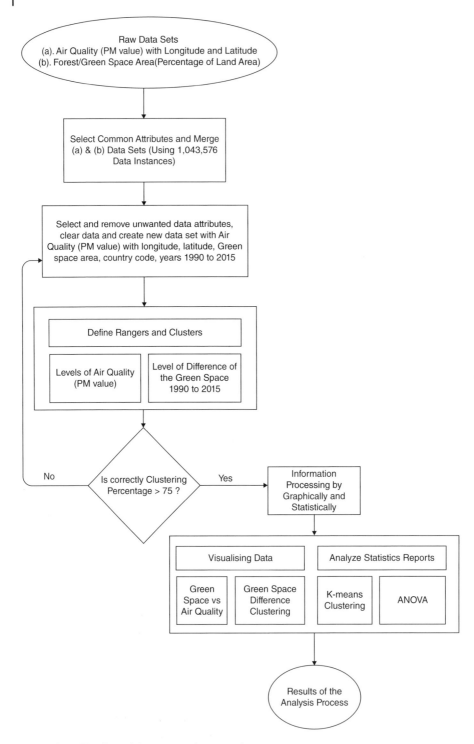

Figure 8.1 The flow of data processing procedure.

other optical technologies, including satellite retrievals of aerosol optical depth and chemical transport models, have been used in this study. This technique covers the estimates of annual exposures of $PM_{2.5}$ levels at high spatial resolution. Air quality data are testified in terms of annual mean concentrations of $PM_{2.5}$ fine particles per cubic meter of air volume (m^3). Routine air quality measurements typically describe such PM concentrations in terms of micrograms per cubic meter ($\mu g\, m^{-3}$) [4, 5]. Approximately, each data point covers $100\, km^2$ of land area ($0.1° \times 0.1°$, which equates to approximately 11×11 km at the equator) [4, 5] on a global scale. According to the definition of the air quality published by the WHO, it can be identified in several categories as follows.

According to WHO guidelines, a PM, an annual average concentration of $10\,\mu g\, m^{-3}$ should be selected as the long-term guide value for $PM_{2.5}$, and this study has followed this guide [4, 5]. This study also characterizes the minor edge range over which important effects on survival were detected in the American Cancer Society's (ACS) study [12]. Finally, air quality data was combined with green space data that have been collected by the United Nations FAO [13]. When collecting these records, it identifies and excludes green spaces that are relevant to the urban parks, forest cultivations below 5 m, and agricultural plantations. After normalizing the data records, we calculated the green space value of differences between each year and categorized them as follows: To recognize the relationship between green space and air quality we performed a static calculation process. For this purpose, we used MS Excel and IBM SPSS software that is selected purposefully for accuracy. Selecting accurate and powerful analyzing tools are prioritized. A Waikato Environment for Knowledge Analysis (WEKA) graph tool was used to identify the approximate geographical changes of the tree areas as well as the level of air quality.

8.2.4 Data Analysis

Using these data, it is expecting to identify whether green space area and air quality have a unique combination or correlation or how they are separated from each other. For this purpose, they are calculated in different tree area percentages of the country land area (df) between each year and identified by significant values and related countries.

$$df_{1990-2015} = \sum_{i=1990\cdots2015} df$$

Difference of the air quality α $df_{1990-2015}$

Then, they categorized those data according to the significance level and created class files accordingly. We analyzed and calculated the static values including a K_{means} cluster using the SPSS tool. We used SPSS and the WEKA tool for a graphical representation of the data using i5 3.0 GHz Windows 10 PC.

8.3 Results

An annual air quality for each country with longitude and latitude coordinated were used. Figure 8.2a shows the air quality ranges in the ground, and Figure 8.2b is a filtered image of Figure 8.2a that considers the intensity of the brightness. The dark areas show the risk

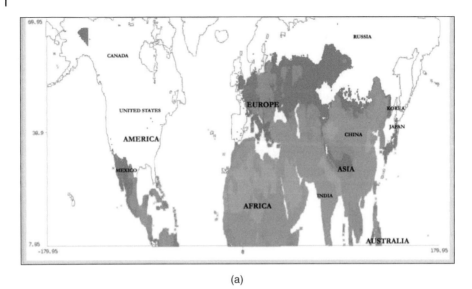

(a)

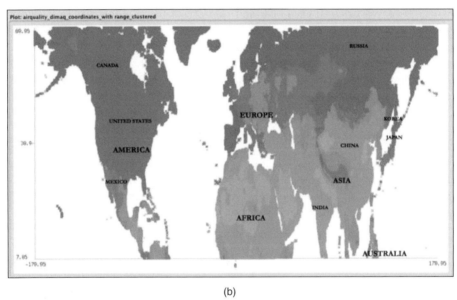

(b)

Figure 8.2 (a) Air quality with land areas in 2014 (using 1 048 576 instances). (b) Air quality value exceeds PM$_{2.5}$ (using 1 048 576 instances).

areas that exceed the PM$_{2.5}$ level and the ranges vary from risk level (above 25 µg m^{-3}) to an extremely high-risk level (70 µg m^{-3} or more) areas. By analyzing this, we can identify Latin America, Europe, and Asia as the low air quality areas. China, Korea, African Countries, Arabia, Germany, Italy, and Mexico belong to the high-risk areas.

In South Asia, China, India, Korea, Pakistan, and Malaysia have more forest areas compared with Africa. Thus, if the air quality is not at a satisfactory level it may affect and

Table 8.2 Categorization of the difference of green space area percentage during 1990–2015.

Difference of green space area percentage during 1990–2015 $(df_{2015} - df_{1990})$	Category
$(-32) < (-16)$	4
$(-16) < 0$	3
$0 < 16$	2
$16 < 32$	1

$df_{2015} - df_{1990}$ negative values present the deforestation.

Table 8.3 Analysis of variance (ANOVA) statistics table.

	Cluster		Error		F	Sig.
	Mean square	Df	Mean square	Df		
Air Quality	248 950 205.386	3	105.143	1 046 112	2 367 721.879	.000
@2014	87 762 175.753	3	87.536	1 046 112	1 002 588.270	.000
Difference	33 494.097	3	11.403	1 046 112	2 937.270	.000

Air Quality: value $(\mu g\,m^{-3})$ [1].
@2014: tree area percentage in 2014 [2].
Difference: tree area percentage variance from year 1990 to year 2014.

promote bad health conditions in conjunction with the absolute adverse effects of urbanization and the rapid development of the manufacturing field. In addition to that, the state of Alaska, northwest of Canada, is the largest and most sparsely populated US state and Latin America (Florida in the United States and Mexico), also poses a risk to the air quality level.

Figure 8.3a took consideration of tree area percentages of the ground in 1990 and Figure 8.3b is relevant to the 2014 with the longitude and relevant latitude. In this figure, intensity is regularly proportional to the tree area percentage of the countries. Some significant changes happened during these 24 years and we can identify some slide differences by comparing these two figures. Figure 8.3c represents the differences between Figure 8.3a and b. Europe, South Asia, and North America have the highest difference of the tree space according to Figure 8.3c.

8.4 Quantitative Analysis

8.4.1 K-Means Clustering

Figure 8.4 summarizes the relation between air quality and forest percentages in 2014 in the countries related to the ranges of air quality defined by WHO. According to Figure 8.4, when

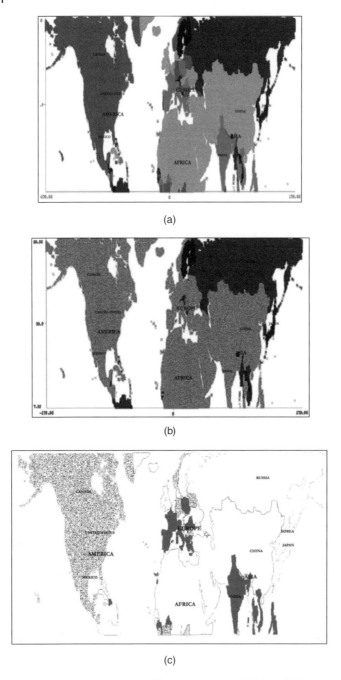

Figure 8.3 (a) Tree area in 1990. (b) Tree area in 2014. (c) Difference of tree area during 1990–2014.

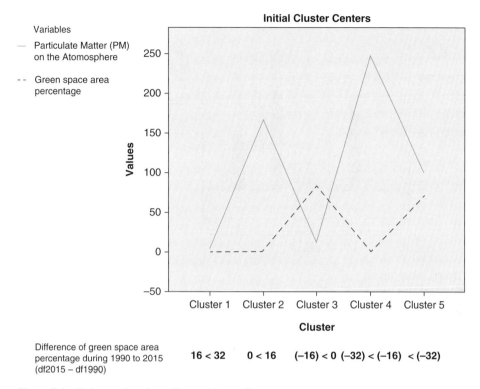

Figure 8.4 Variance of each attribute with coordinates.

the green area is increasing, it has a positive effect and reduces the value of PM in the air and causes good air quality (clustering number 2 to 3 and 4 to 5). Furthermore, decreasing the green space area increases the PM value and results in the increase of bad and risky air quality (clustering number 3 to 4).

By analyzing Figure 8.5, we can identify an inversely proportional combination between air quality and tree area. Figure 8.5 shows a quantitative representation of Figure 8.4. Through Figure 8.5, it can be identified that PM and green space don't have a 1 : 1 proportional relation. But it can be identified as proportional and a green area has a hidden ability to filter air and reduce a considerable portion of particulars. In cluster 2 and cluster 4 there are no considerable tree percentages when compared to the PM values, as it increases the PM value in an uncontrolled manner. Nevertheless, even PM value is high in cluster 3 and cluster 5, as the tree area becomes resistant to increased PM values. Green spaces act as a cofactor regarding air quality.

Figure 8.6 identifies clusters $1(PM_{2.5})$ that represent a smaller number of data that represents good air quality. Cluster 2–6 have represented air quality that exceeds $PM_{2.5}$. The missing pieces can be clearly identified through Figure 8.6, and the value amount of the data set is not a considerable value compared to the amount of the valid instances. Thus, it does not affect the research outcomes in a considerable manner.

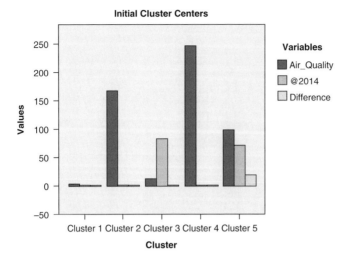

*Air Quality: value (μg/m3)
*@2014: tree area percentage in 2014
* Difference: tree area percentage variance from year 1990 to year 2014

Figure 8.5 Variance of each attribute.

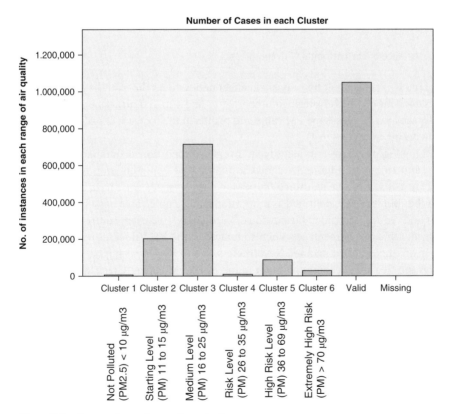

Figure 8.6 Count values of cases in each cluster.

8.4.2   Level of Difference of Green Area

A quantitative amount showing the growth of green space and loss can be identified using Figure 8.7; hence, it helps us to summarize the effect of deforestation on the air. It shows very little growth of green space in comparison with the level of decreasing green space in the past 10 years.

Figure 8.8 shows the combination between PM distribution and green space area percentage in 2014. When the PM value goes down, that is considered a good air quality level and the green space area percentage has a higher percentage value. Hence according to the Figure 8.8, we can indicate that the air quality is proportional to green space percentage.

Figure 8.9 is the further analysis of Figure 8.8 and the distribution of the PM data that follows exponential curve and PM > 70 limit can be identified approximately up to 24% of green space. It means, if a particular country has reduced the green space percentage below 24%, it has a higher probability to have extremely worse air quality level. Also, the same theory is valid for the high-risk level as it belongs to the green space under 65%. Accordingly, when the PM < 10 level, which is recommended by WHO, has a better annual air quality value that can be seen in a few countries that are maintaining at least 84% of their green spaces compared with the whole world.

8.5   Discussion

This study has observed a possible impact on its relation between atmospheric air qualities from 1990 to 2015 with the changes recorded from 190 countries. The green space density

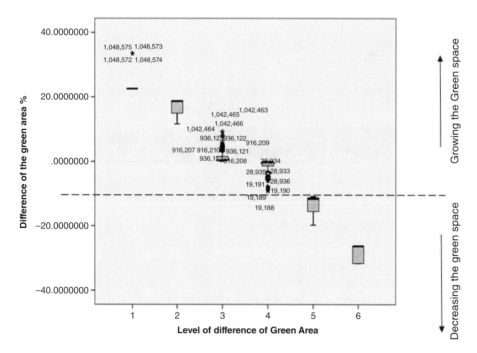

Figure 8.7   Tree area percentage/relation of raw data (difference) and ranges (level of difference).

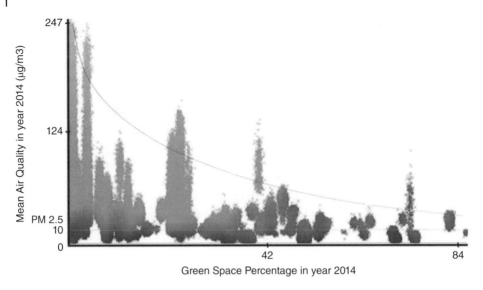

Figure 8.8 Air quality with green space percentage.

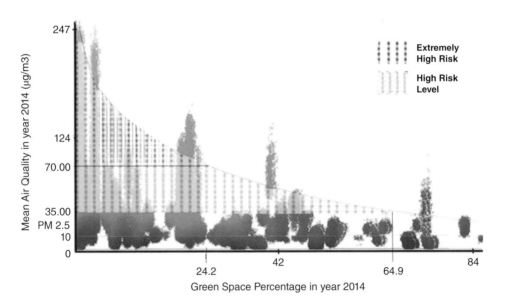

Figure 8.9 Air quality with green space percentage analysis.

has been analyzed and the data set contains more than 1000 K (1 048 000). The effects of deforestation, changing atmospheric chemistry, and air quality can highly impact the nature of the global atmosphere, including the changing life quality of living creatures. Previous studies have exposed and expressed the important effects of natural emissions, deposition, and atmospheric chemistry directly driven by changes in meteorology (such as temperature, humidity, solar radiation) linked with climate modification.

In order to differentiate the effects due to changes in land use/land cover, this chapter has identified green spaces of urban parks, forest cultivations below 5 m, and agricultural plantations, which have seen to be considerably affected by air quality change [14]. Furthermore, variations in troposphere ozone (O_3), nitrogen dioxide (NO_2), and sulfur dioxide (SO_2) in response to green space can also contribute to and affect air quality; however, this criticism is not considered in this chapter. Earlier studies also pointed out that variation in the chemical composition of the atmosphere, such as increasing ozone (O_3) concentrations, can affect vegetation [14] and these effects were also not considered in this chapter. According to some previous studies, in Southeast Asia, smoke from fires related with deforestation, mainly of peat forests, significantly increases existing urban air pollution, especially in El Niño years [15]. Complete tree removal causes an increase in wind speed and air temperature, and a decrease in water vapors, which leads to weakening of the monsoon flow over east China. G. Lee et al. report that observational air surface temperatures are lower in open land than boreal forested areas in Canada and the United States [16]. However, all above studies were limited only either to specific areas of the world or to a specific year. Additionally, some studies have carried out the relation between PM, global air quality, and the climate changes. Therefore, in order to maximize the previous research areas, scientists considered statistic algorithms with a big data set that consists of country-wise air quality values considering PM and country-wise green space values between 1960 and 2015. Further, we used a data set with big data to get more accurate results for this chapter, and in order to reduce anomalies, we used machine learning algorithms. An IBM SPSS modeler is used because it is foremost a data mining software, which can apply several algorithms for data preparation, data rectification, data statistics purposes, data visualization, and for analyzing predictives. A SPSS K-means clustering algorithm is used to cluster and classify those data into manageable groups. The K-means clustering analysis is an excellent approach for knowledge discovery and effective decision-making [17] in huge data sets, which also has seen good results in our previous research [18–20].

8.6 Conclusion

By analyzing air quality data sets of countries and country green space, this chapter identified that there is a relation between the green space density and the atmospheric air quality and for a sustainable prediction, a machine learning algorithm needs to be utilized with the collected results. Since some pollutants such as carbon monoxide (CO), sulfur dioxide (SO_2), nitrogen oxide (NO_2), and ozone (O_3) are not able to be absorbed by tree surfaces; those are not considered in this chapter. PM is considered as the significant pollutant that affects human health. Furthermore, different multivariate tables and graphs have been analyzed in this chapter to gain more accurate results. The outcomes point out that K-means clustering can be used to classify the air quality in each country. Additionally, the same research methodology can be used to carry out air quality, forest area, health effect, and climate-related studies. The gained result can be used to identify the relevant risk level of air pollution for individual countries, and it will help us take urgent actions to prevent future risks. Further, the results with the difference of green space value between each year will lead the relevant governments, World Wildlife Fund (WWF), WHO, etc., to take better actions to protect current green spaces and carry out forestation to prevent further environment changes and to neutralize the effects of air pollutants.

Author Contribution

G.P. and M.N.H. conceived the study idea and developed the analysis plan. G.P. analyzed the data and wrote the initial paper. M.N.H. helped to prepare the figures and tables and finalized the manuscript. All authors read the manuscript.

References

1 Chen, T., Kuschner, W., Gokhale, J. & Shofer, S. (2007). *Outdoor Air Pollution: Nitrogen Dioxide, Sulfur Dioxide, and Carbon Monoxide Health Effects.* [online] Available at: https://www.sciencedirect.com/science/article/abs/pii/S0002962915325933 (accessed 15 May 2017).

2 Kuehn, B. (2014). *WHO: More Than 7 Million Air Pollution Deaths Each Year.* [online] jamanetwork. Available at: https://jamanetwork.com/journals/jama/article-abstract/ 1860459 (accessed 22 Apr. 2017).

3 Irga, P., Burchett, M., and Torpy, F. (2015). *Does Urban Forestry have a Quantitative Effect on Ambient Air Quality in an Urban Environment.* [ebook], 170–175. NSW: University of Technology Sydney. Available at: https://www.researchgate.net/publication/ 281411725_Does_urban_forestry_have_a_quantitative_effect_on_ambient_air_quality_in_ an_urban_environment (accessed 16 May 2015).

4 World Health Organization. (2017). *Ambient and household air pollution and health.* [online] Available at: https://www.who.int/phe/health_topics/outdoorair/databases/en/ (accessed 22 April 2017).

5 World Health Organization. (2017). *Ambient (outdoor) air quality and health.* [online] Available at: http://www.who.int/mediacentre/factsheets/fs313/en/. (accessed 23 May 2017).

6 Adams, E. (2012). *Eco-Economy Indicators - Forest Cover| EPI.* [online] Earth-policy.org. Available at: http://www.earth-policy.org/indicators/C56/ (accessed 15 April 2017).

7 Nowak, D., Hirabayashi, N., Bodine, A., and Greenfield, E. (2014). *Tree and Forest Effects on Air Quality and Human Health in the United States,* 1e [ebook], 119–129. Syracuse, NY: Elsevier Ltd.. Available at: https://www.fs.fed.us/nrs/pubs/jrnl/2014/nrs_ 2014_nowak_001.pdf (accessed 19 April 2017).

8 Nowak, D., Crane, D., and Stevens, J. (2006). *Air Pollution Removal by Urban Trees and Shrubs in the United States,* 4e [ebook] Syracuse, NY: Elsevier, 115–123. Available at: https://www.fs.fed.us/ne/newtown_square/publications/other_publishers/OCR/ne_2006_ nowak001.pdf (accessed 19 April 2017).

9 Pataki, D., Carreiro, M., Cherrier, J. et al. (2011). Coupling biogeochemical cycles in urban environments: ecosystem services, green solutions, and misconceptions. *Frontiers in Ecology and the Environment,* [online] 9 (1): 27–36. Available at: http://doi.wiley.com/ 10.1890/090220 (Accessed 21 April 2017).

10 Setälä, H., Viippola, V., Rantalainen, L. et al. (2013). *Does Urban Vegetation Mitigate Air Pollution in Northern Conditions?* 4e [ebook], 104–112. Syracuse, NY: Elsevier. Available at: http://www.sciencedirect.com/science/article/pii/S0269749112004885 (accessed 23 April 2017).

11 Wu, S., Mickley, L., Kaplan, J., and Jacob, D. (2012). *Impacts of Changes in Land Use and Land Cover on Atmospheric Chemistry and Air Quality over the 21st Century*, 12e [ebook], 2–165. Cambridge, MA: Harvard University's DASH repository. Available at: https://dash.harvard.edu/bitstream/handle/1/11891555/40348235.pdf?sequence=1 (accessed 24 April 2017).

12 Ganzeveld, L. and Lelieveld, J. (2004). Impact of Amazonian deforestation on atmospheric chemistry. *Geophysical Research Letters*, [online] 31 (6), p.n/a-n/a. Available at: https://agupubs.onlinelibrary.wiley.com/doi/full/10.1029/2003GL019205 (accessed 27 May 2017).

13 Data.worldbank.org. (2017). *Forest area (% of land area)* | *Data*. [online] Available at: https://data.worldbank.org/indicator/AG.LND.FRST.Zs (accessed 26 April 2017).

14 Flannigan, M., Stocks, B., and Wotton, B. (2000). *Climate Change and Forest Fires*, 1e [ebook], 221–229. Syracuse, NY: Elsevier. Available at: http://www.environmentportal .in/files/cc-SciTotEnvi-2000.pdf (accessed 30 April 2017).

15 Marlier, M., DeFries, R., Kim, P. et al. (2015). *Regional Air Quality Impacts of Future Fire Emissions in Sumatra and Kalimantan*, 1e [ebook], 2–12. New York: I0P Publishing. Available at: http://iopscience.iop.org/article/10.1088/1748-9326/10/5/054010/meta (accessed 28 April 2017).

16 Varotsos, K., Giannakopoulos, C., and Tombrou, M. (2013). Assessment of the impacts of climate change on European ozone levels. *Water, Air, and Soil Pollution* 224 (6).

17 Wanigasooriya, C.S., Halgamuge, M.N., and Mohammad, A. (2005). The analysis of anticancer drug sensitivity of lung cancer cell lines by using machine learning clustering techniques. *International Journal of Advanced Computer Science and Applications*, [online] 8 (9). Available at: http://thesai.org/Publications/ViewPaper?Volume=8& Issue=9&Code=IJACSA&SerialNo=1 (Accessed 14 May 2007).

18 Halgamuge, M.N., Guru, S.M., and Jennings, A. (2005). Centralised strategies for cluster formation in sensor networks. In: *Classification and Clustering for Knowledge Discovery*, 315–334. New York: Springer-Verlag.

19 Halgamuge, M.N., Guru, S.M., and Jennings, A. (2003). Energy efficient cluster formation in wireless sensor networks. In: *Proceedings of IEEE International Conference on Telecommunication (ICT'03)*, vol. 2, 1571–1576. IEEE Papeete, Tahity, French Polynesia, 23 Feb–1 March 2003.

20 Wanigasooriya, C., Halgamuge, M.N., and Mohamad, A. (2017). The analyzes of anticancer drug sensitivity of lung cancer cell lines by using machine learning clustering techniques. *International Journal of Advanced Computer Science and Applications (IJACSA)* 8 (9).

9

IDA with Space Technology and Geographic Information System

Bright Keswani[1], Tarini Ch. Mishra[2], Ambarish G. Mohapatra[3], Poonam Keswani[4], Priyatosh Sahu[5], and Anish Kumar Sarangi[5]

[1]Department of Computer Applications, Suresh Gyan Vihar University, Jaipur, India
[2]Department of Information Technology, Silicon Institute of Technology, Bhubaneswar, Odisha, India
[3]Department of Electronics and Instrumentation Engineering, Silicon Institute of Technology, Bhubaneswar, Odisha, India
[4]Department of Computer Science, Akashdeep PG College, Jaipur, Rajasthan, India
[5]Student Coordinator, M2M Laboratory, Silicon Institute of Technology, Bhubaneswar, Odisha, India

9.1 Introduction

Data analytics (DA) is a technique where a data set is examined to extract a decision depending upon the contained information using a specialized system, software, and tools. Similarly, in space technology, a huge amount of terrestrial data has to be collected using various techniques and technologies. Primarily, the data contains a huge amount of earth data and space observation data that are basically collected using various spaceborne sensors. These collected data are combined in reference with data from other sources. More likely, a new broad area is coming up with sharp scope of challenges and breakthroughs: computational sensor vis-a-vis space sensor networks that can cover the whole electromagnetic spectrum, such as radio and gamma waves and the gravitational quantum principle. This data analysis can largely contribute to the universe and can also enhance life on earth. Figure 9.1 shows a generalized data collection scheme from various sources from space. Implementation of big data in space technology can provide a boost to this area by providing a common platform to various scientific communities and users. In addition to this, a broad effort is also awakening the public by offering new opportunities to uplift the individuality of a person in a society [1].

In space exploration, space communication and networking has emerged as a new research arena. In earlier days, the communication in space was done using radio signals that were blasted toward the antenna of a spacecraft within range. Moreover, the software platform used in different missions was not versatile enough and was absolutely different for every mission, which forced each mission to be mutually exclusive. A standardized, futuristic technology and interconnected novel network that can support space communication is a solution to this problem. The interplanetary internet (IPN) is a solution to this kind of problem.

Intelligent Data Analysis: From Data Gathering to Data Comprehension,
First Edition. Edited by Deepak Gupta, Siddhartha Bhattacharyya, Ashish Khanna, and Kalpna Sagar.
© 2020 John Wiley & Sons Ltd. Published 2020 by John Wiley & Sons Ltd.

Figure 9.1 Data collection from various sources from the space.

The IPN is capable of deep space exploration. It has a wireless backbone, even if the links are error prone, and only a slight delay range varying from minutes to hours whenever there is a connection. The basic principle on which the IPN works is called a store-and-forward network of internets. There are phenomenon, such as dust devils, magnetic interference, and solar storms that can disrupt communication among spacecraft. In addition to this, the farther spacecraft are from earth, the more likely they are to have obsolete technologies compared to the spacecraft launched very recently. Hence, the existing terrestrial and internet protocol suite will not be effective to overcome constrictions compounded by such extremities. Only a deep space network will be able to handle such extreme conditions. Protocols used in earth-based internet and planetary network can connect to the IPN backbone through mobile satellite gateways, and can seamlessly switch among protocols used [2], for example, a delay tolerant network (DTN), which is capable of integrating a bundle layer on heterogeneous lower layers. NASA has its own open-source, customizable overlay network called interplanetary overlay network (ION) that can implement DTN, which can be used for heterogeneous routing protocols.

The term big data from space has got its own meaning. It primarily refers to the amalgamation of data from spaceborne sensors and ground-based space observer sensors. Big data has three attributes, called three Vs, namely velocity, variety, and veracity. The volume of data generated by these is enormous, for example, the archived data are of an exabyte scale. The velocity (the rate at which new data are collected) is very high; the variety (data delivered from heterogeneous sources with various parameters such as frequency and spectrum) is extraordinary. The velocity and variety have veracity (the accuracy validated from the collected data). Hence, big data in space has got two dimensions, the capability to collect data and the ability to excerpt information out of the collected data [3].

Landsat was the very first revolution in earth observation (EO) in which the largest historical collection of geodata was made public in the United States in 2008, which is the largest collection of earth imagery. Similarly, the European Space Agency (ESA) of the European Union (EU) has created its own mark in EO through it sentinel twin-satellite constellation, the Copernicus program. ESA is indeed making EO truly an era of big data by producing up to 10 TB of data while gearing up to the full capacity of its program [3]. Additionally, inclusion of the data relay satellite of Europe (EDRS-A) to the orbit is a significant step to creating a space data highway. EDRS is on its way to creating Europe's own space communication network that can transmit data at a significantly higher rate (more than 1 Gbps), and in turn, will generate a staggering amount of data on the ground [3].

In this context, the United States has its own way of providing open access to satellites and data in association with the National Oceanic and Atmospheric Administration (NOAA) and various industries. Data, specifically earth imagery, from April 2016 generated by advanced spaceborne thermal emission and reflection radiometer (ASTER), a Japanese remote sensing instrument operating aboard NASA's Terra spacecraft, are free for use. Similarly, the Japanese Space Agency (JSA) provides a similar model from data generated by the advanced land-observing satellite "DAICHI" (ALOS-1) at no charge, which is used by various countries to use unlimited open-access data. Public and private sectors are the major investors in achieving this goal of building and launching collections of small satellites to orbit. Apart from satellites, other techniques used in geospatial studies also add volumes of data, which adds value to all the three Vs of big data, and furthermore, can be used by anyone.

Other countries around the globe are also taking the path to revolutionizing the EO in line with space observation. This, however, has boosted various communities to come up with a common umbrella of big data. The main aim of big data and analytics is to nurture innovation, research, science, and technology and other such related developments that are capable of handling challenges in EO as well as promoting extracting useful data in space science.

Similarly, the geographic information systems (GIS) is an information technology–enabled tool that can analyze, manipulate, visualize geographic information, and store data, usually in a map. In the past years, advancements can be seen in the development and implementation of tools that can attain, analyze, share, and store geotopological data. Technically, GIS illustrates two primary tuples <x,y> where x is a location including temporal dimension on earth's surface and y is an attribute of that location. These technologies have widespread application and are designed for data processing, data analysis, data modeling, and data storage. Using GIS information handling, such as systems for acquiring imagery from aircraft or space, otherwise known as remote-sensing systems, can also be accomplished. GIS started to show its presence in the late nineteenth century, and is of a greater significance in today's industrial sector.

GIS is primarily a collection of voluminous geographic data and documents such as images, graphic charts, tables, multimedia, and text along with the related exploration data and developments. The acute phenomenon is GIS analysis of the spatial geographic data along with storing and analyzing the data in line with the operation being performed on the data. GIS, however, is capable enough to interconnect geographic data with databases, manage data effectively, and overcome the data-storing issues involved.

Ever since its inception in the mid-twentieth century, GIS has been applied in problems related to spatiotemporal and decision-making [4]. With the increasing advancement in information technology, the GIS as a software platform is being witnessed as a steady evolution. The evolutions of GIS in the last several years are cited in Figure 9.2, which basically signifies several representative architectures of GIS [5–7].

Desktop GIS. Here, all information and programs are accomplished by central software; basically GIS application software installed in a stand-alone machine. All the geospatial data are stored in a stand-alone system and the computations are also managed in the same system.

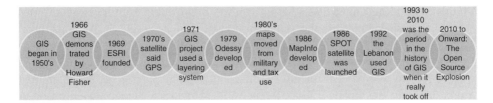

Figure 9.2 GIS evolution and future trends.

Client-Server GIS. Web technologies applied in traditional GIS model results in a client-server GIS. Implementations such as web-GIS and internet-GIS are adopted in this model. With reference to numerous GIS utilities to a client or a server, the methods can be considered as "thin-client, fat-server" and "fat- client, thin-server" [8, 9].

Component-based GIS: This sanctions GIS software to be fabricated by the integration of recyclable codes from multiple developers, which in turn can overcomes the disadvantage on overhead in code creation, such as various coupling strategies to link GIS with other software packages [1].

Service-oriented GIS: This is a combination of GIS services over the web services, along with the service computing technologies, which have been widely employed in the past decade and can support cyber spatial data infrastructure.

With the advent of big data technologies in GIS it has been possible to encounter the need to store massive geospatial data by processing and analyzing the data and visualization. However, the intense growth of spatiotemporal data, such as data captured from remote sensing, GPS, and many more, along with the three V, exerts new challenges to GIS. There is a need of integrating GIS and sensor data along with volunteered geographic information (VGI). Also there is a requirement of creating new and effective algorithms and data models to manage dynamic, unstructured, and multidimensional data. The computing requirements such as real-time analysis, information gathering from very large data set, and stream processing must be taken care of. Moreover, there is more research to be done toward data mining and data analysis of achieved data and information. Innovative methods are to be applied for mapping and visualization of data and displaying, analyzing and simulating geo-phenomenon.

9.1.1 Real-Time in Space

The real-time event detection in space has always been an area of interest. So, a basic question arises here as to the existence of the real-time entity in space. A spacecraft is programmed to accomplish a predefined task prior to sending it to space and any other spatial location. However, it is practically not possible to plan all possible situations, in order to make the best out of a spacecraft; each craft is provided with certain level of autonomy. Dust devils are an uncontrollable inflow of micro meteoroids and are an unpredictable but frequent phenomenon found out in detailed studies of space. The dust devil is a great challenge to encounter in space. Earlier. the dust devil was thought to be untraceable because of the limited resources of knowledge. However, it is difficult to know the occurrence of a meteoroid shower in real time due to the time taken in speed of light to earth. The real-time detection is possible by applying change detection mechanism on the spacecraft [10]. This mechanism, however, allows a spacecraft to pause observation tasks and focus on the meteoroid shower and capture the motion sequence of the dust devil. The captured motion sequence of the dust devil will alert scientists about the complete sequence back on earth. The required autonomy in a spacecraft is achieved by deploying more computational capability and analytic capability on the spacecraft or rover and authorizing the system to have an autonomous response system.

In the era of exponential advancement in information and communication technology (ICT), big data analytics has its own place as a cutting edge technology. Big data analytics is exceptionally backed by real-time earth observation system (EOS). It may be thought that the amount of data generated by a single satellite in EOS is not voluminous enough, but the inclusive data generated by various satellites can produce a sheer volume of data, maybe even in exabytes. Here the major challenge in EOS is to extract useful information from such a high volume of data. Challenges such as aggregation of collected data, storing, analysis of data, and managing remotely collected data are impending. This requires scientists to achieve a universal and integrated information system of EOS with associated services such as real-time data processing. In addition to this, there is a requirement of an ecosystem consisting of data acquisition, data processing, data storage, and data analysis and decision-making (explained in Section 9.3), which will provide a holistic approach that can ruminate data flow from a satellite to a service using big data architecture. This kind of system can provide a real-time mechanism to analyze and process the remotely captured EOS data.

There is a great deal of interest in ICT and advent because information technology has boosted the exponential growth in data volume and data analysis [11, 12]. The last two years have witnessed the generation of 90% of the whole data, as per a report from IBM [13]. This resulted in an appraisal of the big data concept as trendy and a cutting edge technology. This also has added a huge number of research challenges pertaining to various applications such as modeling, processing, data mining, and distributed data repositories. As discussed earlier, big data encompasses three Vs, namely volume, velocity, and veracity of data, making it difficult to collect, process, analyze, and store data with the help of currently used methodologies.

In a real-time scenario, huge amounts of data are to be collected for processing and analysis; this represents the volume. Additionally, high-speed processing and analysis of real-time data, such as data collected from online streaming (real-time remote sensing is another aspect); this represents the velocity. Finally, the validity of data collected from vastly different structures, such as internet of things (IoT), machine to machine, wireless sensor networks, and many more, is a very vital aspect to be considered in a real-time scenario; this represents the veracity.

However, existing services, such as web services and network devices, are also part of generating extensive data and it can also be expected that in a real-time environment the majority of data will be generated from different sensors. According to Oracle, sensors will soon generate data in scale of petabytes [14].

In a way, the progresses in big data and IT have transformed the fashion in which remote data are collected, processed, analyzed, and managed [15–17]. The continuous data streams are generated by integrating satellites to various EOS and other platforms. These continuous streams of data are generally termed as "big data" that are putting a day-to-day challenge to the real-time environment [18]. This results to a critical task, as it involves a scientific understanding of the remotely sensed data [19]. Moreover, because of the rate of increase in the volume of real-time data, shortly there will be a demand of a potential mechanism that would efficiently aggregate, process, and store the data and sources in real time.

Data acquisition is the first step in remote sensing, where a satellite generates a huge amount of raw data by monitoring the earth continuously. The data captured in this phase are raw data of no interest that are to be cleaned at different orders of magnitude. Filters used in the data cleaning process don't discard useful information. Also, a generation of a précised metadata is equally important that primarily describes the data composition and the data collection and analysis mechanism. Analysis of such metadata is of course a challenging task that requires us to understand the source of each data in remote sensing.

9.1.2 Generating Programming Triggers

In order to achieve the programming triggers, the spacecraft uses the onboard systems that runs a variety of algorithms about priorities, resource management, and functional capabilities and also interrupts its ongoing activity. Again, one question still remains unanswered: what this trigger is? It can be understood that a system of such capabilities and response has to be programmed exhaustively. The real issue here is how to make a trigger generalized so that it can be applied to multidimensional scenarios [20]. However, an event detection and response undoubtedly requires multidisciplinary amalgamation, such as atmospheric scientists, planetary geologists, and other experts who can explain events.

Moreover, a robust enough technique and contemporary algorithm have to be devised to detect such an event and apply a significant trigger herein. The architecture needs to be designed with an onboard facility that can allocate resources and required capabilities to activate a trigger when an event is occurring. Numerous action items would be required to achieve this response. Taking a camera as an example: it needs to focus at a correct direction and its power must be available. Safety of the vehicle is also to be considered. These are primarily the building blocks of detecting an event and designing the response framework [21, 22].

9.1.3 Analytical Architecture

The sources used in big data are varied, such as social networks, web services, satellite imagery, senior data, and many more. Irrespective of its evolution, various literatures would be found pertaining to big data analytics architecture. In this context, the remote-sensing big data architecture has got its own significance to boost the efficiency of data collection mechanism used in space science [23]. Figure 9.3 shows "n" number of satellites that captures the earth observatory big data imagery, along with sensors and cameras using these images that are stored with the help of radiations. In order to produce conventional and thematic maps, resource surveys' cutting edge technologies are applied to interpret remote sensing imagery. In the whole process, the big data architecture is divided into three aspects, namely, remote sensing data acquisition unit (RSDAU), data processing unit (DPU), and data analysis and decision unit (DADU).

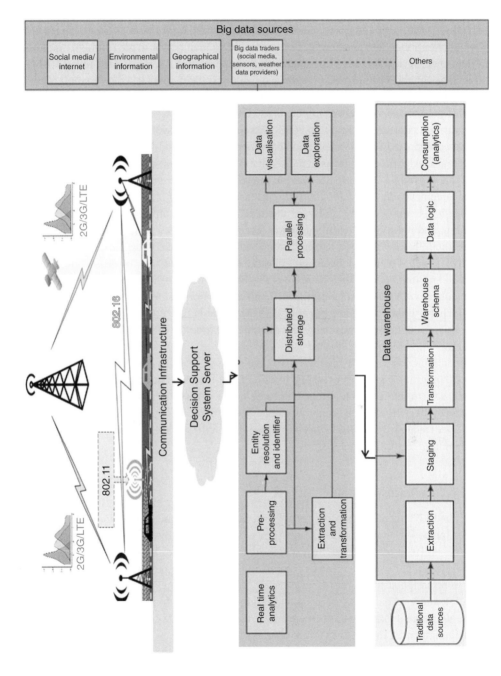

Figure 9.3 Remote sensing big data architecture.

9.1.4 Remote Sensing Big Data Acquisition Unit (RSDU)

To satisfy the computational requirement, remote sensing can be used as an earth observatory system that is a cost-effective parallel data acquisition system. This is an approved solution, by the earth and space science society, as a standard for parallel processing in EO [24]. Conventional data processing cannot provide enough power to process high amounts of data because a satellite is integrated with a high amount of sophisticated units, for EO, so as to be capable of improved big data acquisition. Indeed, there is a requirement of parallel processing of such huge data and this can be effectively analyzed in big data. So, the RSDAU proposes a remote-sensing big data architecture that is capable of collecting data from various satellites around the globe. There are possibilities that the received data are corrupted/distorted by scattering and absorption by gasses and dust. The remote sensing satellite uses algorithms such as SPECAN or Doppler to extract an image format from the raw data, presuming the satellite has the capability to correct the erroneous data [25]. In order to achieve effective data analysis, decrease storage cost, and improve accuracy, the satellite is able to preprocess and integrate data from various sources. Data integration and cleaning, and redundancy elimination are few of the followed relational data preprocessing methodologies [26]. Data is then transmitted to the ground station through a downlink channel, which is usually done using a relay satellite in a wireless atmosphere with suitable antenna and link. Taking various reasons for distortion in the data, such as movement of the platform with reference to earth, earth curvature, nonuniform illumination, dynamic sensor characteristic, and platform attitude, various methods are applied to correct the data. After this, the data are forwarded to the earth base station for further processing via direct communication link.

The data processing is divided into two stages, first, a real-time big data processing and second, an offline big data processing. In case of real-time big data processing, the data are directly sent to the filtration and load balance server (FLBS) because if the incoming data are stored, this will degrade the performance of the real-time data processing. In the latter case, the data are transmitted by the earth base station for storage to a data center. The data are then used for future analysis.

9.1.5 Data Processing Unit

Here, the FLBS primarily has two tasks. First, data filtration and second, load balancing of processing power. In data filtration only the useful data for analysis are identified, as it only allows useful information, and the rest of the data are discarded. This, in turn, enhances the overall system performance. The load balancing of processing power facilitates the filtered data to be assigned to various processing units in chunks. The algorithm applied toward filtration and load balancing varies according to the analysis done. For example, if system has a requirement to analyze only the sea wave temperature data, only these data are filtered out and segmented into chunks and forwarded for processing.

The processing server has its own implementation of algorithm to process the incoming chunk from the FLBS. A processing server makes statistical, mathematical, and logical calculations to produce results for each chunk of incoming data. The servers carry out tasks mutually exclusive and work parallel, the performance of the system is highly enhanced, and the results of each chink of data are generated in real time. The results from each processing server are forwarded to an aggregation server for compilation, organization, and storing for further processing.

9.1.6 Data Analysis and Decision Unit

An aggregation and compilation server, result storing server, and decision-making server are the major servings of this unit. The DPU sends only partial results to the aggregation and compilation server, which are unorganized and are not compiled. So, there is a requirement that the related results be aggregated, organized in a proper shape for processing, and stored. Moreover, the aggregation and compilation server supports various algorithms that can compile, organize, store, and transmit results. However, the algorithms vary depending on the type of analysis being performed. The aggregation and compilation server, with a purpose that the result be used by any other server anytime, stores the compiled and organized results in resulting storage. Moreover, the aggregation server forwards a copy of the result to the decision-making server to carry out a proper decision made on the given results. The decision-making server supports various decision-making algorithms that can inspect different aspects from the results and consequently help to make decisions, such as analysis of land, sea, and ice, and findings, such as fire, tsunami, and earthquake. The algorithms are, however, robust and efficient enough to produce results that are capable to extract unknown things, which is helpful for decision-making. Moreover, the decision-making server is of high significance because it has a low error margin; even the smallest error in decision-making will degrade the efficiency of the overall data analysis. The decision is finally broadcast so as to be used by various applications and services, such as business software, community software, and many more, in real time to data analysis and other such related developments.

9.1.7 Analysis

The primary objective of remote sensed image analysis is to cultivate robust algorithms that would support the discussed architecture and monitor various geographical areas, such as land and sea, and additionally, can be able to appraise the system. The satellite data are captured from various satellite missions that monitor the earth from almost 1000 km from the earth's surface. For example, the ENVISAT data set, captured by the ENVISAT mission of the ESA using advanced synthetic apertures radar (ASAR) satellite sensors, monitors various geographical areas, such as deserts, forest, city, rivers, roads, and houses, of Europe and Africa.

9.1.8 Incorporating Machine Learning and Artificial Intelligence

Machine learning (ML) is a software application that provides accurate prediction outcomes without being explicitly programmed. The basis of ML is to build algorithms that can provide predictive and accurate output and updates the outputs on the arrival of new data. With the inclusion of AI, the ML system is given an ability to automatically learn and improvise

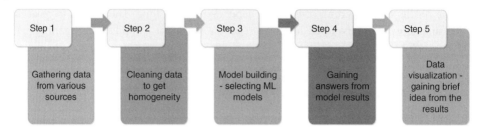

Figure 9.4 The machine learning process.

from prior experience. The common process for the machine learning model is shown in Figure 9.4.

So, ML can be inferred as an essential component in this process. Examples of events are identified in the training phase. To make it possible for the computing system, a generalized pattern is determined through various techniques. A data set for similar events are then found out as an additional training example. This overall process is referred to as a supervised learning model, as at every stage the model is to be trained.

Depending on the noisiness of the system and how the system responds, the training sets vary to achieve success. In general, a system may have more than 100 example training sets. But, the solution solely depends on the context of the problem and behavior of the data set. However, training a model as per the requirement is a very tedious job and subject to expertise.

9.1.8.1 Methodologies Applicable

The term big data has created a need for parallel and scalable analytics for technologies, such as image processing and automated classification. The primary goal of remote sensing from satellite is to comprehend the experimental data and categorize (classification) significant features of different geographical entity types. Support vector machines (SVMs), a parallelization technique, can be applied so as to accelerate the categorization (classification) of these geographical entities [27]. SVMs are one of the widely applied methods in remote sensing. Algorithm implementation, the analysis of remote sensing data sets, primarily poses four scalability subjects, namely (i) number of different data classes, (ii) number of features, (iii) parallel computing nodes, in term of available processing and memory, used and (iv) actual volume of data and input output throughput. Out of these four subjects, data classes (i) and features (ii) are driven by software such as data set and parallel computing (iii) and I/O throughput (iv) are driven by hardware technology.

9.1.8.2 Support Vector Machines (SVM) and Cross-Validation

The SVM algorithms segregate training samples as those belonging to different classes by tracing the maximum margin hyperplanes in the space where the samples are mapped. In general, complex classes are not linearly separable in the novel feature space. So, to generalize the samples to a nonlinear decision function, a kernel trick such as radial basis function (RBF) kernel is used [28]. Moreover, a cross-validation mechanism is used for SVM model selection to attain the SVM and kernel factors systematically.

9.1.8.3 Massively Parallel Computing and I/O

The massively parallel computing is alternatively termed as high performance computing (HPC) and has two primary significances, namely, (i) faster and reliable process intercommunication and (ii) parallel input and output (I/O). Message passing interface (MPI) is used in order to achieve a faster and reliable process intercommunication because MPI is standardized to perform process intercommunication across parallel computing nodes with distributed memory [29]. Similarly, binary hierarchical data format (HDF) standard supports parallel I/O by storing data and metadata in the same file [30]. The parallel I/O is useful when a parallel file system (GPFS) is encountered making it compatible with big data sets in parallel environment. HDF with parallel I/O reduce the time complexity in reading the input data in a cluster.

9.1.8.4 Data Architecture and Governance

The source of the data set is the primary element in making a system reproducible. The source of the data set in turn depends on the system that generates it. The need for search and analysis can be handled by modern and sophisticated data models and taxonomies. Implicit data architecture was in use for a long period of time. Gradually, a data dictionary was developed, which was used to manage the changes. Prior to the evolution of a sophisticated software system it was very difficult to link data architectures, to contextualize observation data, and to capture and archive data [31]. Turning into a nonimplicit data design, recognizing the prominence of data architecture and software classifications are linked so that a model-driven architecture can be evolved.

Eventually, to describe the planetary data that may include mission data, instrument data, observation data, and all the types of data, a sophisticated system can be designed. The data definition can be integrated with a big data system so that the software can adopt any ontological changes in the system. The relationship between common applications and various machine learning models is depicted in Figure 9.5. Hence, the data can be generated and validated at the ontology level and can be shared across missions.

From the futuristic information network point of view, remote sensing via satellites, mobile communication technologies, and space exploration are the upcoming services with applications in various scenarios such as city, mountain, ocean, and forest.

When users on the ground access the user access system of the space network, the channel becomes more perplexed than that of the conventional space communications and earthly wireless network. The use of big data in remote sensing is shown in Figure 9.6. The channel diverges with a variation of relative angle and attitude between the space nodes and ground terminals. That the fading is slow in the channel is a result of propagation and shadowing and the channel fading is fast as a consequence change in attitude and movement of vehicles. In case of low elevation of the satellite the channel selects a frequency and when the satellite elevation is high the channel frequency becomes Gaussian. Shadowing is absent on the ocean, however, the sea surface has a stronger reflection. In urban areas and mountains, the channel is much worse. In the case of a longer broadcast distance, a weak signal is received with a larger Doppler shift. Gaining and processing such a weak signal is a perplexing task [32]. Moreover, another challenge here is to improvise the energy efficiency and spectral efficiency of the space nodes because the power and energy available to each space node is limited; techniques such as green radio may be considered

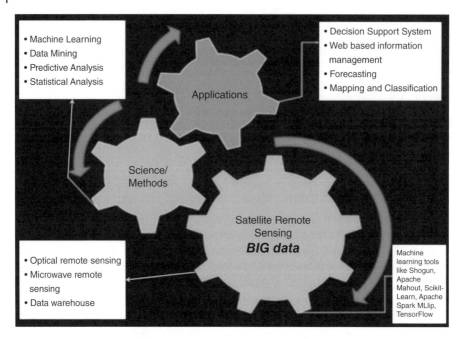

Figure 9.5 Big data in remote sensing.

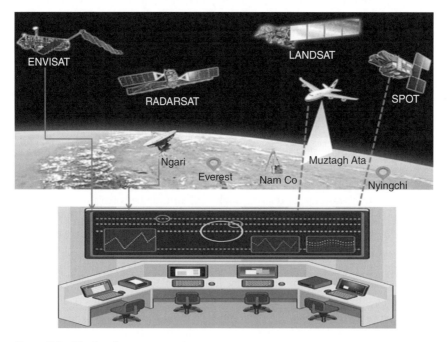

Figure 9.6 Big data in remote sensing.

to address such issues. To incorporate coordination among multiple space nodes, cooperative techniques are implemented to achieve a diversity gain to resolve the channel fading.

A burst of service is a primary feature provided by mobile communication and the user trend is varied, with a plethora of services available. So, there is a burning requirement of space information network to address such emerging trends. For example, in a city a user can be blocked and shadowed using a space information network. Moreover, the space information network must be in line with an earthly wireless system to overcome the challenge of a dynamic, weak signal by providing service to handheld devices [33].

Ground stations receive real-time sensing data transmitted from remote sensing satellites using the space information network, thereby reducing the overhead satellites flying over ground stations. In this scenario a larger data rate is what is more important than that of the delay. Here, the channel is capable of selecting a frequency to operate, and multiple point transfer methods are implemented to achieve a data rate [34].

9.1.9   Real-Time Spacecraft Detection

Real-time event detection and data collection such as volcanic eruption, ice-sea breakup, forest fire, and so on... are even closer to the periphery of earth rather than the fact that these phenomenon take place on earth. However, integrating the data analytics in cross sequence along with the data life cycle is a very critical process. In this process the data is created, captured, processed, managed, and finally distributed. Various techniques, namely ML, AI, computational intelligence, and statistical algorithms are to be applied to the data life cycle so as to integrate distributed data with scientific archive data.

As a limitation in the number of ground tracking stations resulting in low coverage percentage of terrestrial space, responsive nodes of space system are primarily responsible for tracking the spacecraft, interchange data, and command information and sending the information back to the tracking station via the space information network. This tracking service requires less delay in transmission with higher reliability in accessing the space system of information [33].

Variation of attitudes between the responsive nodes and the spacecrafts targeted and Doppler shifts caused by relative motion make the channel, which is considered as Gaussian, fast fading thereby making anti-interface capability as a requirement for the channel [35]. To facilitate remote sensing of satellite and tracking a spacecraft, specific ground stations are positioned appropriately. Methods to access the space information system are designed considering the available distribution and scale and the futuristic fashion of services primarily focusing on sensitive and disastrous areas.

In the course of spacecraft detection mechanism, it requires specific technical requirements, specification, and parameters of each service, which includes data rate, time delay, bit error rate, etc. Moreover, it also takes the channel fading, propagation distance, and characteristic study of various geolocations into consideration. The channel fading requires a bit of modeling and simulation with design significance in modulation scheme, power, and radio access and to allocate proper resource priority of each space service is determined [36].

Hence, there is a requirement of such physical interfaces that can be set up on the basis of development, which must be futuristic and must satisfy the regulation of both space and terrestrial network, and nonetheless, must be compatible with current and future wireless communication systems. The implementation, if it has to be homogeneous with current and future terrestrial wireless communication systems, requires a unified physical link interface. The development should be based on international regulations for space and terrestrial networks and should be compatible with future trends.

9.1.9.1 Active Phased Array

Frequency division and space division multiplexing are realized using narrow point beam by multiple beam satellites, which suppress interference and increases throughput. Understanding the distribution of ground users is a major factor, in order to schedule the beams effectively. The distribution varies widely in different cities and in different geographical areas. A precise and efficient mathematical distribution model can help beaming to be more flexible and efficient. Performance improvements can be achieved by adopting MIMO based on a cross-layer optimization and antenna diversity [37].

9.1.9.2 Relay Communication

Ground terminals are easily shadowed and the line of sight is blocked in city and mountain areas. Ground relay nodes in association with space nodes can increase the coverage area to minimize the problem. A heterogeneous communication network is formed when integrated with a terrestrial mobile communication system, thus helping ground relay nodes to work in path diversity and packet retransmission. The scenario can give rise to a spectrum coordination problem, which can be addressed with the help of a cognitive radio and cross-layer approach and which can be applied to get seamless integration and optimized resource allocation.

9.1.9.3 Low-Latency Random Access

Latency is a service-dependent phenomenon. Out of these services, the spacecraft tracking is the hardest to achieve and it requires an exclusive channel. Moreover, the telemetry part of the channel is not too extreme and depends on circumstances. A substantial number of users will burst and participate in the channel access in the growing communication era.

9.1.9.4 Channel Modeling and Prediction

In a space-terrestrial hybrid network, the access channel has multiple paths and is time-varying. Signal fading is caused by environmental factors, as well as the relative altitude change of the antenna. For an increased channel capacity, the delay of channel feedback should be less than its correlation time. A channel that varies at a faster rate with respect to time is always hard to predict. A fast time-varying channel is hard to predict. Network coding is implemented to reduce the latency and enhance the system throughput in order to improve this network. An innovative model is to be applied when multi-access inference of different terminals makes the channel no longer a Gaussian one, even between two space nodes.

9.2 Geospatial Techniques

Various geospatial techniques are followed, but this can be classified into three broad classifications. Position metrics calibrate a geographical position to a location, data collection mechanisms collect GIS data, and data analysis models analyze and use the collected data [9]. A pictorial view of the various geospatial techniques is shown in Figure 9.7.

9.2.1 The Big-GIS

Big-GIS is a consolidation of big data technology with GIS and big data in GIS or the Big-IS and has a very high significance over the data characteristic. Moreover, it is an ample grouping of knowledge discovery with data analytics. Big-GIS handles challenges related to this sheer volume of data using cutting edge information science, IoT, distributed and HPC, web and cloud GIS, and cyber GIS.

The primary purpose of Big-GIS is to sense, process, integrate, visualize, and store a large volume of geospatial data. Keeping the traditional methods of GIS as well as technologies used and data management in view, Big-GIS provides solution to make GIS more efficient. The basic characteristics of Big-GIS are coordinated observation, HPC, parallel processing, knowledge discovery, distributed storage, and efficient geospatial analysis. Big-GIS will definitely overcome issues related to voluminous data management, data processing and visualization, and data analytics in the field of geospatial study. Big-GIS makes it possible to extract information from varied sources of data by efficiently controlling a sheer volume of spatiotemporal data, thereby influencing problem solving and decision-making.

9.2.2 Technologies Applied

An efficient data model, information and network infrastructure, and computing models are to be improvised to meet the challenges compounded by big data. Various technologies, such as NoSQL, cloud computing, stream processing, and big data analytics, are available and those can address data management, parallel processing, data analysis, and visualization challenges.

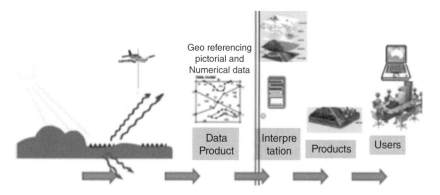

Figure 9.7 Geospatial techniques.

9.2.2.1 Internet of Things and Sensor Web

The advancement witnessed in IoT and web of things (WoT), where a huge amount of sensors produce voluminous data, are the prime contributors to sources in big data. Now, these two sensor-based technologies (IoT and WoT) influence high-speed data capture and provide ease of access to data, and as a result, influence the big data trend [38]. These technologies will continue to deliver colossal interpretations of data, as the technologies are omnipresent and can be easily deployed anywhere.

9.2.2.2 Cloud Computing

This is a new trend in parallel and distributed computing, HPC, and grid computing. Cloud computing affords efficient and cost-effective computing and storage needs [39]. This technology is undoubtedly scalable and effective for huge data storing, parallel processing, and data analytics. The services provided by cloud computing are infrastructure as a service (IaaS), software as a service (SaaS) [40], platform as a service (PaaS), network as a service (NaaS) [41], and database as a service (DBaaS) [42].

9.2.2.3 Stream Processing

This technology is highly appreciated for processing streaming data. Unlike conventional computing methods, this technology emphasizes reading the data directly from software services and from sensors (sources of stream data) instead of reading data from a database. It is presumed that the input data are an infinite order of sequence. However, algorithms in stream processing work on finite and up-to-date data. In this way, stream processing backs real-time data analytics. Examples of stream computing are storm and simple scalable streaming system (S4) [43].

9.2.2.4 Big Data Analytics

With the advent of new technologies and demand of data increasing day by day, the data generation is growing at an exponential rate. The conventional method of data analytics is not capable to work on this high volume of raw data generated. This, however, has motivated science to generate powerful algorithms for data analytics, utilizing methods and software that can extract useful information from a big data set of various applications. The big data analytics comprise data mining, natural language processing (NLP), predictive analytics, social computing, statistical modeling, machine learning, text analytics, web analytics, network analytics, in-memory databases, and advance visualizations [44].

9.2.2.5 Coordinated Observation

Data are collected from varied sources, such as air, land, and ocean-based sensors, and are used further for information mining and knowledge discovery. The coordinated observation mechanism is vital so as to keep synchronization among the data collected. Untamed issues prevalent in global climatic shift, orbital processes, and response to disasters can now be addressed with interactive usage of EO data by implementing coordinated observation. However, there are scopes for further studies in integrated sensor systems pertaining to space, air, land, and ocean, optimized model for data visualization, multisensor coordination mechanism, interactive processing methodologies, and heterogeneous sensor data assimilation.

9.2.2.6 Big Geospatial Data Management

Significant aspects of geospatial data such as spatiotemporal data models, spatiotemporal indexing, and distributed data management are to be addressed so as to meet the strain impounded by big data management. Figure 9.8 shows a roadmap for the geospatial big data management.

- Spatiotemporal data model
  There must be a model that would be capable to support spatial and attribute data and unstructured spatiotemporal data such as sensor data, GPS data, and stream data. This innovative model must take care of spatiotemporal data, scale, and semantics of big data and must organize the object, state, and data hierarchically.
- Distributed data management
  In order to implement distributed data management in data storage and data management, patterns in data division need to be explored as per the data types and spatiotemporal adjacency. Big-GIS, to deal with the real-time data analysis, backs internal and external storage methods.
- Spatiotemporal indexing
  To achieve rapid retrieval of data and proper indexing, methods such as spatiotemporal query optimal algorithms, multidimensional joint query, batch indexing methods, and dynamic updating algorithms are required.

9.2.2.7 Parallel Geocomputation Framework

This framework is a parallel processing framework and is used for processing a sheer volume of data in a big data environment. This framework provides a platform that is useful and essential for a geocomputation modeling tool, for geoprocessing tasks, and that incorporates a multigranularity parallel model, and a parallel geocomputation interface and algorithms library. Parallel algorithms are carried out using varied hardware such as multicore central processing unit (CPU)/graphic processing unit (GPU)/clusters depending on the geoprocessing convolution. Hybrid parallel architecture, parallel programming model, and cluster architecture (such as YARN and Mesos) make the parallel algorithm and computing resources more efficient.

9.2.3 Data Collection Using GIS

Data collection using GIS mostly pertains to imagery data. Mounting high-end equipment on a spacecraft can provide much finer high-resolution imagery from space so as to

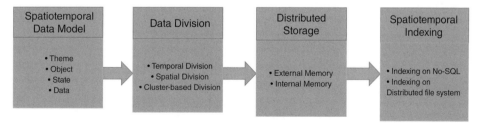

Figure 9.8 A roadmap for geospatial big data management.

overcome submeter resolution imagery by remote sensing from satellites [8]. These optical wavelength, infrared sensor–based images are found to be useful in many applications, a temperature map of the world, for example. Usage of active sensors, capturing transmission signals, radar-based remote sensing, and laser transmission such as LiDAR can also provide a very high precision of data [45].

9.2.3.1 NoSQL Databases

Due to constantly growing demands in exponential data storage and query, as well as high-volume and large-scale data environments, it is inadequate for the relational database management systems (RDBMSs) to meet these requirements. Another challenge for the RDBMS is to handle unstructured data. The possible solution to this kind of problem is the NoSQL database. The NoSQL databases are progressively in use in big data applications because it is simple, scalable, and possesses high performance. As stated in Table 9.1 the NoSQL database comprises four types such as (i) key value database (Redis), (ii) column-oriented database (HBase, Google Bigtable), (iii) document database (MongoDB, CouchDB), and (iv) graph database (Neo4j) [46].

9.2.3.2 Parallel Processing

A high volume of data can be processed in a timely fashion in parallel processing. The parallel algorithm pertains to efficient computation over distributed processors [47]. Various models, such as open multiprocessing (OpenMP), MPI, Map Reduce, and GPU-based systems with multicore processing capability, and cluster computing are used in order to achieve parallel computing. Out of these, the Map Reduce is a very simple and robust platform that can efficiently handle large data in a distributed system and has been implemented in a Google file system (GFS) and Hadoop distributed file system (HDFS) by Apache [48]. However, Map Reduce in the Apache Hadoop (also available in the cloud as Microsoft Azure, Amazon Elastic Compute Cloud, and Simple Storage Service S3) is an open-source application created using Java programming language.

9.2.3.3 Knowledge Discovery and Intelligent Service

The key parameters in knowledge discovery are integrating dissimilar data, domain-related knowledge, and related semantics and algorithms for parallel data mining. As stated in

Table 9.1 NoSQL database types.

Data model	Mainstream databases
Key-value	Dynamo, Redis, Riak
Column-oriented	Bigtable, Cassandra, HBase, Hypertable
Document	CouchDB, MongoDB, XML database
Graph	AllegroGraph, Neo4J, InfiniteGraph

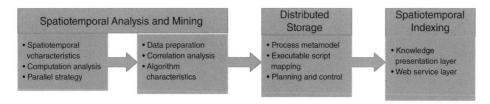

Figure 9.9 A roadmap knowledge discovery and service.

Figure 9.9, it is evident to implement data mining, process modeling, and service technologies for achieving an efficient knowledge discovery and geospatial analysis. In general, a data mining algorithm is designed for small applications where the scope of data mining is very limited and implemented in a stand-alone computer. On the contrary, big data mining requires aggregation of a wide range of varied data and the processing of data using parallel computing. In dynamic streaming, the traditional knowledge discovery process will not be effective enough. Similarly, process modeling technologies needs to provide services such as script mapping, process planning and control, and its evaluation and source, and spatiotemporal temporal meta-model over big data [49].

9.2.3.4 Data Analysis

The primary goal of GIS is data analysis; for example, analyzing a land and dividing it by area. Currently, lot of techniques have been devised to mine the data, match patterns, make inference, test hypothesis, simulate, and predict future changes in surface of earth (implementing the digital representation of land), etc. [1]; it can be inferred that GIS is capable of any conceivable operation on geographic information, from the modeling of the earth's surface to emergency route planning for a spacecraft [10].

Fog Computing and FogGIS

Cisco Systems in 2012 has created the concept of fog computing, which refers to a model where an interface is closely associated to the device that captures data [50]. The fog computing processes data locally, thereby reducing data size, lowers the latency, increases the throughput of the system, and stimulates the power efficiency of a cloud-based system. Fog computing applications can be found in smart cities and health care [51, 52]. Basically, a fog device is an embedded computer that acts as an interface between the cloud and any mobile device; for example, smartphones and mobile GIS [53]. Fog computing is a low-power gateway that boosts the throughput of the system, reducing the latency around the edge of the geospatial clients. Moreover, for the geospatial big data in the cloud, the storage requirement also diminishes. As stated in Figure 9.10, these lead to a reduction in transmission power, thereby increasing the performance of the whole system.

Lossless Compression Techniques

In a fog layer of fog computing, various compression algorithms are used to reduce the data size. Similarly, compression techniques are also used in GIS technologies such as network GIS and mobile GIS [54]. However, the compression techniques used in mobile GIS can be

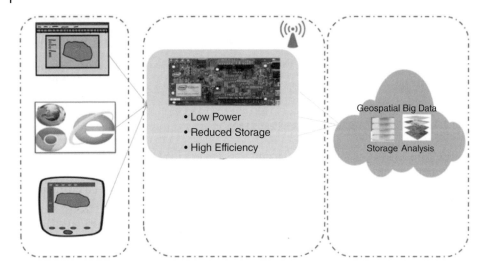

Figure 9.10 Conceptual diagram of the proposed fogGIS framework for power-efficient, low latency, and high throughput analysis of the geospatial big data.

translated to a corresponding fog layer [55]. In the similar context, any geospatial data can be compressed on the fog computer and can be transmitted to a cloud layer later. In the cloud layer the data is either compressed or decompressed before processing and finally analyzed and visualized.

9.3 Comparative Analysis

Various concepts, challenges, requirements, and scopes are discussed in this chapter. A primary problem of meteoroid shower, also known as the dust devil, is discussed, and possible solutions are presented. Table 9.2 depicts various aspects of integrated data representation, storage model, computation, and visual analysis. Table 9.3 justifies various data models with supported data types and scalability with examples of various products that use a corresponding data model.

9.4 Conclusion

Various concepts, challenges, requirements, and scopes are discussed in this chapter. A primary problem of a meteoroid shower, also known as the dust devil, is discussed, and possible solutions to overcome the dust devil are proposed. In addition to this, amalgamation of AI and ML with respect to space study and GIS are discussed, which will definitely help to calibrate space technology to the next level. Moreover, involvement of big data to remote sensing and GIS data collection techniques, such as Big-GIS and fog GIS, can be used so as to make the data collection easier and prediction more accurate.

Table 9.2 NoSQL database types.

Integrated representation	
Integrated representation based on tensor	Data characteristics are represented as tensor orders
	Tensors can be joined to an identical tensor through the tensor extension operator
	Elements of the identical order are accumulated together
	Extract valuable core data through dimensionality reduction
	The association between tensor orders, namely data features, is worthy of further research
Integrated representation based on graph	i. Graph-based text representation model, which could represent texts with weighted directed graphs
	ii. Utilize graph to indicate the geographic information
	iii. More information and supports for the analysis of land space structure could be provided
	iv. Widely used for unstructured data such as text and image
Integrated storage model	
Key-value storage model	i. Stores data as key-value pairs or maps, in which a data object is represented as a value, while key is the only keyword for mapping data to value by the hash function
	ii. Key-value model include real-time processing of big data, the horizontal scalability across nodes in a cluster or a data center
	iii. Can quickly query, but the defect is querying a continuous block of key
	iv. Ordered key value model and key column and key-document can be used to overcome the defect
Column storage model	i. Data are organized as a one-dimensional array in physical memory
	ii. Same column data could be saved continuously
	iii. Scalability of column databases is more in distributed extension
Document database model	i. Stored in a document in XML, YAML, and JSON format
	ii. Organized into different collections for data management
	iii. Every field in a document is a unique key, and the content indicates the value
	iv. Low performance of querying
Graph database model	i. Manages data in a network structure
	ii. Quickly process the JOIN operation in massive data, without any pre-defined models
	iii. Can store and manage the semi-structured data
Integrated computation	i. Deep belief networks, convolutional neural networks are applied to integrated big data computation
	ii. Restrict the network architecture with local connectivity and weight sharing
	iii. This is a synthesis of a variety of big data computing methodologies
Integrated visual analysis	i. Extracts more complete, intuitive, and hidden information
	ii. Can provide more comprehensive, intuitive, and targeted data presentation
	iii. Top-down method is used to organize data into three categories: numeric data, texts, and binary waveforms and images
	iv. After the categorization, they visualize the data by the combination of the data table, trend chart, timeline, thumbnails, and keywords

Table 9.3 NoSQL database types.

Data model	Data type support			Scalability		Map reduce support	Products
	Structured	Unstructured	Semi-structured	Horizontal	Vertical		
Key-value	Available	Available	Available	Available	Available	Available	Voldemort, Redis
Column oriented	Available	Available	Available	Available	Available	Available	Hypertable, Hbase
Document databases	Available	Available	Available	Available	Available	Available	MongoDB, CouchDB
Graph databases	Available	Available	Available	Not-available	Available	Not-available	Neo4j, Infinite Graph

References

1 Maliene, V. (2011). Geographic information system: old principles with new capabilities. *Urban Design International* 16 (1): 1–6.

2 Mukherjee, J. and Ramamurthy, B. (2013). Communication technologies and architectures for space network and interplanetary internet. *IEEE Communication Surveys and Tutorials* 15 (2): 881–897.

3 Marchetti, P.G., Soille, P., and Bruzzone, L. (2016). A special issue on big data from space for geoscience and remote sensing. *IEEE Geoscience and Remote Sensing Magazine* https://doi.org/10.1109/MGRS.2016.2586852.

4 Coppock, J.T. and Rhind, D.W. (1991). The history of GIS. *Geographic Information System: Principles and Applications* 1 (1): 21–43.

5 Abel, D.J., Taylor, K., Ackland, R., and Hungerford, S. (1998). An exploration of GIS architectures for internet environments, computer. *Environment and Urban Systems* 22 (1): 7–23.

6 Yue, P., Gong, J., Di, L. et al. (2010). GeoPW: laying blocks for geospatial processing web. *Transactions in GIS* 14 (6): 755–772.

7 Zhao, P., Foerster, T., and Yue, P. (2012). The geoprocessing web. *Computers & Geosciences* 47 (10): 3–12.

8 Cheng, T. and Teizer, J. (2013). Real-time resource location data collection and visualization technology for construction safety and activity monitoring applications. *Automation in Construction* 34: 3–15.

9 Cheng, E.W., Ryan, N., and Kelly, S. (2012). Exploring the perceived influence of safety management practices on project performance in the construction industry. *Safety Science* 50 (2): 363–369.

10 Pradhananga, N. and Teizer, J. (2013). Automatic spatio-temporal analysis of construction site equipment operations using GPS data. *Automation in Construction* 29: 107–122.

11 Agrawal, D., Das, S., and Abbadi, A.E. (2011). Big data and cloud computing: current state and future opportunities. In: *EDBT*, 530–533. New York: ACM.

12 Dean, J. and Ghemawat, S. (2008). Mapreduce: simplified data processing on large clusters. *Communion of the ACM* 51 (1): 107–113.

13 IBM, Armonk, NY, USA, Four Vendor Viewson Big Data and Big Data Analytics: IBM [Online], 2012. htt://www-Ol.ibm.comlsofware/in/data/bigdata

14 Big Data (2011). *Business Opportunities, Requirements and Oracle's Approach*, 1–8. Cambridge, MA: Winter Cororation.

15 Schowengerdt, R.A. (1997). *Remote Sensing: Models and Methods for Image Processing*, 2e. New York, NY: Academic Press.

16 Landgrebe, D.A. (2003). *Signal Theory Methods in Multispectral Remote Sensing*. New York, NY: Wiley.

17 Richards, J.A. and Jia, X. (2006). *Remote Sensing Digital Image Analysis: An Introduction*. New York, NY: Springer.

18 Plaza, A., Benediktsson, J.A., Boardman, J. et al. (2009). Recent advances in techniques for hyperspectal image processing. *Remote Sensing of Environment* 113: 110–122.

19 Christophe, E., Michel, J., and Inglada, J. (2011). Remote sensing processing: from multicore to GPU. *IEEE Journal of Selected Topics in Applied Earth Observation and Remote Sensing* 4 (3).

20 Mell, P. and Grance, T. (2011). The NIST definition of cloud. *Computing*, 800-145: 1–3.

21 S. Azhar, BIM for facilitating construction safety planning and management at jobsites, In: Proceedings of the CIB-W099 International Conference: Modeling and Building Safety, Singapore.

22 Park, C.S. and Kim, H.J. (2013). A framework for construction safety management and visualization system. *Automation in Construction* 33: 95–103.

23 Samuel Marchal, Xiuyan Jiang, Radu State, Thomas Engel, A Big Data Architecture for Large Scale Security Monitoring, IEEE International Congress on Big Data, 2014. DOI: https://doi.org/10.1109/BigData.Congress.2014.18

24 Cohen, J., Dolan, B., Dunlap, M. et al. (2009). Mad skills: new analysis practices for big data. *PVLDB* 2 (2): 1481–1492.

25 Envi Sat, A. S. A. R. product Handbook, European Space Agency, Issue 2.2 Feb 2007.

26 Angrisani, L., D'Arco, M., Ianniello, G., and Vadursi, M. (2012). An efficient pre-processing schemeto enhance resolution in band-pass signals acquisition. *IEEE Transactions on Instrumentation and Measurement* 61 (11).

27 Cortes, C. and Vapnik, V. (1995). Support-vector networks. *Machine Learning* 20 (3): 273–297.

28 Scholkopf, B. and Smola, A. (2002). *Learning with Kernels*. MIT Press.

29 Skjellum, A., Gropp, W., and Lusk, E. (1999). *Using MPI: Portableparallel Programming with the Message-Passing Interface*. MIT Press.

30 HDF Group, Hierarchical data format, version 5, Website, 2015, URL http://www.hdfgroup.org/HDF5.

31 Armbrust, M. (2010). A view of cloud computing. *Communications of the ACM* 53 (4): 50–58.

32 Farserotu, J. and Prasad, R. (2000). A survey of future broadband multimedia satellite systems, issues and trends. *IEEE Communications Magazine* 6: 128–133.

33 Chuang, J. and Sollenberger, N. (2000). Beyond 3G: wideband wireless data access based on OFDM with dynamic packet assignment. *IEEE Communications Magazine* 37: 78–87.

34 Kul Bhasin, Jeffrey Hayden, Developing Architectures and Technologies for an Evolvable NASA Space Communication Infrastructure, National Aeronautics and Space Administration, Washington, DC, 2004.

35 AOS Space Data Link Protocol. Recommendation for Space Data System Standards, CCSDS 732.0-B-2, Blue Book, Issue 2, Washington D.C., July 2006.

36 Taleb, T., Kato, N., and Nemoto, Y. (2006). An efficient and fair congestion control scheme for LEO satellite networks. *IEEE/ACM Transactions on Networking* 14: 1031–1044.

37 Castro, M.A.V. and Granados, G.S. (2007). Cross-layer packet scheduler design of a multibeam broadband satellite system with adaptive coding and modulation. *IEEE Transactions on Wireless Communications* 6 (1): 248–258.

38 Lohr, S. (2012). The age of big data. *New York Times* 11: 11–23.

39 Buyya, R., Yeo, C.S., Venugopal, S. et al. (2009). Cloud computing and emerging IT platforms: vision, hype, and reality fordelivering computing as the 5th utility. *Future Generation Computer Systems* 25 (6): 599–616.

40 Peter, M. and Grance, T. (2009). *The NIST Definition of Cloud Computing (Draft)*, vol. 53, pp. 50, pp. 1216-1217. National Institute of Standards and Technology.

41 Lehner, W. and Sattler, K.U. (2010). Database as a service (DBaaS). In: *2010 IEEE 26th International Conference on Data Engineering (ICDE)*, 1216–1217. IEEE.

42 Costa, P., Migliavacca, M., Pietzuch, P., and Wolf, A.L. (2012). NaaS: networkas-a-service in the cloud. In: *Proceedings of the 2nd USENIXconference on Hot Topics in Management of Internet, Cloud, and Enterprise Networks and Services, Hot-ICE*, vol. 12, 1–13. New York: ACM.

43 Neumeyer, L., Robbins, B., Nair, A., and Kesari, A. (2010). S4: distributed stream computing platform. In: *IEEE International Conference on Distributed Systems and Modelling*, 170–177. IEEE.

44 P. Russom, Big data analytics, TDWI Best Practices Report, Fourth Quarter, IEEE, 2011.

45 Baertlein, H. (2000). A high-performance, high-accuracy RTK GPS machine guidance system. *GPS Solutions* 3 (3): 4–11.

46 Chang, F., Dean, J., Ghemawat, S. et al. (2008). Bigtable: a distributed storage system for structured data. *ACM Transactions on Computer Systems (TOCS)* 26 (2): 4–15.

47 Schadt, E.E., Linderman, M.D., Sorenson, J. et al. (2010). Computational solutions to large-scale data management and analysis. *Nature Reviews Genetics* 11 (9): 647–657.

48 Dean, J. and Ghemawat, S. (2010). MapReduce: a flexible data processing tool. *Communications of the ACM* 53 (1): 72–77.

49 Di, L., Yue, P., Ramapriyan, H.K., and King, R. (2013). Geoscience data provenance: an overview. *IEEE Transactions on Geoscience and Remote Sensing* 51 (11): 5065–5072.

50 Bonomi, F., Milito, R., Zhu, J., and Addepalli, S. (2012). Fog computing and its role in the internet of things. In: *Proceedings of the First Edition of the MCC Workshop on Mobile Cloud Computing*, 13–16. ACM.

51 Hancke, G.P. and Hancke, G.P. Jr., (2012). The role of advanced sensing in smart cities. *Sensors* 13 (11): 393–425.

52 Dubey, H., Yang, J., Constant, N. et al. (2015). Fog data: enhancing telehealth big data through fog computing. In: *SE BigData & Social Informatics*, vol. 14. ACM.

53 Yi, S., Li, C., and Li, Q. (2015). A survey of fog computing: concepts, applications and issues. In: *Proceedings of the Workshop on Mobile Big Data*, 37–42. ACM.

54 Monteiro, A., Dubey, H., Mahler, L. et al. (2016). FIT: a fog computing device for speech tele-treatments. In: *IEEE International Conference on Smart Computing (SMART-COMP), St. Louis, MO*, 1–3. https://doi.org/10.1109/SMARTCOMP.2016.7501692.

55 F. Chen, H. Ren, Comparison of vector data compression algorithms in mobile GIS, 3rd IEEE International Conference on Computer Science and Information Technology (ICCSIT), 2010. DOI: https://doi.org/10.1109/ICCSIT.2010.5564118

10

Application of Intelligent Data Analysis in Intelligent Transportation System Using IoT

Rakesh Roshan[1] and Om Prakash Rishi[2]

[1] *Institute of Management Studies, Ghaziabad, India*
[2] *Department of Information Technology, University of Kota, Kota, India*

10.1 Introduction to Intelligent Transportation System (ITS)

As indicated by the UN overview in 2014, the greater part of the total populace is presently living in urban zones and expanding day by day, without a doubt cautioning city organizers. Associated urban areas develop when internet of things (IoT) advancements and socially mindful system frameworks are used as total organizations over an entire associated metropolitan area. When considering associated urban zones, one may consider innovative urban communities that have the noticeable front-line advances for their residents. Be that as it may, little private networks have likewise been profiting from interfacing people, organizations, city foundations, and administrations. This chapter gives the detail of city transportation issues and a segment of the troubles that are included with creating across-the-board IoT systems. The alliance of world-class IoT change foresees working with everyone of these brilliant urban networks that empower residents to make innovation use more sensible, versatile, and manageable. Numerous urban areas and towns around the world are swinging to socially keen gadgets to take care of urban issues, for instance, traffic issues, natural pollution, medicinal services, and security observations to improve the expectations for everyday comforts for their overall population. Brilliant sensors have been introduced all through a city, in vehicles, in structures, in roadways, in control checking frameworks, security reconnaissance, and applications and gadgets that are used by people who are living or working there. Conveying data to the general population using these cutting-edge services, provide brilliant urban community openings. The enormous information can be used to settle on how open spaces are arranged, how to make the best usage of their benefits, and how to pass on authoritative notices all the more capable, suitable, and proper.

With the evolution of smart cities transforming towns into digital societies, making the life of its residents hassle free in each facet, the intelligent transport system (ITS) turns into

Intelligent Data Analysis: From Data Gathering to Data Comprehension,
First Edition. Edited by Deepak Gupta, Siddhartha Bhattacharyya, Ashish Khanna, and Kalpna Sagar.
© 2020 John Wiley & Sons Ltd. Published 2020 by John Wiley & Sons Ltd.

the quintessential component among all. In any urban area, mobility is a key difficulty; be it going to office, school, and college or for another motive for residents to use a transport facility to journey within the city. Leveraging residents with an ITS can save them time and make the city even smarter. ITS pursuits to achieve optimization of traffic performance by means of reducing traffic troubles is a worthy goal. It alerts drivers with previous data about the traffic, local convenience concurrent running statistics, seat availability, etc. This reduces the travel time of commuters and improves their security and comfort.

The utilization of ITS [1] is broadly acknowledged and utilized in numerous nations today. The utilizationisn't simply restricted to movement of traffic control and data, yet is used additionally for street well-being and effective foundation use. In view of its unlimited potential outcomes, ITS has now turned into a multidisciplinary conjunctive field of work, and along these lines, numerous associations around the globe have created answers for giving ITS applications to address the issue [2].

The whole utilization of ITS depends on information accumulation, examination, and utilizing the consequences of the investigation in the tasks, control, and research ideas for movement administration where an area assumes an imperative job. Here the use of sensors, global positioning system (GPS), communication protocols, messages, and data analysis plays an integrated role for the implementation of:

a. *Traffic management system (TMS).* TMS is the part of intelligent transportation to improve the flow of different types of vehicles in city or highway traffic. There are different sensors and communication protocols that can be used to implement the TMS.

b. *Traffic information system.* Whatever data receive by the different sensors can be managed through the traffic information system and gives the accurate information as per the requirement of the TMS.

c. *Public transportation system.* Different sensors, GPS, and communication protocols can be used in the public transportation system for smooth running, and real-time information can be shared with the public so that there will be less waiting line on the road. The public can track the vehicle as well as pay their fare online.

d. *Rural transportation system.* There are different types of rural transportation, like local bus, truck, tractor, etc. Sensors, GPS, and communication protocols can be used to track the vehicle and identify the environment.

e. *Road accident management system.* By using the sensors, GPS, and communication protocols, road accidents can be easily identified and quick help can be initialized to the injured person so that causality will be decreased. It will help a lot in the highway because so many injured persons in accident end up deceased because they run out of time waiting for emergency services.

f. *Commercial vehicle control system.* The rules for the commercial vehicle are different in every country, including India. There is a separate permit for the vehicle and products are required for different states. So much paperwork can be removed by utilizing the sensors, communication protocols, and message passing.

g. *Vehicle tracking system.* GPS can be used to track every vehicle. Now, in India, there is no mechanism to track every vehicle, whether it be a public, private, or commercial vehicle. But using the intelligent data analysis (IDA), the vehicles can be easily tracked and identified by the concerned authority.

10.1.1   Working of Intelligent Transportation System

A traffic control center (TCC) is an important unit of ITS. It is essentially a specialized framework directed by the transportation expert. Here, all information is gathered and dissected for further activities and control administration and data of the movement and whereabouts of the neighborhood transportation vehicle [3].

An efficient and capable TCC relies on automatic information accumulation with exact area data to produce precise data to transmit back to explorers. The entire workings of ITS are as follows [4]:

- *Data collection through smart devices or sensors.* This requires exact, broad, and specific information gathering with real-time observation for intelligent planning. So the information here is gathered through devices that lay the base of further ITS capacities. These smart devices are for the identification of vehicles, location of vehicles, sensors, cameras, and so forth. The equipment chiefly records the information, such as movement check, reconnaissance, travel speed, and travel time, area, vehicle weight, delays, and so forth. These hardware smart devices are associated with the servers that for the most part store a lot of information for further investigation.
- *Data sharing with the vehicle, traffic control center (TCC), and authorities.* Data should be communicated to the vehicle, TCC, and authorities in real time. The efficiency of data transmission to the stakeholders of intelligent transportation system (ITS) is the key factor for the system. Here, communication protocol plays an important role for the real-time data transmission to all stakeholders. By data sharing, many problems can be easily removed, and it will also help to prevent critical situations.
- *Analysis of the gathered data.* The information that has been gathered and received at Traffic Management Center (TMC) is processed further in numerous steps. These steps are blunders rectification, data cleansing, data synthesis, and adaptive logical analysis. Inconsistencies in information are diagnosed with specialized software and rectified. After that, the record is further altered and pooled for evaluation. This mended collective data is analyzed further to expect a visitor's situation in order to deliver appropriate information to users. There are so many tools available, especially for the analysis of data.
- *Send real-time information to the passengers.* Passenger information system (PIS) is utilized to illuminate transportation updates to the passengers. The framework conveys ongoing data like travel time, travel speed, delay, mishaps on streets, change in course, preoccupations, work zone conditions, and so on. This data is conveyed by an extensive variety of electronic gadgets like variable message signs, highway warning radio, website, SMS, and customer care (Figure 10.1).

10.1.2   Services of Intelligent Transportation System

ITS and its services [5] are the concept of how mobile information systems can be applied in the transportation sector. With ITS, the services should give additional information to the passengers while the same information for operators should improve transportation and travel performance. In mid-sized vehicles, ITS is used to help drivers navigate, avoid traffic

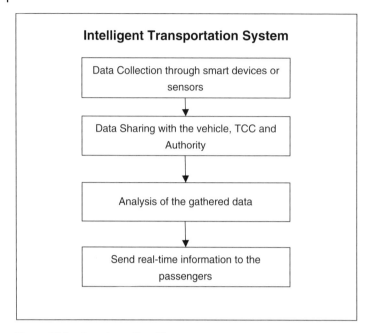

Figure 10.1 Overview of intelligent transportation system.

snarls, and collisiòns. ITS on buses and trains are used to manage and adapt fleet operations and to provide automated ticketing and real-time traffic information to passengers. ITS on the roadside is used to coordinate traffic signals, detect and manage events, and display information for passengers, drivers, and pedestrians.

There are various types of services that can be given to the users (Figure 10.2). The number of services can be changed (added or removed) from time to time according to the situation and demand by the public. Some of the important services for the user are [6]:

a. Traffic information system
b. Accident management system
c. Electronic payment
d. Automatic payment to toll plaza
e. Live information about the public transportation
f. Commercial vehicle operations
g. Intelligent parking.

In the current scenario, most of the countries have a partial ITS, that is, not all the services are implemented properly. For instance, India has services like automatic payment toll plaza (which is not yet implemented in all parts of the country), electronic payment, traffic information system (which is again, not yet implemented in all part of the country). But there area lot of developments that should be required, such as accident management system, live information of public transport, intelligent parking, etc. Parking is the common problem in all the major metros or big cities, and people do not have any information about the vacancy of parking available in nearby areas. Most of the traffic increases are due to the

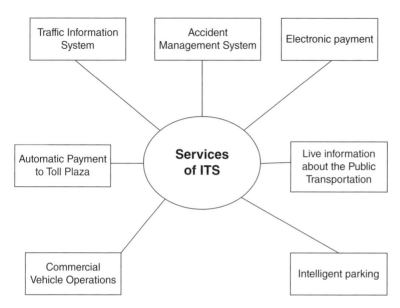

Figure 10.2 Services of intelligent transportation system (ITS).

parking problem, so intelligent parking plays a vital role in resolving the traffic problem, which is essentially congestion from lack of parking space.

10.1.3 Advantages of Intelligent Transportation System

ITS combines a wide range of data and correspondent technology to make a system of frameworks that connect devices in order to ensure smooth traffic in the city, safety of vehicles, etc. As an ever-increasing number of parts of our transportation moves toward becoming arranged, ITS will modify the way drivers, organizations, and governments manage street transport. These propelled frameworks can help enhance transportation in a few different ways.

- *Health benefits.* Formation of traffic-free zones and low emission zones in urban areas decrease contamination and unexpected losses. Coordination of vehicle frameworks with portable correspondences and progressed mapping innovation gives a potential fuel sparing of 10–20%. ITS improves safety as well as capacity. Normal speed cameras convey substantial upgrades through decreases in loss of life, improvement of congestion, and enhanced environment.
- *Improvement of traffic safety.* Regarding speeding, risky climate conditions, and overwhelming mass movement would all be able to prompt mischance and the death toll, thus, ITS helps with these. Ongoing climate checking frameworks gather data on deceivability, wind speed, precipitation, and street conditions and that's only the tip of the iceberg, permitting an activity controller's up-to-date data on every conceivable driving situation. In completely arranged frameworks, this data would then be able to be utilized to refresh caution signs and even speed confines when the need emerges, keeping drivers alert to the situations around them. Emergency vehicles can react rapidly to accidents,

as they are continuously updated by real-time traffic monitoring system. ITS movement control occupies activity far from occupied or hazardous zones, counteracting congested roads in addition to decreasing the danger of crashes.

- Benefits of public transport. Use of electric and hybrid buses or public transport vehicles always reduces pollution. Understanding of online data to transports, train, and their travelers makes a superior educated traveler and administrator. E-ticketing empowers faster, easier travel by public transport, and gives administrations data to make upgrades in the vehicle's framework. Smart cards for the public transport vehicles save time and money of the passengers.

- *Loss of infrastructure reduced.* Heavy vehicles can put too much stress on the road network, especially when they are overloaded. The weight of vehicles in weight stations and other old forms of weight control is reduced but with the cost of time and delayed traffic is increased. Weight-speed-speed system measures the size and weight of the vehicle while transmitting the collected data returned to the central server. The overloaded vehicles can be recognized and suitable measures taken, which result in high compliance between the passengers and the driver and the reduction in lengthy road routes. Not only do these systems simplify the enforcement, they can reduce the expenses of repairing the road, so that it can be allocated somewhere else.

- *Traffic and driver management benefits.* Online and street-side data will be available to drivers, and vehicles are equipped with driver help frameworks to enhance the proficiency and security of road transport. Administration of vehicle armadas, both cargo and open transport, through online data and two-way correspondence among chief and driver helps to limit confusion [7, 8].

- *Reduced parking problem.* Invalid parking contributes to busy streets and makes a city hazardous, and causes problems for other drivers; city vehicles and others need spaces in which to park. Stopped vehicles cause time-consuming traffic to grow in crowded areas because visitors find these places in which it is very difficult to park. Manual parking management systems can be expensive and ineffective; they can also add more people to the crowded areas. Cities should explore smart parking systems, which scans parked vehicles and communicates with the parking meter to search and record unauthorized parked vehicles. Rather than requiring a human parking enforcement officer, smart parking systems let drivers know that they will be automatically identified for illegal or extended parking. These automated system drivers help to improve traffic flow by enhancing conformity to parking rules and keeping ample parking spaces available for use.

10.2 Issues and Challenges of Intelligent Transportation System (ITS)

ITS is a comprehensive management system of transportation services, aimed at providing innovative ideas in various ways of transport management. ITS integrates high technology and improvements in the information system, including communication, sensor, controller, and highly developed mathematical methods to dovetail with traditional systems of transport infrastructure, and it is one of the most important characteristics of ITS. When the transport system is integrated into the infrastructure and the vehicle itself, these techniques can get rid of the crowded areas, improve safety, and increase productivity.

Clogs in urban rush hour gridlock systems are an extraordinary danger to any economy, and the regular habitat, and adversely affects the nature of individual life. Traffic jams call for extraordinary systems to diminish travel period, queue, and unsafe gas emanations, particularly unused time spent sitting during extended traffic lights. Traditional activity administration arrangements fail to confront numerous restrictions primarily because of the absence of sufficient correspondence between vehicles themselves and additional vehicle-to-infrastructure connection. In the past couple of years, there have been endeavors to build up an assortment of vehicle automation and communication systems (VACS) [9] for enhancing traffic capability and well-being. Additionally, there have been significant advancements in intelligent automatic vehicles for this reason.

10.2.1 Communication Technology Used Currently in ITS

Communication technology assumes a noteworthy job in the viability of ITS. Presently, the majority of networking and communication technology is composed in ITS applications [9, 10]. The important communication technologies currently used in ITS include the following:

- RFID uses the vehicle for communication of infrastructure, and it has proved useful in identifying vehicles, in order to facilitate parking with automatic toll collection. Still, if the range of communication is limited in RFID, ITS can be supported from this technique in different ways. An RFID application that has not been widely implemented has to fix RFID tags for traffic signals. This can be accessed/read by vehicles passing nearby.
- Bluetooth [11] is utilized primarily to accumulate the MAC address of mobile devices when passing through vehicles to build source-destination chart. This information can be utilized for ITS applications for the prediction of traffic flows in the future. Dynamically, Bluetooth can be used to portray traffic masses of specific areas of the city. In comparison to RFID, the correspondence distance of Bluetooth is very restricted.
- Wi-Fi is broadly utilized in ITS experiments for correspondence technology, particularly the 802.11a variation that works properly at 5.8 GHz. The explanation behind this is the accessibility of the ad hoc mode and the correspondences range, which is roughly 50 m. It is equipped for duplex correspondence and locally supported TCP/IP.
- 2G, 3G, 4G, and 5G mobile communications has been widely used in vehicles for decades to achieve location-based services such as real-time maps and internet. These mobile communications can be used to connect various devices to vehicles as well as vehicles to infrastructure.
- GPS are used in vehicles to get the accurate position of the vehicle. Location-based services are not possible without the help of GPS. GLONASS and Galileo are other options for this, and these services are the main competitors of GPS.
- WiMAX was initially anticipated to be included in VANETS due to its long range of 1 km. Be that as it may, it had a moderate start, and a few of the communication organizations quit utilizing it. This issue has prompted a shortage of hardware utilizing it and any experimentation generally stayed on a simulation level.

- ISM RF is free from license. It is a duplex wireless communication that can cover the area from 20 m to 2 km, but it depends upon the equipment fitted with it, like antennas and RF power. Scientists can use any wireless protocol with it for the research work. It is difficult to blend and match segments from various producers of devices, since there is no normal convention. The frequencies used are 868 MHz, 915 MHz, and 2.4GHz.

10.2.2 Challenges in the Implementation of ITS

There are lots of challenges in the implementation of ITS with respect to communication technology, connection issues, management of device and sensors, and awareness to the public about the technology, as well as the opportunity of new inventions like the self-driving car, public transport, etc., after the successful implementation of ITS. Some of the challenges in the implementation of ITS include the following:

- *Management of the communication of the two devices in a long-distance pairing.* The distance between any two devices that are communicating plays a vital role, whether communication is reliable or unreliable, continuously without connecting and disconnecting, and whether the amount of data that can be sent without any noise. In one-hop communication, a message can start traveling only when direct communication from source to destination is established. In multi-hop communication protocols, the long distance is covered by the intermediate nodes. However, the complication of these protocols means that more research and testing is required to check that a message can travel from the source of the desired distance to the destination. Implementations of some multi-hop protocols use the infrastructure as intermediate devices.
- Effect of bandwidth and MAC (medium access control). The bandwidth has to do with the amount of information that can be passed to any message or message on a communication network from source to destination if the information is changed to bits per second. Due to concentration on wireless devices, many devices are the worst at the same frequency, so a MAC protocol is used to avoid collision, and as a method to identify dropped information. There is a requirement of a different bandwidth to work by different services. Emergency or urgent messages are generally small and require continuous bandwidth, whereas standard VANET information is required to have high bandwidth and large messages.
- *Internet on vehicle.* Current traffic operational strategies are based mostly on TCC or any centralized control room. When the scenario is complex, it will cause a significant amount of computing resources. With the help of the internet on vehicle technology, we can distribute part of the vehicle's decision work. In increasing information exchange between vehicles, many decentralized solutions can be proposed, which can provide efficient traffic operations and ensures system scalability.
- *Handling of emergency situations.* The number of times the vehicle interacts with smart traffic lights is important for the efficiency of optimization algorithms and bandwidth usage. A very few number of messages and traffic lights will be obsolete or inoperative, and many messages and bandwidth will have multiple conflicts, and traffic light controllers may not have sufficient time to process information from time to time for the use of customizable algorithms. In case of emergency, real-time information is important,

such as an accident that has already happened, an inevitable collision in the future, violation of red light, involvement of pedestrian (movable device), and bicycle traffic being light. Apart from this, there is also a priority of emergency vehicles, heavy vehicles, and transit/busses on traffic light schedules.

- *Efficient use of data received from multiple sources.* Compared to two single sources, multiple source supplements can provide data, and multiple stores of data fusion can reduce the uncertainty related to individual sources, creating a better understanding of the observed condition. Apart from this, it is quite expensive to install and maintain sensor suits. If due to some failure it is shown that one or more sensors are deteriorating, it will only reduce the entry rate and will have less effect on the performance of the system. With the help of ITS, we can get real-time information in many sources of traffic data. Consequently, this will allow us to manage the transport system more efficiently.
- *Verifying of model in ITS.* In spite of the fact that countless models have been proposed, there is no appropriate method to demonstrate their worth. On one hand, the transportation framework is refined and reproduction cannot give a comparable forecast about this present reality. Then again, a reality check on an extensive scale is incomprehensible. In the future, with the improvement of ITS, we might have the capacity to utilize more information gathered to remove the gap between simulation and the real world.
- *Security and privacy of gathered data or information.* Wireless communication is telling us that eavesdropping is notorious for ease of ease. Due to cheap storage and big data analysis or data science, the information or data collected from a passing car and recorded communication can disclose enough information to know a lot about a person, for example, a social site's data, the person's schedule, and their shopping habits. This information can be sold for advertisement and can be used as a way to hijack their normal routes. A more important concern is the active malicious communication node that spreads wrong information or rejects complete communication. This problem is extremely risky because it casts doubt on the system used for emergency, and doubt of such a system will make it obsolete.

Any new communication mechanism such as VANET could be used by "malicious users" to use the technology for fun or with the intention to harm the particular user using the technology. In addition to that, modifying the hardware to send the wrong information will cause the local government, manufacturers, and other middlemen to take the risk of entering the central computer attack. The performance of the vehicle can reduced its chances for attack by using the VANET as a gateway (Figure 10.3).

10.2.3 Opportunity for Popularity of Automated/Autonomous/Self-Driving Car or Vehicle

With the development of ITS-related technology and communication protocols, there is an opportunity for the frequent running of self-driving cars/vehicles in the city. By the use of sensors and IOT, the commercial automated driving vehicle is coming soon. Researchers must innovate the frameworks so that the self-driving vehicles can reduce the congestion and consumption of fuel. The levels of autonomy in vehicles include the following:

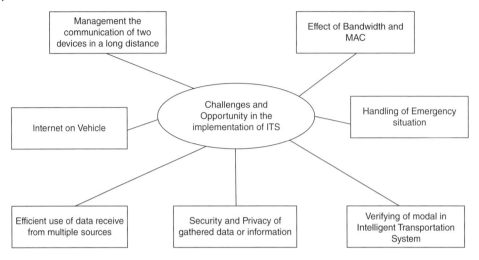

Figure 10.3 Challenges and opportunities in the implementation of ITS.

- *Connected vehicle*. Full control by driver.
- *Automated car/vehicle*. Automatically performing the function of driver by integration of mechanical and electronic. Automated technology contains an emergency break, alert before collisions, different warning system, etc.
- *Autonomous car/vehicle*. Autonomous car is also called a self-driving car/vehicle. This is equipped with camera, radar, and sensors.

10.3 Intelligent Data Analysis Makes an IoT-Based Transportation System Intelligent

10.3.1 Introduction to Intelligent Data Analysis

The beginnings of intelligent data analysis (IDA) are from different disciplines. Statistics and machines are two important fields to learn for data analysis. Of these, statistics are old machines that can expect to learn that they are not useful for long. But the only fact of the machine learning is not that it has its own culture, it is that it has its own emphases, purpose, and objectives, which are not always reliable with the data. There are differences in these two topics in the heart of IDA, i.e., there is a creation tension, which has benefited the development of data analytical tools [12].

In contrast, the origins of the machine learning community are very high in computer practice. It gives birth to practical orientation; there is a desire to test some to see how well it works without waiting for the proper proof of effective formality.

Modern figures are more or less entirely inspired by the concept of the model. It is a fixed structure, or an estimate for the structure, which can be due to the data. There are different types of models available. In fact, the data analysis tools have multifaceted interconnections, and this is not just a collection of different techniques. There are some extraordinary examples within the statistics of schools or philosophies that are not model-driven, but they are notable for lacking in innovation and for their general isolation. These exceptions are common with the approach of data analysis developed in the machine learning community compared to traditional data. Instead of emphasizing the statistical emphasis on the model, the machine emphasizes learning algorithms.

Data always play an important role in the journey of scientific research. Data/information are important in demonstrating, analyzing, and tackling logical issues, and data mining is the most famous apparatus to process information appropriately. In the beginning of data mining, the data sets that should be taken care of are generally spotless ones of moderately little scale. With the improvement of data collection techniques, the wellsprings of data are getting to be more extravagant and more pervasive, which straightforwardly prompts the increase in the data sum. In the meantime, the gathered data have additionally demonstrated a solid character of continuous and heterogeneous data, with a specific extent of messy data blended inside too. In this manner, the customary information mining is standing up to extraordinary difficulties. With the end goal to conquer the difficulties, IDA strategies with higher preparing speed, higher precision, and higher effectiveness are created. In the development process of IDA, different researcher contributes the revised or updated algorithms from time to time to solve a particular problem in a different field. The past trends in the development of IDA include the following:

- The concept of IDA becomes smarter
- The data set is bigger day by day
- The organization of data set is transformed from structured data set to unstructured/heterogeneous data set.

First, the coverage area of IDA was climate, geography, medical science, physics, image processing, system engineering, etc., but now the application of IDA is spread to e-commerce, public sectors, government sectors, social networks, etc. Some of the IDA algorithms have done great progress but still there exists some challenges in data analysis. These challenges include the following:

- *Selection characteristic of disproportion data set.* Although the most important features of multisource asymmetric data set by data preprocessing, this feature can create an unbalanced data set. Conventional IDA algorithms all hope that the generalization should be given so that the unbalanced data set will be neglected as a minority. However, as a result of fault diagnosis, the minorities often include valuable information. Therefore, a special feature selection algorithm for developing unbalanced data needs to be developed.
- *Data analysis distribution.* In view of dispersed registering, the high performance computing innovation can upgrade the capacity of each and every hub, with the goal that the ability and handling speed of the whole disseminated framework can be additionally reinforced. One of the principal methodologies of high performance computing is to exploit graphic processing units' (GPU) capacity to do gigantic dreary monotonous counts, and extra more central processing unit (CPU) assets to accomplish more complex computational work, or, in other words arrangement without any changes on equipment.
- *Big data analytics.* Although distributed knowledge analysis platforms could provide an answer to cope with data of huge size, the information is based mostly on modeling in big data where the atmosphere is still challenged. There are two effective solutions to the big data analytics problem: (a) using machine learning design, an algorithm of deep learning to processing of huge multidimensional data, and (b) converting the whole data set into subsets, where submodels can be developed, and then get the whole model by logically integrating the submodels.

10.3.2 How IDA Makes IoT-Based Transportation Systems Intelligent

IDA modules provide basic information about data analysis for user identities, behavioral identities, and service creation in an AI module. That is, the technique can create a meaningful mapping layer through different models, including the use of models, i.e., device model, logic model, and knowledge model [13].

There are many coexistent IOT frameworks for device access, such as Amazon Alexa, Google Home, and so on. In particular, companies are committed to setting up global standards for IOTs and its application in the areas of medical, energy/power management, security, entertainment, etc. However, in reality, the international model for smart device management is still to come. The smart devices from different manufacturers adhere to IOTstandards, including device details, useful instructions, and control interfaces. Different standards have flexible and costly IOT configurations, and they drop the development of standard IOT services. Therefore, to ensure the data analysis of different IOT platforms, these different IOT platforms must be delivered to the same IDA module. Then, several services of the IOT platform can be enjoyed by the people for the same purpose. In addition, IDA technologies enable the device to understand the intentions of humans. The latest technologies, such as natural language processing (NLP), describe the living environments of people, which is also the key to machine learning. In traditional intelligent service solutions, speech recognition is implemented and people can give instructions by voice commands to receive services. However, these voice commands and instructions are categorized and matched only by an underlying instruction library. The traditional intelligent service machine is not based on NLP, and is not as intelligent as required. To understand the user's activities for IDA and to predict user intentions, everything must be connected. IDA can be implemented through meaningful matching, including association computing and similarity computing. Association computing determines a series of people's needs. Equality computing determines alternative services for people's intentions. For example, when a service is out of service, the other service can give the same results.

10.3.2.1 Traffic Management Through IoT and Intelligent Data Analysis

Everybody hates traffic jams, from residents, planners, and visitors, and the crowd of a city can be a huge danger. With a large amount of information and the combination of net cloud platforms, interconnected cars, and networks of sensors, traffic management is becoming smart and very economical.

The concept of big data and IoT means that the quantity of traffic that should be allowed to roll out at a particular time can be regulated. Information can be collected in real time with the use of sensors, cameras, wearable gadgets, and smart devices. With the rapid surge of connected vehicles, agencies and startups are using data analytics and mobile networks to help cities manage traffic more efficiently.

Let's see how city transport planners are using IDA to solve traffic challenges:

- *Decision support system for traffic.* The smart-traffic call network is born before the establishment of huge information analytics. It breaks down data into chunks and helps to share this data across multiple systems and departments. The answer collects all types of traffic data mistreatment sensors to facilitate time period traffic observance and management. In addition, the solutions feature the power to predict traffic trends and support realistic simulations and mathematical models. The aim of all these methods is to help decision-makers with reliable and scientific data.

- *Addressing traffic congestion and parking.* Fifty percent of the world's population lives in cities, and the population of cities is growing at roughly 2% each year. Whereas incremental growth is nice for the city's economic health, this increase usually causes more problems.

  Data analysis isused to assist town planners to establish the causes of the group's growth. The planner will currently inspect the origin of the visitors, and experiences encountered during their visit, as well as the final purpose of the travel, and may make sure that there's available parking downtown. Town planners will use knowledge analytics to search out the popular parking slots preferred by most drivers.

- *Long-term traffic.* American census data shows that average American workers spend 20% more today than they used to do in the 1980s. It is a positive step for cities to attract visitors and businesses. However, it is never easy to scale the road potential to maintain additional traffic. Large data and IOTs can be used to determine the length of the journey, to decide the long journey, where the journey begins and finally ends, and who is doing it. The planner can access when those passengers have unlimited access to practical driving options. Analytics can show the location of the longest travel time and final miles traveled along with the interval. This data helps to identify alternate routes that can encourage drivers to use the options.

Traffic has a significant impact on people staying in cities and on the efficiency of cities during busy times. Despite the population growth, efficient use of data and censors will help manage the traffic efficiently. The goal of smart traffic management is to make urban driving more intuitive and efficient. As smart cities develop, services and infrastructure will be further integrated. As time goes on, issues such as traffic, waste management, and energy conservation will greatly benefit from the concept of things and the internet of big data.

That is, the entire process of analysis has been depicted in Figure 10.4. When the device is used, they will register on the forum and their basic information will be availablefor a meaningful model search. Then their examples will be stored in the local database. That is, the combination will analyze these examples and implement meaningful annotation and meaningful association building according to the knowledge model. When service requirements are detected, meaningful reasoning and analysis will be called to find related services. Finally, users' models will be analyzed and proper services will be generated.

IDA transforms data into working knowledge, allowing transport users to make informed decisions to ensure the safe and efficient use of facilities. For example, in this type of system, each passenger has access to the most reliable and up-to-date position of almost all transport modes from any point on the transport network. Passenger devices use information about traffic from smartphones, tablet computers, and roadside information. Then they can choose the mode and route that will provide them with minimum travel time and dynamic adjustments with real-time information.

10.3.2.2 Tracking of Multiple Vehicles

Tracking of vehicles is the integral part of ITS, and includes the monitoring of traffic, ground surveillance, and driver support systems. To improve central object tracking performance in

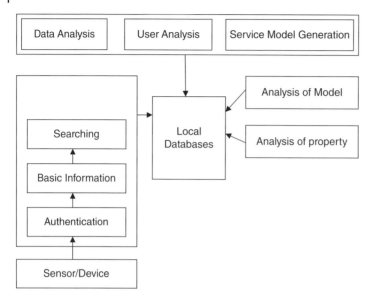

Figure 10.4 Process of intelligent data analysis.

vehicle tracking research, planners must exploit any additional prior information. As vehicles drive on roads, on-road hurdles or road map information can be considered as preexisting. Information for tracking any and all on-road obstacles is considered and was directly applied to form a soft hindrance by filtering the directional process noise. As a sequence of linear segments by modeling the problem of vehicles with on-road obstacles, the tracking is considered as a state with linear equality barriers, and it has been demonstrated that the projection of optimum solution is the penman filter. Other customizable-based filters are also enabled to fix the on-road goal tracking issue. Recently, road map information is included in the vehicle tracking in road coordination. The constrained state estimation problem then becomes a standard unrelated problem. In [14], a single vehicle is tracked on a one-dimensional road, a coordination system in which only longitudinal speed is considered in a vehicle; thus, this approach is only applicable in single-lane cases. This method is extended to the multidimensional by adding an extra dimension in order to model several alleys [15] in the case. Then, exact longitudinal and lateral motions of the vehicle are individually estimated.

10.4 Intelligent Data Analysis for Security in Intelligent Transportation System

With the IoT, modern vehicles and transport systems make the backbone of the future ITS. The associated autonomous vehicles are becoming reality; Google's public commitment to self-driving cars and some other companies are some examples of industry attempts in this direction. Developers of automated driving platforms, which help in delivering their technologies, are working to supports its use by various types of automobile companies.

Figure 10.5 Three-dimensional model for security in ITS.

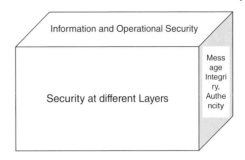

Securing ITS is the foundation for ITS, since ITS systems should be secure before they will improve the effectiveness of the surface installation. ITS as an associate data system should be protected so that ITS applications are trusty, reliable, and obtainable. In summary, we tend to describe the safety within the realm of three-dimensional models with multiple layers, as shown in Figure 10.5.

- ITS ensures the security at all layers of the networking system, such as physical layer security, data link layer security, network layer security, transport layer security, and application layer security.
- Information security and operational security are the other dimensions for the security in ITS. Whatever the information is that is transmitting from one vehicle to another vehicle, vehicle to control center, vehicle to devices/sensors during the operations must be secure and not accessed by unauthorized persons. So, ensuring the security during information sharing and operations is the crucial part of security in ITS.
- Privacy: The contents of the transit information transmitted in the ITS are true and confident, and only the sender and the desired receiver understand the transmitted message. Because it is a content-based network, the information is the truth and also the confidence key. Message integrity requires that this data is transmitted between the sender and the receiver, which is unchanged in the transmission for any reason. Availability means that we have to make sure that the network communication is working unnecessarily without interruption. Disclaimer means that the sender and the receiver are both capable of identifying the other party involved in the communication. Authentication ensures message integrity and adds the sender's digital signature with the message.

While the software is the main area of innovation and value in modern vehicles, additional complexity comes at the cost. Experts estimate that today the cost of software and electronics in some vehicles is already 35–40%, in the future, some types of vehicles can reach 80%. In addition, more than 50% of car warranty costs can be attributed to electronics and their embedded software. In February 2014, after the discovery of faulty software in the car's hybrid control system, the carmaker Toyota recalled 1.9 million hybrid cars worldwide. Software disturbances can stop the hybrid control system while driving, resulting in a lack of electricity, and the vehicle suddenly comes to a stop.

Moreover, security-related incidents are a dangerous threat. Exploiting vulnerabilities within the vehicle's natural state will permit the remote control of car components, whereas an assaulter will put off the lights or perhaps control the brakes while on the move. Additionally, quite recently, attacks on vehicles exploited their vulnerabilities, enabling hackers to use the brakes, kill the engine, and steer the vehicle over the internet; the company urged car owners to update their cars' software packages to patch the known vulnerabilities. There is currently no standard policy on security for transportation. In addition, methods and tools are rare to estimate the combination of physical and cyber risks, and provide only limited guidance for the transport sector on the way to assess these risks. Nevertheless, such a combination of risks is expected to increase. It has highlighted the need for the development of transport-specific equipment to help in the analysis of joint physical attacks and cyberattacks, especially those that are extremely high risks, on interdependent and dependent land transport systems. Steps should be taken to provide a method for multi-risk analysis in the ITS mutually dependent/dependent system. This analysis should be based on a detailed and reliable vulnerability assessment, pointing out how joint attacks can provide cascading effectiveness and how ITS can spread the notice of these threats during deployment. Apart from this, the effectiveness of these efforts should be enhanced by introducing a method for analyzing the necessary level of detail and the problem range and a scalable framework that can be extended to include additional systems and threat scenarios.

IDA in ITS enables intelligent decisions for better transport security and traffic effectiveness through sensors fitted in vehicles, which helps incooperation and wireless communication. Similar to other control systems, ITSs should work properly to limit losses due to system failure, such as invasion or communication breakdown.

In the communication system of the vehicle, the network points broadcast beacon messages to the vehicle and roadside unit (RSU) senders to share information related to the situations, environment, traffic situations, and incidents of the period. The beacon can also be transmitted, sporadically, sent to one or several vehicles. Generally, one beacon message is sent by a vehicle every hundred to one thousand milliseconds. Such communication is generally commented on as a vehicle-to-vehicle (V2V) or vehicle-to-basic infrastructure (V2I). For information shared between participating organizations, messages can also be sent through unicast, multicast, or geocast. Roadside-aided routing is connected to finish the IoT infrastructure through different communication channels. All vehicles share data about themselves by utilizing signals, for example, speed, area, street condition occasions, mischance area, change path/blend movement cautioning, brake light cautioning, crisis vehicle cautioning, and so forth. RSUs send cautioning messages principally about street and weather conditions. Again, the cloud-based services givento the vehicle, for example, vehicle upkeep update or e-toll, may require a session to be set up between the vehicle and a RSU utilizing unicast messages.

IDA plays an important role in the security of the ITS. By smart data analysis, the system can easily identify the congestions, traffic, accidents, and intentions of drivers easily and take the corrective action. For example, suppose that the car is moving in a road abnormally and the cameras are able to identify the abnormal or unwanted activity in the road by the vehicle to the control center or server, so that the corrective action can be initiated. These abnormal activities may be due to speeding, drinking and driving, criminal activity, or due to a health problem. The following steps should be taken for the security of transportation system by the help of IDA:

- *Identify the real usage of ITS.* First, list the realistic ITS usage and applications required that reflect a series of resale processes. Most of these applications can be encouraged or motivated by specific financial needs, barriers to expect social impacts, jurisdiction, or even native issues.

- *Identify specific functional requirements.* This practice requires the adoption of the problem so that it can fully understand what is needed as well as obstacles, and how the tasks are expected to interact. The result in this phase is the motivation for advancing those solutions that solve the appropriate problems.
- *Define the important security issues.* Every functional de-crest means that it uses unique security architecture. This is because the opponents try to attack the system by their open interface, and all functional decomposition shows an exclusive grouping/combination of interfaces. In all the actual systems, the attacker will benefit from some of the accessible interfaces, and to reduce these potential exploitations, requirements of security are cautiously defined as guided targets. In a real system, the reduction of all vulnerabilities can be cost-prohibitive.
- *Critical analysis of data by considering all of the above issues of security.* Apply IDA to the collected data (from heterogeneous devices) by considering all the security issues critically. The benefit of IDA is the quick outcome or decision.

10.5 Tools to Support IDA in an Intelligent Transportation System

Companies that are not taking advantage of data analytic tools and techniques will be left behind. Since the data analytics tool captures products that automatically embrace and analyze the data, as well as provide information and predictions, you can improve prediction accuracy and refine the models. This section will discuss four data analytics tools for success. Organizations can analyze the data and remove active and commercially relevant information to boost performance. There are many extraordinary analytical tools available and you can take advantage of it to enhance your business and develop skills. The properties of the good IDA tools are:

- Visual representation (2D, 3D, or multidimensional)
- Regression analysis
- Correlation analysis
- Results embedded
- Data exploration
- Version records
- Integration with other tools
- Scalability.

Some of the tools that support the IDA and that are the best for the analysis of ITS are:

- *See5.* See5/C 5.0 is designed to analyze millions of records and hundreds of databases in numerical, time, date, or nominal areas. To accelerate the See5/C 5.0 analysis, one or more CPUs (including Intel hyper-threading) also takes advantage of the computer up to eight cores. In order to maximize interpretation, See5/C 5.0 classifiers are expressed as a set of decision trees or immediate rules, which are generally easier to understand than neural networks. This tool is available for Windows as well as for the Linux operating system. RuleQuest provides the C source code of this tool so that See5 can be embedded in any system of organization.

- *Cubist*. This tool has some feature of See5, such as, it can also analyze millions of records and hundreds of databases at a time. To maximize interpretation, Cubist models are expressed as a collection of rules, where each rule has a multi-special linear model. Whenever a situation matches the conditions of the rule, then the approximate value of the respective model is used to calculate it. This tool also available for Windows as well as for the Linux operating system. Cubist is easy to learn and advanced knowledge of statistics or machine learning is not necessary.
- *Inductive learning by logic minimization (ILLM)*. This tool innovates the classification models in the form of different rules, which shows the hidden relationship among the data.

The following programming language can support IDA for the ITS:

- *Python programming language*. This programming language is the procedural language as well as the object-oriented programming language. Python works easily in different platform like Windows, Linux, and Mac operating systems. This language has a very rich and comprehensive library, and is easy to learn and understand. Python programming language supports the hardware and sensors that are used in ITS. Python programming language is free to all because it is an open-source software managed by the Python Software Foundation. Currently, Python programming language is one of the first choices for implementing the smart systems, especially for data science, data analysis, and machine learning. The main data types that work with Python are number, string, list, tuple, and dictionary (Figure 10.6).
- *R programming language*. R programming language is the free software environment for research purposes, solving the statistical problems and predictions. R provides a variety of statistical techniques like linear and nonlinear modeling, time-series analysis, classical statistical test, clustering, and regression tests. Different types of graphical representation

Figure 10.6   Data types of Python.

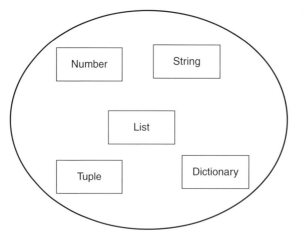

Table 10.1 Objects, statistical techniques, and graphs supported by R programming.

R objects	Statistical techniques supported	Type of graph supported
• Vectors	• Linear and nonlinear modeling	• Histogram
• Lists	• Time-series analysis	• Bar charts
• Matrices	• Classical statistical test	• Piecharts
• Arrays	• Clustering	• Box plots
• Factors	• Regression test	• Line graph
• Data frames		• Scatter plots

can be easily done by the R programming like histogram, bar charts, piecharts, box plots, line graphs, and scatter plots. R programming can be especially useful for the IDA because ITS produces a large amount of homogenous and heterogeneous data daily. For instance, there is a requirement of deep analysis to identify dangerous driving in real time, so that the information about that vehicle can be sent to the concerned authorities immediately for quick action (Table 10.1).

References

1 Rodrigue, J.P., Comtois, C., and Slack, B. (2013). *The Geography of Transport Systems*. New York: Routledge.

2 Sussman, J. (2005). *Perspectives on Intelligent Transportation Systems (ITS)*. Boston, MA: Springer Science + Business Media.

3 Cobo, M.J., Chiclana, F., Collop, A. et al. (2014). A bibliometric analysis of the intelligent transportation systems research based on science mapping. *IEEE Transactions on Intelligent Transportation Systems* 15 (2): 901–908.

4 Diakaki, C., Papageorgiou, M., Dinopoulou, V. et al. (2015). State-of-the-art and -practice review of public transport priority strategies. *IET Intelligent Transport Systems* 9: 391–406.

5 Eriksson, O. (2002). Intelligent transport systems and services (ITS). In: *Information Systems Development* (eds. M. Kirikova, J. Grundspenkis, W. Wojtkowski, et al.). Boston, MA: Springer, Springer.

6 Y. Lin, P. Wang, and M. Ma, Intelligent Transportation System (ITS): Concept, Challenge and Opportunity. IEEE Conference on Big Data Security on Cloud (2017).

7 Chen, B. and Cheng, H. (2010). A review of the applications of agent technology in traffic and transportation systems. *IEEE Transactions on Intelligent Transportation Systems* 11 (2): 485–497.

8 Barmpounakis, E.N., Vlahogianni, E.I., and Golias, J.C. (2015). A game theoretic approach to powered two wheelers overtaking phenomena. In: *Proceedings of the Transportation Research Board 94th Annual Meeting, Washington, DC, USA*, 1–4.

9 A. Maimaris & G. Papageorgiou, "A Review of Intelligent Transportation Systems from a Communication Technology Perspective," IEEE International Conference on ITSC, Brazil 2016.

10 Cheng, W., Cheng, X., Song, M. et al. (2012). On the design and deployment of RFID assisted navigation systems for VANETs. *IEEE Transactions on Parallel and Distributed Systems* 23 (7): 1267–1274.

11 Liu, Y., Dion, F., and Biswas, S. (2005). Dedicated short-range wireless Communications for intelligent transportation system spplications:state of the art. *Transportation Research Record: Journal of the Transportation Research Board* 1910: 29–37.

12 F. Victor & Z. Michael, "Intelligent Data Analysis and Machine Learning:are they Really Equivalent Concepts?" IEEE conference RPC, Russia, (2017).

13 Lu, H., Zhiyuan, S., and Wencong, Q. (2015). Summary of big data and its application in urban intelligent transportation system. *International Journal on Transportation Systems Engineering and Information* 15 (05): 45–52.

14 Ulmke, M. and Koch, W. (2006). Road-map assisted ground moving target tracking. *IEEE Transaction Aerospace Electronic system* 42 (4): 1264–1274.

15 Chen, Y., Jilkov, V.P., and Li, X.R. (2015). Multilane-road target tracking using radar and image sensors. *IEEE Transaction Aerospace Electronic System* 51 (1): 65–80.

11

Applying Big Data Analytics on Motor Vehicle Collision Predictions in New York City

Dhanushka Abeyratne and Malka N. Halgamuge

Department of Electrical and Electronic Engineering, Yellowfin (HQ), The University of Melbourne, Australia

11.1 Introduction

11.1.1 Overview of Big Data Analytics on Motor Vehicle Collision Predictions

Due to population growth, there are more traffic accidents, which have become a global concern. In the past, researchers have conducted studies to find out the common cause for motor vehicle collisions. Log-linear models, data mining, logical formulation, and fuzzy ART are some of the methods widely used to perform research [1]. Even with the use of these methods, data analysis is a complicated process. However, with the improvements in technology, data mining is defined to be a highly accurate method for the analysis of big data.

With the use of big data application, few researchers have focused mainly on understanding the significance of vehicle collisions. A research performed by Shi et al. [2] explains the significance of identifying the traffic flow movements on highways to minimize the impact on vehicle collisions. This research has used a time series data approach under clustering analysis to comprehend traffic flow movements. It was converted using the cell transformation method.

A similar study by Yu et al. [3] considers traffic data mining to be a big data approach. The author further carries out the study using past traffic data with the application of common rules via data mining, which is based on a cloud computing technique. Traffic trend prediction and accident detection within a MAP-reduce framework are used. However, the data sets have many missing pieces and were redundant with values that cannot be normalized [1].

Data mining is a widely used computing technique adapted to determine unfamiliar patterns in a large data set [4]. For prospective use, data mining is a comprehensive field that presents meaningful patterns. These techniques are classified into three categories; classification, prediction, and data clustering, in order to trace seasonal trends and patterns [4]. Classification is a commonly used method that forecasts unknown values acknowledged to generated mode [5].

This chapter proposes a framework that uses a classifying technique for collision predictions that has been undertaken by Python programming language. Python is an effective

Intelligent Data Analysis: From Data Gathering to Data Comprehension,
First Edition. Edited by Deepak Gupta, Siddhartha Bhattacharyya, Ashish Khanna, and Kalpna Sagar.
© 2020 John Wiley & Sons Ltd. Published 2020 by John Wiley & Sons Ltd.

programming language that is used in the theoretical and mathematical analysis for large data sets [6]. Python programming language supports various data mining techniques and algorithms that mainly clusters and classifies. As this option has many beneficial features, it is one of the most suitable tools for making scalable applications. Thus, it can be utilized for the framework of big data analysis in wide motor vehicle collision data sets to obtain reliable results.

A specific data set has been obtained for this chapter about vehicle collision in a large city in the United States (New York City). A data mining technique is exercised to perform further analysis on the data set. Based on recent news reports, New York City (NYC) roads are believed to have an increase in motor vehicle collisions. The National Safety Council has conducted a preliminary survey that confirms that the year 2016 had the deadliest accidents on NYC roads over several decades [7]. Thus, there is a need to predict and assess the association of vehicles involved in a collision and their attributes found in the data set.

11.2 Materials and Methods

This chapter uses an analytical approach to data mining, which forecasts relevant attributes corresponding to the source of other related attributes. Analysis of variance (ANOVA) table generated from k-means clustering, k-nearest neighbor (kNN), naïve Bayes, and random forest classification algorithms are used in this chapter to understand the association of statistical data collected.

11.2.1 Collection of Raw Data

In this chapter the data is collected from an open data website of the New York Police Department (NYPD) for the period 2012–2017. The data set contains information about motor vehicle collisions occurred in NYC with all collision-related attributes.

11.2.2 Data Inclusion Criteria

This raw data set consists of 26 attributes and 1 048 575 cases filtered down using vehicle type code. All the unknown vehicle type codes and blank values have been removed. Filtered data includes 14 attributes. Date attribute was further expanded to generate four attributes (day of the week, day, month, and year). As an example, 1/1/2017; day of the week = Sunday, day = 1, month = Jan, and year = 2017. Finally, the selected collisions data set contains 998 193 and 17 attributes of data (Table 11.1).

11.2.3 Data Preprocessing

The study of Sharma and Bhagat [8] recognizes that the raw data gathered contains noisy data, missing values, data dependency, and multiple sources of data. Thus, the preprocess method identified as the initial process of data mining is recommended to make raw data more reliable. It has three stages: data cleaning, data integration, and data transformation. Primarily, data cleaning excludes missing values and errors while recognizing attributes of

Table 11.1 Illustration of data set attributes.

Attributes	Description
Day of the week	Day of the week collision occurs (Mon–Sun)
Day	Day of the month collision occurs (1–31)
Month	Relevant month collision occurs (Jan–Dec)
Year	Relevant year collision occurs (2012–2017)
Time	Relevant time of the day collision occurs
Borough	Location where the collision happened.
Latitude	Geographical location
Longitude	Geographical location
Number of persons injured	Number of people were injured in the collision.
Number of persons killed	Number of people were killed in the collision.
Number of pedestrians injured	Number of pedestrians were injured in the collision.
Number of pedestrians killed	Number of pedestrians were killed in the collision.
Number of cyclists injured	Number of cyclists were injured in the collision.
Number of cyclists killed	Number of cyclists were killed in the collision.
Number of motorists injured	Number of motorists were injured in the collision.
Number of motorists killed	Number of motorists were killed in the collision.
Vehicle_Group	Group of vehicles which caused the collision

Table 11.2 Categorized vehicle groups.

Vehicle group	Vehicle type-codes
Large_Vehicle	Bus, fire truck, large commercial vehicle
Medium_Vehicle	Pick-up truck, small commercial vehicle, van, livery vehicle
Small_Vehicle	Passenger vehicle, sort-utility/wagon, taxi, pedicab
Very_Small_Vehicle	Motorcycle, scooter, bicycle

raw data. Subsequently, data integration links the data into a reliable structure. In the last part of preprocessing, data is converted into acceptable forms for data mining [8].

During the data preprocessing, the vehicle type-code is further categorized into four groups, depending on vehicle passenger capacity and size of the vehicle. This developed attribute is then added for analysis into the data set. Table 11.2 illustrates categorized groups.

11.2.4 Data Analysis

In this chapter, data has been analyzed using the Python programming language. Python programming language presents significant support for the process of experimental

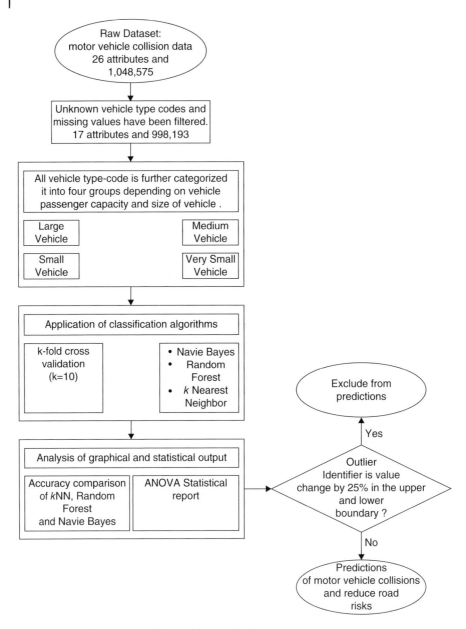

Figure 11.1 Overall methodology of data analysis process.

data mining, combined with categorizing data inputs, for measuring statistical learning schemes. Relevant statistical data is generated rapidly and is precise. Further, it visualizes the input data and learning outcome for a large set of data that becomes clearer [6]. Figure 11.1 illustrates the overall data analysis process of this chapter.

Table 11.3 Description of classification algorithms and functionalities.

Classifiers	Description
Naive Bayes	Naive Bayes is a probabilistic classifier that uses a statistical approach for classification [9].
*k*NN	*k* nearest neighbor is a simple, widely known and efficient algorithm for pattern recognition. It classifies samples considering the class of nearest neighbor [10].
Random Forest	Random forest is a combination of tree-structured predictors via tree classification algorithm. It is identified as an automatic compensation mechanism with the advantages of speed, accuracy, and stability [11].

11.3 Classification Algorithms and K-Fold Validation Using Data Set Obtained from NYPD (2012–2017)

11.3.1 Classification Algorithms

Table 11.3 demonstrates the description of the applied classification algorithms and its functionalities.

11.3.1.1 k-Fold Cross-Validation

The k-fold cross-validation is identified as a well-known statistical experimental technique where k disjointed blocks of objects are produced by the database with the random division. Later, the data mining algorithm is formulated using k−1 blocks and the remaining block is applied to test the function of algorithm; this is a repetitive process by k times [12]. K-fold (K = 10) cross-validation method is used to analyze data in this chapter.

This chapter follows three main steps for data analysis. The three steps are as follows:

1. Data selection
2. Data preprocessing
3. Data analysis using classification methods

All three algorithms are systematically carried out for data analysis using K-fold validation.

Algorithm 11.1 Data Selection

```
LOAD NYPD_collision_data
STRING [] VTC= SELECT Vehicle_Type_Code
READ VTC
IF VTC = " ", "Unkown", "Other"
     DELETE
ELSE
```

```
        ADD
END IF
SAVE filtered_collision_data [n=998,193]
THEN
LOAD filtered_collision_data
STRING [] Date= SELECT Date
SEPARATE Date = "Day_of_ the_ week", "Day", "Month", "Year" manually
SAVE Seperated_Filtered_vollision_Data [n=998,193]
```

Algorithm 11.2 Data Preprocessing

```
LOAD Seperated_Filtered_collision_Data
STRING [] VTC= SELECT Vehicle Type Code
READ VTC
FOR n = 1 to 998,193
IF VTC= "Bus, Fire Truck, Large Commercial Vehicle" THEN
SAVE VTC as "Large Vehicle"
END IF
IF VTC= "Pick-up Truck, Small Commercial Vehicle, Van, Livery Vehicle" THEN
SAVE VTC as "Medium Vehicle"
END IF
IF VTC= "Passenger Vehicle, Sort-Utility/Wagon, Taxi, Pedicab" THEN
SAVE VTC as "Small Vehicle"
END IF
IF VTC= "Motorcycle, Scooter, Bicycle" THEN
SAVE VTC as "Small Vehicle"
END IF
SAVE Grouped_Filtered_ Collision_Data [n=998,193]
```

Algorithm 11.3 Classification Analysis Using k-Fold Cross-Validation

```
LOAD Grouped_Filtered_ Collision_Data
INT [] k= 10 (number of tests)
Step1:
SELECT k-NEAREST NEIGHBOUR
FOR k= 1 to 10 DO
Learn k nearest Neighbour based on predictions
END FOR
RETURN k = Accuracy% (k1), Accuracy%(k2), Accuracy% (k3),......, Accuracy% (k10)
Step2:
SELECT RANDOMFOREST
FOR k= 1 to 10 DO
Learn RandomForest based on predictions
END FOR
RETURN k = Accuracy% (k1), Accuracy%(k2), Accuracy% (k3,......, Accuracy% (k10)
```

```
Step3:
SELECT NAVIEBAYES
FOR k= 1 to 10 DO
Learn NavieBayes based on predictions
END FOR
RETURN k = Accuracy%  (k1), Accuracy%(k2), Accuracy%  (k3) ,......., Accuracy%  (k10)
```

11.3.2 Statistical Analysis

Statistically, the data is analyzed by using a Python one-way ANOVA table. In fact, Python programming allows users to understand significance value (*p*-value) in a data set variable. In Python, the scipy library inherits functionality to carrying out one-way ANOVA tests called scipy.stats.f_oneway. One-way ANOVA is a statistical implication test that allows the comparing multiple groups simultaneously [13]. This helps to find out that each vehicle group mean value in a given period differs from each other depending on the time period.

11.4 Results

11.4.1 Measured Processing Time and Accuracy of Each Classifier

This chapter has classified using algorithms using k-fold across-validation test option and k-means clustering are the main techniques used to achieve results. Accuracy and processing times were computed by using *k*NN, random forest, and Naive Bayes classification. Table 11.4 displays a comparison of each classifier results.

The high accuracy recorded 95.03% and 94.93% in random forest and *k*NN classifiers. Subsequently, Naive Bayes indicates the accuracy of 70.13% in contrast to *k*NN and random forest classifiers that comprise maximum inaccurate instances of 29.87%.

The maximum processing time is consumed by *k*NN with 682.5523 to construct the model subsequently with random forest of 507.98 seconds. Naive Bayes has determined the minimum time to build the model = 5.7938 seconds.

Figure 11.2 demonstrates a graphical representation of *k*NN, random forest, and Naive Bayes classifiers accuracy output comparing to k-fold value. The result explains the average accuracy of data set as constant. However, *k*NN and random forest outperformed Naive

Table 11.4 Comparison of classifier results.

Classifier	Processing time	Accuracy %
Random forest	507.98316	95.03
*k*NN	682.5523	94.93
Naive Bayes	5.7938	70.13

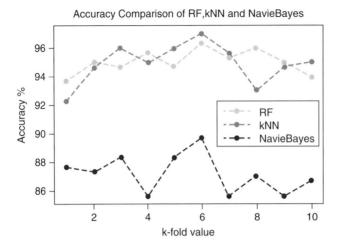

Figure 11.2 Accuracy comparison of RF and *k*NN.

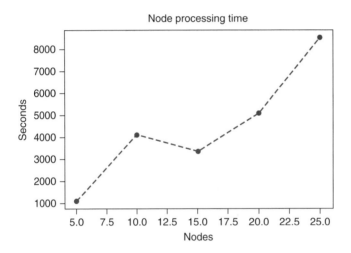

Figure 11.3 Random forest node processing time.

Bayes accuracy. The highest accuracy has been recorded k6 = 97% in *k*NN while the lowest is recorded in k7 = 85.66% in Naive Bayes. Further, Figure 11.1 indicates k6 instance as the highest accuracy for all three classifiers. Nonetheless, comparing all above results, it is evident that random forest and *k*NN prediction related to vehicle groups will be highly accurate.

Figure 11.3 explains that the total processing time of total nodes is in a linear growth where there is evidence of the accurate frequency of data set in a random forest classifier. However, 10–15 node processing time has decreased in the classifier.

Figure 11.4 illustrates the constant accuracy of 95.033% data comparing to each total number of nodes. It is further evident constant high accuracy data prediction in random forest is spread among each node.

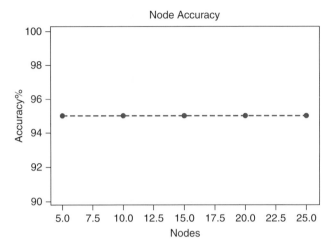

Figure 11.4 Random forest node accuracy.

11.4.2 Measured *p*-Value in each Vehicle Group Using K-Means Clustering/One-Way ANOVA

Python one-way ANOVA is carried out considering three groups of time periods in the data set. The relevant python algorithm of one-way ANOVA process is shown.

Algorithm 11.4 Data Preparation of Vehicle Group

```
LOAD Grouped_Filtered_collision_Data
STRING [] VG= SELECT Vehicle_Group
READ VG
FOR n = 1 to 998193
IF VG= "Large Vehicle" THEN
SAVE VG as "Test 1"
END IF
IF VG= "Medium Vehicle" THEN
SAVE VG as "Test 2"
END IF
IF VG = "Small Vehicle" THEN
SAVE VG as "Test 3"
END IF
IF VG = "Very Small Vehicle" THEN
SAVE VG as "Test4"
END IF
SAVE pvaluetest_ Collision_Data [n=998193]
```

Algorithm 11.5   Data Preparation of Year Group

```
LOAD pvaluetest_Colision_Data
STRING [] Period= SELECT Year
SEPARATE Period = "2016-2017", "2014-2015", "2012-2013" manually
SAVE pvaluegrouptest_Collision_Data [n=998193]
```

Algorithm 11.6   *p*-Value Test

```
LOAD pvaluegrouptest_Collision_Data
INT [] VG = Select Vehicle Group
STRING [] Period= SELECT Year group
Step1:
SELECT ONE-WAY ANOVA
FOR VG = "Test 1" DO
Learn pvalue based on mean
END FOR
Step2:
SELECT ONE-WAY ANOVA
FOR VG = "Test 2" DO
Learn pvalue based on mean
END FOR
Step3:
SELECT ONE-WAY ANOVA
FOR VG = "Test 3" DO
Learn pvalue based on mean
END FOR
Step4:
SELECT ONE-WAY ANOVA
FOR VG = "Test 4" DO
Learn pvalue based on mean
END FOR
RETURN INT [] p-value = (p₁, p₂, p₃, p₄)
```

SAVE p_1, p_2, p_3, p_4

According to the above analysis of algorithms, it is observed that different vehicle groups are compared against a period of years. In the following four tests p-value <0.05 considered as statistically significant (Table 11.5).

The tests output yields every p-value of $p < 0.001$. Its justified p-value is significant at a 99.99% confidence level while indicating a high significance between different means of each vehicle group in given periods.

Figure 11.5 Heat map of large vehicle collisions.

11.4.3 Identified High Collision Concentration Locations of Each Vehicle Group

Visualization of collision heat maps for each vehicle group are generated using random forest classifier. It represents the locations that have more collision occurrence. The heat maps illustrate a meaningful pattern that precisely confirms the association between number of collisions and location for each vehicle group. Latitude and longitude attributes are used to represent this information on the heat maps (Figures 11.5 and 11.6).

11.4.4 Measured Different Criteria for Further Analysis of NYPD Data Set (2012–2017)

Figure 11.7 trends indicate between the years 2013 and 2016 the numbers of collisions and number of persons injured has been increasing year-wise while killed persons were stable.

Table 11.5 Analyzed p-value test results.

Test no	Vehicle_Group	p-Value
Test 1	Large vehicle	1.777e-159
Test 2	Medium vehicle	0.001
Test 3	Small vehicle	0.001
Test 4	Very small vehicle	4.798e-98

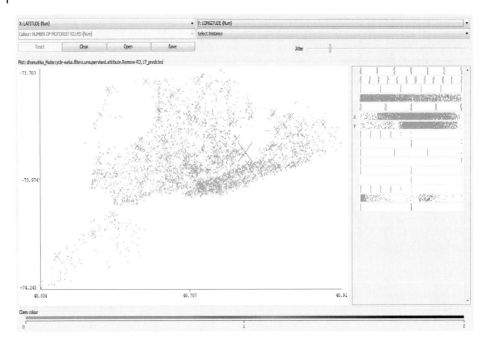

Figure 11.6 Heat map of very-small vehicle collisions.

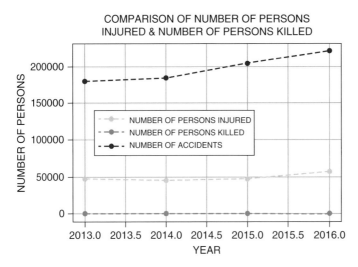

Figure 11.7 Comparison of number of collisions, persons injured, and persons killed year-wise.

Further, this is evident by numbers recorded for 2017 until the month of April. Within four months of 2017, the number of collisions, injured, and killed persons recorded were 122 245, 31 057, and 105, respectively, which is an approximately 50% growth comparing to the previous year. Therefore, the year 2017 was more fatal than 2016 and could be recorded as the deadliest year for NYC roads in decades.

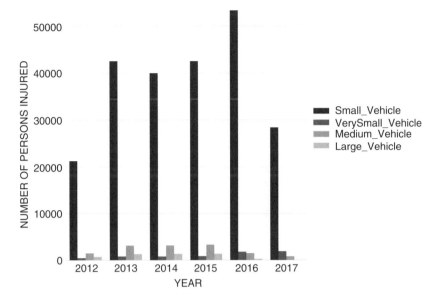

Figure 11.8 Number of persons injured based on vehicle groups.

Figures 11.8 and 11.9 show that most of the fatal collisions are recorded from medium-sized vehicles. The reason for this could be that the increasing number collisions occurred by passenger vehicles in NYC. In 2017, it recorded rapid growth compared to other vehicle types. Further Figure 11.8 shows the number of persons killed in very small vehicles is considerably higher than other groups of vehicles. This could be due to less safety in very small vehicles.

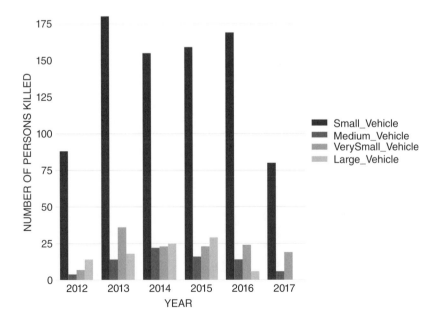

Figure 11.9 Number of persons killed based on vehicle groups.

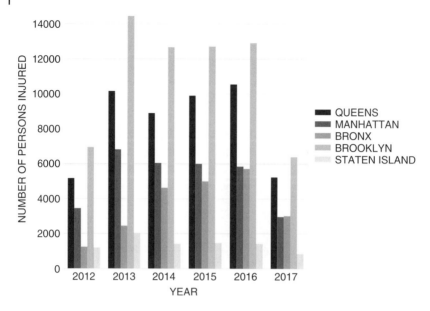

Figure 11.10 Number of persons injured based on borough.

Figures 11.10 and 11.11 shows the number of persons injured and killed in each borough. Queens and Brooklyn can be identified as the areas where many fatal crashes occur. This could be due to decreased road safety in both of these areas as well as high traffic conditions.

After analyzing 998 193 motor vehicle collisions (reported during 2012–2017) in NYC, there is a probability of extreme cases. Therefore, distribution of collision severity of

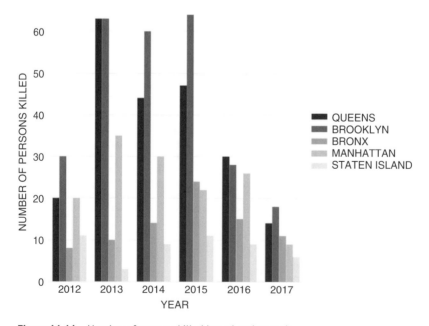

Figure 11.11 Number of persons killed based on borough.

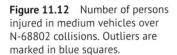

Figure 11.12 Number of persons injured in medium vehicles over N-68802 collisions. Outliers are marked in blue squares.

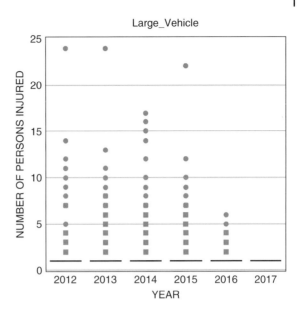

each vehicle group requires testing the normality of their distribution. The full data set (n = 998 193) were used. Mean value is calculated for "injured persons" and "killed persons" attributes in data set and outliers were identified.

The outlier is a critical representation of the spread of data, as value change by 25% in the upper and lower boundary, which do not affect any prediction. However, if data is from a normal distribution, the outlier is considered inefficient compared to the standard deviation [14]. With the purpose of estimating outliers, one person injured, and another killed one have been used. Therefore, any value below or above one is defined as the outlier. Figures 11.12–11.19 show the outliers in each year injured and killed people are recorded based on the vehicle group.

In Figures 11.12–11.19, significant numbers of outliers for the severity of collisions in each vehicle group can be observed. Most of the outliers are visible in the injured numbers. This could be due to the growth of injured numbers over the reported killed. However, the forecasts were carried out by excluding the extremely severe cases, which were found during the outlier analysis process. In these predictions, taking into consideration the outliers of severe collision cases is significant and very critical.

11.5 Discussion

In recent years, the growth of motor vehicle collisions has become critical. The number of road incidents has increased leading to more injuries, disabilities, and mortality on a global level. On a daily basis, more people will experience collision as a result of traffic congestion, which causes delays for vehicles passing through areas with lane closures.

The outcome of this chapter will predict emerging patterns and trends in motor vehicle collisions that may reduce the road risks. In fact, it will help to predict patterns of collision

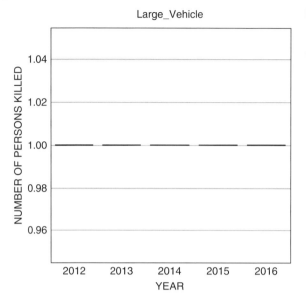

Figure 11.13 Number of persons killed in medium vehicles over N-68802 collisions. Outliers are marked in blue squares.

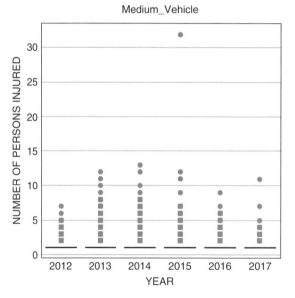

Figure 11.14 Number of persons injured in large vehicles over N-27508 collisions. Outliers are marked in blue squares.

and severity engaged with each type of vehicle. This chapter has used 998 193 large gen-uine data sets from NYPD as the source for data analysis. Therefore, the analyzed patterns were very reliable for overcoming road risks. The results of this chapter can even be used by NYPD to identify and prevent road risk on NYC roads. This chapter has used machine learn-ing classification algorithms of k-NN, random forest, and Naive Bayes, as these received good results in our previous research [15, 16]. Using these three accuracy classifiers can predict the different vehicle groups and identify particular risk groups.

Figure 11.15 Number of persons killed in large vehicles over N-27508 collisions. Outliers are marked in blue squares.

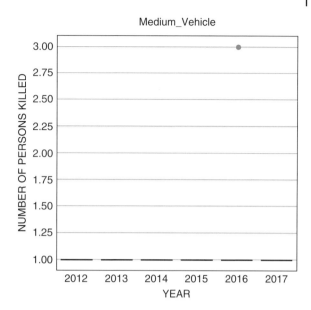

Figure 11.16 Number of persons injured in small vehicles over N-892174 collisions. Outliers are marked in blue squares.

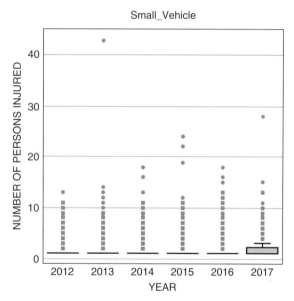

Among three classifiers' data sets, the random forest generated the highest data prediction accuracy results. Random forest is a tree-structured algorithm that is used for pattern recognition [11]. The main reasons for using random forest as a classification technique is due to its nature of producing accurate and consistent predictions [9]. The random forest algorithm previously used in several studies demonstrates predictions. For instance, random forest has been used for a data-driven model for crossing safety by predicting collisions in railway and roadway crossings [17].

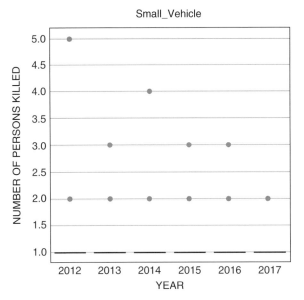

Figure 11.17 Number of persons killed in small vehicles over N-892174 collisions. Outliers are marked in blue squares.

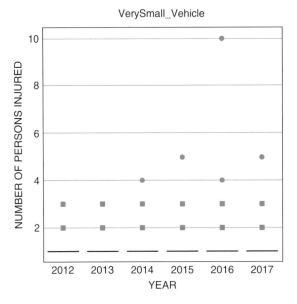

Figure 11.18 Number of persons injured in very small vehicles over N-9705 collisions. Outliers are marked in blue squares.

On the other hand, studies carried by several researchers [18, 19] suggested Naive Bayes classification as the accurate classification technique for collision analysis. These studies have used only numerical inputs for relevant prediction analysis. Therefore, Naive Bayes as a statistical pattern recognizer has produced high accuracy than other classifiers. Nevertheless, according to this chapter, random forest provides the most accurate prediction over the selected data set.

Additionally, the data set was statistically analyzed using one-way ANOVA. It shows high significance between different means of each vehicle group in the given periods. This is

Figure 11.19 Number of persons killed in very small vehicles over N-9705 collisions. Outliers are marked in blue squares.

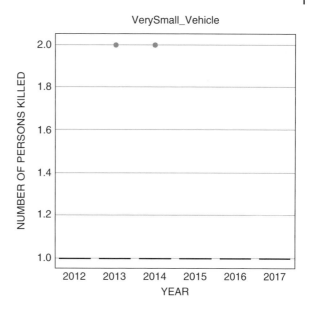

evident that vehicle groups are highly significant for collision patterns. Therefore, collision severity has been analyzed comparing to the vehicle group. It is evident that small vehicles were the reason for the high collision severity. This chapter reveals that between the years 2012–2017 motor vehicle collisions have increased in NYC with severity. However, it is observed that some extreme values are present in the data. Hence, this chapter excludes extreme values using outlier analysis. Further graphical representation of location using latitude and longitude confirmed the pattern of collision for each vehicle group. Brooklyn and Queens boroughs are identified as locations with the highest severe collisions.

The main limitation of this chapter occurred in the data analysis phase. During the classification carried out using Python programming, Naïve Baiyes did not capture the data in $k = 4$ and $k = 7$ k-fold instances. However, this did not impact the overall result since Naïve Bayes generated lesser value for other k-fold instances comparing to kNN and random forest.

The results obtained from this chapter confirmed the significance of vehicle groups in motor vehicle collisions and road risks. Identified vehicle groups could accurately predict the location and severity of the collisions. Therefore, further studies can consider these identified vehicle groups for future road accident–related research. This academic chapter acknowledged patterns through vehicle collision analysis and converted this knowledge for relevant road safety authorities and law enforcement officers to minimize motor vehicle collisions.

11.6 Conclusion

The data mining classification techniques are widely used for data analysis to obtain valuable findings. In this chapter, there are three main classifiers that have been used to generate accurate prediction over motor vehicle collision data. In total 998 193 data collisions were tested among these three classifiers. The analysis in this chapter shows that the random forest classifier has the highest data prediction accuracy algorithm with 95.03%, following kNN at 94.93%, and naïve Bayes at 70.13%. Additionally, the analysis of random forest node processing time and accuracy further confirms it is the most suitable prediction classifier of this collision data. The findings of this chapter show that there has been an increase in motor vehicle collisions on NYC roads during 2012–2017. Considering only the recent years 2016–2017, there has been a growth of approximately 50% in vehicle collisions. Among these collisions, the small vehicle group recorded the highest. It is evident that the identification of motor vehicle groups has a significant impact on the severity of collisions on NYC roads. Brooklyn and Queens boroughs are identified as locations with the highest and most severe collision rates. However, this chapter generates results for collisions in NYC only; it cannot be universally applied to collisions occurring in other parts of the world. Further, these results can be used to improve road safety and minimize any potential road risks. Finally, it highlights valuable information for road authorities and law enforcing bodies to manage inner-city traffic in an efficient manner.

Author Contribution

D.A. and M.N.H. conceived the study idea and developed the analysis plan. D.A. analyzed the data and wrote the initial paper. M.N.H. helped preparing the figures and tables, and in finalizing the manuscript. All authors read the manuscript.

References

1 Abdullah, E. and Emam, A. (2015). Traffic accidents analyzer using big data. In: *2015 International Conference on Computational Science and Computational Intelligence (CSCI)*, 392. Las Vegas: IEEE https://doi.org/10.1109/CSCI.2015.187.

2 Shi, A., Tao, Z., Xinming, Z., and Jian, W. (2014). Unrecorded accidents detection on highways based on temporal data mining. *Mathematical Problems in Engineering* 2014: 1–7. https://doi.org/10.1155/2014/852495.

3 Yu, J., Jiang, F., and Zhu, T. (2013). RTIC-C: a big data system for massive traffic information mining. In: *2013 International Conference on Cloud Computing and Big Data*, 395–402. IEEE https://doi.org/10.1109/CLOUDCOM-ASIA.2013.91.

4 Sharma, S. and Sabitha, A.S. (2016). Flight crash investigation using data mining techniques. In: *2016 1st India International Conference on Information Processing (IICIP)*, 1–7. IEEE https://doi.org/10.1109/IICIP.2016.7975390.

5 Chauhan, D. and Jaiswal, V. (2016). An efficient data mining classification approach for detecting lung cancer disease. In: *2016 International Conference on Communication and Electronics Systems (ICCES)*, 1–8. IEEE https://doi.org/10.1109/CESYS.2016.7889872.

6 Ince, R.A.A., Petersen, R., Swan, D., and Panzeri, S. (2009). Python for information theoretic analysis of neural data. *Frontiers in Neuroinformatics* 3 https://doi.org/10.3389/neuro.11.004.2009.

7 Korosec, K., 2017. 2016 Was the Deadliest Year on American Roads in Nearly a Decade [WWW Document]. Fortune. http://fortune.com/2017/02/15/traffic-deadliest-year (accessed 10.8.17).

8 Sharma, S. and Bhagat, A. (2016). Data preprocessing algorithm for web structure mining. In: *2016 Fifth International Conference on Eco-Friendly Computing and Communication Systems (ICECCS)*, 94–98. IEEE https://doi.org/10.1109/Eco-friendly.2016 .7893249.

9 Witten, I., Frank, E., and Hall, M. (2011). *Data Mining Practical Machine Learning Tools and Techniques*, 3e. Boston: Morgan Kaufmann.

10 Shi, A., Tao, Z., Xinming, Z., and Jian, W. (2014). Evolution of traffic flow analysis under accidents on highways using temporal data mining. In: *2014 Fifth International Conference on Intelligent Systems Design and Engineering Applications*, 454–457. IEEE https://doi.org/10.1109/ISDEA.2014.109.

11 Chen, T., Cao, Y., Zhang, Y. et al. (2013). Random Forest in clinical metabolomics for phenotypic discrimination and biomarker selection. *Evidence-based Complementary and Alternative Medicine* 2013: 1–11. https://doi.org/10.1155/2013/298183.

12 Salvithal, N. and Kulkarni, R. (2013). Evaluating performance of data mining classification algorithm in Weka. *International Journal of Application or Innovation in Engineering and Management* 2: 273–281.

13 Rao, J., Xu, J., Wu, L., and Liu, Y. (2017). Empirical chapter on the difference of Teachers' ICT usage in subjects, grades and ICT training. In: *2017 International Symposium on Educational Technology (ISET)*, 58–61. IEEE https://doi.org/10.1109/ISET.2017.21.

14 Halgamuge, M.N. and Nirmalathas, A. (2017). Analysis of large flood events: based on flood data during 1985–2016 in Australia and India. *International Journal of Disaster Risk Reduction* 24: 1–11. https://doi.org/10.1016/j.ijdrr.2017.05.011.

15 Halgamuge, M.N., Guru, S.M., and Jennings, A. (2005). *Centralised Strategies for Cluster Formation in Sensor Networks, in Classification and Clustering for Knowledge Discovery*, 315–334. Cambridge, UK: Springer-Verlag.

16 Wanigasooriya, C., Halgamuge, M.N., and Mohamad, A. (2017). The analyzes of anti-cancer drug sensitivity of lung cancer cell lines by using machine learning clustering techniques. *International Journal of Advanced Computer Science and Applications (IJACSA)* 8 (9): 1–12.

17 Trudel, E., Yang, C., and Liu, Y. (2016). Data-driven modeling method for analyzing grade crossing safety. In: *2016 IEEE 20th International Conference on Computer Supported Cooperative Work in Design (CSCWD)*, 145–151. IEEE https://doi.org/10.1109/ CSCWD.2016.7565979.

18 Al-Turaiki, I., Aloumi, M., Aloumi, N., and Alghamdi, K. (2016). Modeling traffic accidents in Saudi Arabia using classification techniques. In: *4th Saudi International Conference on Information Technology (Big Data Analysis) (KACSTIT)*. Riyadh: IEEE https://doi.org/10.1109/KACSTIT.2016.7756072.

19 Li, L., Shrestha, S., and Hu, G. (2017). *Analysis of Road Traffic Fatal Accidents Using Data Mining Techniques*, 363–370. IEEE https://doi.org/10.1109/SERA.2017.7965753.

12

A Smart and Promising Neurological Disorder Diagnostic System: An Amalgamation of Big Data, IoT, and Emerging Computing Techniques

Prableen Kaur and Manik Sharma

Department of Computer Science and Applications, DAV University, Jalandhar, India

12.1 Introduction

Neurological disorders are critical human disorders that are generally related to problems of the spinal cord, brain, and nervous system. In general, the structural disturbance of neurons in the human body leads to these disorders. Lifestyle, hereditary, infection, nutrition, environment, and major physical injuries are some of the important causes of neurological disorders. There are different signs and symptoms for each neurological disorder. Therefore, it is difficult to mention the complete list. However, in general, the change in behavior, emotions, and physical appearance are some signs of these critical disorders. In general, neurological disorders lead to the impairment of the brain, spinal cord, nerves, and several neuromuscular functions in the body [1–3]. These disorders are the deformities that ensue in the central nervous system of the human body. The central nervous system consists of nerves inside the brain and spinal cord, whereas the nerves outside the brain and spinal cord are part of the peripheral nervous system [4]. The brain controls different body functions such as breathing, hormone release, heart rate, body temperature, movements, sensations, desires, and emotions in the body, etc. The spinal cord carries the brain signals to peripheral nerves. And the peripheral nerves connect the central nervous system with the other body organs [5]. The diseases related to the spinal cord and brain are considered central nervous system disorders. Meningitis, brain tumor, stroke, seizures, epilepsy, Alzheimer's disease, Parkinson's disease, and multiple sclerosis are some of the major central nervous system disorders, whereas the diseases related to nerves outside the spinal cord and brain are known as peripheral nervous system disorders. Some of the examples of peripheral nervous system disorders are Guillain-Barre syndrome, carpal tunnel syndrome, thoracic outlet syndrome, complex regional pain syndrome, brachial plexus injuries, sciatica, neuritis, and dysautonomia. The brief list of central and peripheral nervous system disorders is depicted in Figure 12.1 [6, 7].

12.1.1 Difference Between Neurological and Psychological Disorders

Neurology and psychology have a close relationship with each other. The study of the human mind and its functioning is called psychology [8], whereas neurology is related to

Intelligent Data Analysis: From Data Gathering to Data Comprehension,
First Edition. Edited by Deepak Gupta, Siddhartha Bhattacharyya, Ashish Khanna, and Kalpna Sagar.

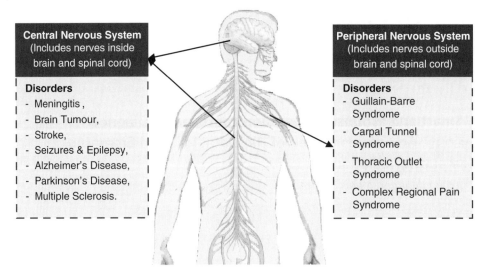

Figure 12.1 Types of neurological disorders.

neurons in the nervous system of the human body [9]. The neurological disorders in human beings generally affect the functioning of brain and behavior. As stated earlier, generally, the disorders related to the brain, nervous system, and the spinal cord in human beings are categorized as neurological disorders [3]. On the other hand, the disorders related to an ongoing dysfunctional pattern of thought, emotion, and behavior that causes significant distress are known as psychological disorders [10]. Some of the major differences among neurological and psychological disorders are presented in Table 12.1

Table 12.1 Difference between neurological and psychological disorders.

Neurological disorders	Psychological disorders
Malfunction or damaged neurons in nervous system.	The dysfunctional pattern of thought, emotion, and behavior.
Patients have the realization that they are not well.	Patients have loss of touch with reality.
Depends upon biological, cognitive, and behavioral factors.	Depends on the psychological, biological, social and emotional, and behavioral factors.
Affects walking, speaking, learning and moving capability of the patient.	Affects thinking and behaving capability of the patient.
Stroke, brain tumors, Alzheimer's, Parkinson's, meningitis, multiple sclerosis, and epilepsy are some neurological disorders.	Depression, anxiety, panic attacks, schizophrenia, autism, and post-traumatic stress, etc. are some psychological disorders.
Can be diagnosed at an early stage by observing clinical symptoms.	Difficult to diagnose at early stage.
A neurological examination can be performed by some tests such as x-ray, fluoroscopy, biopsy, CT scan, MRI, and PET, etc.	The psychological examination can be performed by conduction interviews and counseling sessions between patients and health care experts.

According to the University of California [11], there exist about 600 known neurological disorders that can be diagnosed by a neurologist [12]. Some neurological disorders can be treated, whereas some require a serious prognosis before starting treatment. The diagnostic test includes regular screening tests such as blood or urine test, genetic tests such as chorionic villus sampling, uterine ultrasound, etc., and other neurological examinations, such as x-ray, fluoroscopy, biopsy, brain scans, computed tomography (CT scan), electroencephalography (EEG), electronystagmography, electromyography, magnetic resonance imaging (MRI), thermography, polysomnogram, positron emission tomography (PET), etc. [13]. All these methods are time-consuming and costly. Moreover, several computational models have also been designed to efficiently diagnose neurological disorders. The objective of this research work is to devise a smart and promising neurological diagnostic framework using the amalgamation of different emerging computing techniques.

12.2 Statistics of Neurological Disorders

In the last 25 years, the burden of neurological disorders has increased and become one of the leading causes of disability. Some prevalent neurological disorders are migraine, dementia, stroke, and Alzheimer's, which may adversely affect the behavior and overall performance of human beings [14]. The prevalence of neurological disorders in human beings, along with the associated death rate in 2015, has been depicted in Figure 12.2 [15].

A severe headache is the most prevalent disorder in human beings. Fortunately, it does not lead to death. However, there are other neurological disorders, viz. dementia, stroke,

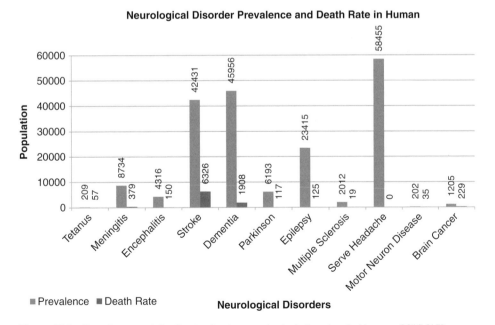

Figure 12.2 Prevalence and death rate due to neurological disorders in the year 2015 [15].

Prevalence of Neurological Disorders in Different Countries

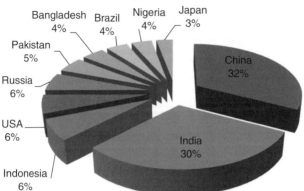

Figure 12.3 Prevalence of neurological disorders in different countries [15].

and epilepsy, which were highly prevalent in human beings. Also, death rate due to stroke and dementia is very high. It is observed from Figure 12.2 that the percentage of death rate according to prevalence is high in patients suffering from tetanus, brain cancer, motor neuron disease, and stroke, i.e., 27%, 19%, 17%, and 15%, respectively. The risk of neurological disorders has spread widely all over the world. Figure 12.3 shows the top 10 countries with the highest prevalence of neurological disorders.

In 2015, China and India were the highest populated countries and also have shown a high prevalence of neurological disorders in their population. Other than China and India, Indonesia, the United States, Russia, Pakistan, Bangladesh, Brazil, Nigeria, and Japan are also countries with a high prevalence of neurological disorders.

12.3   Emerging Computing Techniques

12.3.1   Internet of Things

Internet of things (IoT) was firstly coined by Kevin Ashton in 1999. Itis an emerging computing network composed of different computing or communicative devices, physical objects or things, and is capable to transmit data without requiring human intervention [16]. The use of IoT makes the objects to communicate with the external environment and human beings without their intervention. This interaction can be performed by embedding multiple sensors on the human body or by using smart and wearable devices. IoT systems can be categorized as local and remote IoT systems. For local systems, different objects can be hooked up by using Wi-Fi local area network only, i.e., no internet connection is required. However, for remote systems, the internet is mandatory as the data transmission between the different IoT devices will be done using an internet connection. IoT is used for the devices that would not be expected to have an internet connection, such as an air conditioner, door, lights, and refrigerators, which can be automatically switched on or off with the help of smartphone applications [17]. This can be done by introducing smart sensors on devices.

Primarily, IoT was used in manufacturing industries and business to predict new opportunities with analytics. Nowadays, it is applicative with different sectors such as banking, health care, finance, learning, transportation, communication, entertainment, etc. with smart devices. The effective use of IoT devices seems to be very beneficial in the health care sector. It can improve the diagnostic process, treatment efficiency and hence the overall health of the patients [18]. Real-time monitoring helps to save patients' life in emergency events such as worse heart conditions, seizures, and asthmatic attacks [19]. A huge amount of data (structured and unstructured) is generated by IoT devices from different sources such as a hospital, laboratory, clinics' embedded health care devices, and research centers. This data requires large storage, integration, assortment, security, and analysis for remote health monitoring and medical assistance. These issues can be solved by using big data analytics to diagnose various chronic diseases such as diabetes, cancer, cardiac disorders, and psychological and neurological disorders, in order to deliver precise data to perform potential research, etc.

Due to uncontrollable data growth, the effective utilization and analysis of data are imperative. This effective data analysis can be practicable in sale forecasting, economic analysis, disease diagnosis, social network analysis, business management, etc. Some organizations erstwhile use analytics on the systematic data in the form of reports. The consequences related to data size and computing powers can be figured out by introducing the concept of big data [20].

12.3.2 Big Data

The primitive idea of big data was broached in the paper, "Visually Exploring Gigabyte Datasets in Real Time," published by the Association for Computing Machinery in 1999 [21]. In 2001, Doug Laney depicts the characteristics of big data in four Vs, i.e., volume, velocity, variety, and veracity. The volume represents an analysis of the huge quantity of data (in terabytes) [22, 23]. Variety represents data incorporated from distinct sources and in different formats (structured, unstructured, semi-structured) [22, 23]. Velocity represents fast data processing to promote the decision-making process [22, 23]. And veracity represents data credibility and appropriateness for users [22, 23]. Hadoop is one of the dominant frameworks used to analyze huge and unstructured data [24]. Big data offers vast opportunities in the health care sector to improve operational efficiency, advanced patient care, detection of health threats, early diagnosis of diseases, real-time alerts, and patient care, reduction of treatment cost and readmission rates, telemedicine, and preventative visits to health care experts [25]. The relationship between the characteristics of IoT and big data is presented in Figure 12.4.

12.3.3 Soft Computing Techniques

Soft computing is an optimization technique that is used to provide solutions to imprecise, uncertain, approximate, and real-life complex problems. In the last few years, various soft computing techniques have been designed and employed to solve a number of real-life applications related to business, finance, and engineering as well as health care. In general, these techniques are divided into evolutionary and machine learning techniques [26]. Figure 12.5 presents some of the major subcategories of these techniques.

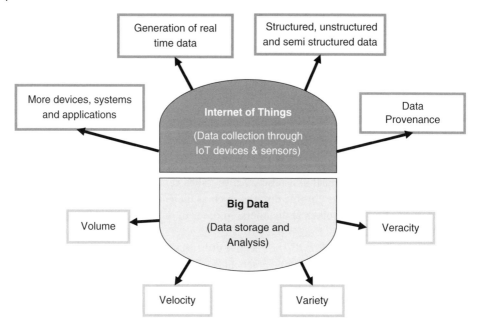

Figure 12.4   IoT and big data.

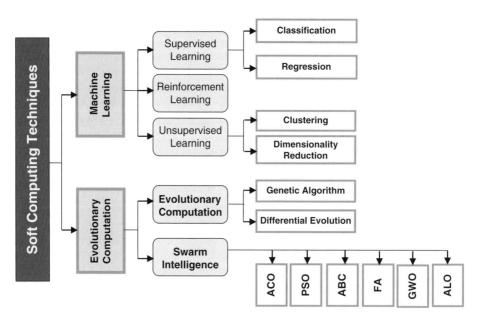

Figure 12.5   Soft computing techniques.

Figure 12.6 The process to generate an optimal solution [76, 77].

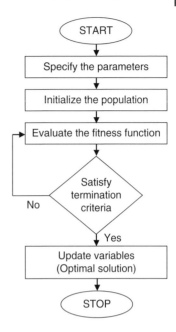

Evolutionary computations are used to solve nonlinear, complex, optimization problems that are difficult to solve using deterministic techniques [26]. The idea of evolutionary computation is taken from biology, physics, chemistry, and genetics. Genetic algorithm and swarm intelligence are coming under the category of evolutionary computation. A genetic algorithm was introduced by John Holland in the 1970s [27]. It is a search algorithm intends to generate new offspring by combining genes of the population. Basically, it depends upon the principle of natural biological process. It is a stochastic technique to produce a quality solution with less time complexity. It asserts a population consisting of different individual chromosomes to derive under selection rules to breed a state for objective function optimization. Genetic algorithm (GA) lucratively works on the population of solutions rather than on single solution and employs some heuristics such as "selection," "crossover," and "mutation" for improved solutions [28]. Swarm intelligence is another innovative optimization method inspired bythe biological behavior of birds, animals, birds, and nature [29]. The encouragement comes from the herding, swarming, and flocking behavior in vertebrates. Swarm intelligence is basically used to employ the self-organizing behavior of agents through interaction with the nearest neighbor.

Figure 12.6 shows the general process to generate the optimal solution using swarm intelligence.

Some swarm intelligence algorithms are

- ant colony optimization (ACO) [30],
- particle swarm optimization (PSO) [31],
- artificial bee colony (ABC) [32],
- firefly algorithm (FA) [33],
- ant lion optimizer (ALO) [34],
- gray wolf optimizer (GWO) [35],

- crow search algorithm (CSA) [36].
- whale optimization algorithm (WOA) [37]
- honey bee (HB) [38],
- cuckoo search (CS) [39],
- water wave algorithm (WWA) [40],
- ray optimization (RO) [41],

All of the above-mentioned soft computing techniques have been used to solve the optimization problems related to operation site allocation, disease diagnosis, agriculture, stock analysis, engineering, and many more [42–46].

Machine learning is a mature and well-recognized branch of computer science that is concerned with innovations of new algorithms and models with data. It helps to perform different tasks and to build models that unearth connections and help organizations to perform better decisions without any human intervention [47]. Traditional approaches to programming rely on hard-coded rules, which set out how to solve a problem, step by step. In contrast, machine learning systems set tasks and give a large amount of data to use as an example of how this task can be achieved or from which to detect patterns. The system then learns to show how best achieve the desired output. The learning element also creates systems that can be adaptive, and that continue to improve the accuracy of their results after they have been deployed. Nowadays, machine learning techniques have been successfully employed to solve several problems from different domains [48]. The rate of machine learning based articles related to different applications has been depicted in Figure 12.7.

Machine learning techniques are divided into three categories, i.e., supervised, unsupervised, and reinforcement learning. Supervised learning is the predictive method used to induce the estimation or prediction about the future outcomes. Techniques come under this category are classification and regression. The process of discovering rules to assign new items into a predefined category is called classification [49–51]. Decision tree, support

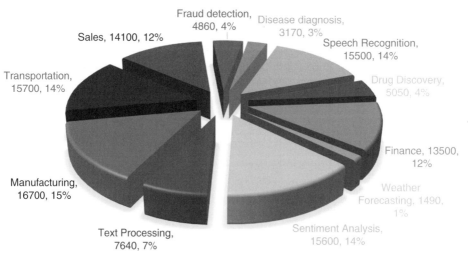

Figure 12.7 Machine learning applications.

vector machine (SVM), random forest, naive Bayes, and ID3 are some of the dominantly used classifiers. Regression is the process of imperative data mining used to determine a trend between dependent and independent variables. The aim of the regression is to predict the value of the dependent variable based on the independent variable [52].

Contrarily, unsupervised learning techniques are employed when the data is not labeled. These techniques are normally suited for transactional data. These techniques can also be beneficial in finding facts from the past data. Clustering and dimension reduction are two well-known unsupervised learning techniques. Clustering is a process of data segmentation used to divide large data sets into similar groups. Some methods for clustering are hierarchical, partitioning, grid-based, density-based, and model-based methods [53, 54].

12.4 Related Works and Publication Trends of Articles

Data is the consequential component of data mining. It is required to collect patients' data from different sources, viz. medical devices, IoT, hospitals, neurological clinics, online repositories, research centers, etc., for the diagnosis of several neurological disorders. This study emphasizes the diagnosis of several neurological disorders using different emerging techniques in computer science. This section briefly summarizes the study of different researchers for diagnosing different neurological disorders using soft computing and machine learning techniques.

Emina Alickovic et al. [55] introduced a model to detect and predict seizures in patients by classifying EEG signals. Freiburg and Children's Hospital Boston- Massachusetts Institute of Technology (CHB-MIT) EEG databases were used to conduct the study. The proposed model was defined by four components: noise removal, signal decomposition, feature selection, and classification of signals. Multi-scale principle component analysis was used for noise removal; decomposition of signals was performed using empirical mode, discrete wavelet transform, and packet decomposition; feature selection was performed using statistical methods; and signal classification was done using machine learning techniques (random forest, SVM, multi-layer perceptron [MLP], and k-nearest neighbor [KNN]). It was observed that the best results have shown by the study using wavelet packet decomposition with SVM. The performance of the proposed model was evaluated and given by accuracy (99.7%), sensitivity (99.6%), and specificity (99.8%).

Pedro P. Rebouças Filho et al. [56] proposed a feature extraction method called analysis of brain tissue density (ABTD) to detect and classify strokes. The study was applied over the data set of 420 CT images of patients' skulls. Data preprocessing, segmentation, feature extraction, and classification steps were included to devise the method. MLP, SVM, KNN, optimal path forest (OPF), and Bayesian classifiers were used to classify the data set. The best diagnostic accuracy (i.e., 99.3%) was shown by OPF with a Euclidean distance method.

Abdulhamit Subasi et al. [57] devised a hybrid model using SVM with GA and PSO for the detection of epileptic seizures in patients. A data set of five patients with 100 EEG for each was used to conduct the study. Discrete wavelet transformation was used for feature extraction. It was observed that hybrid SVM with PSO has shown better classification accuracy (i.e., 99.38%) as compare to simple SVM and hybrid SVM with GA.

U. Rajendra Acharya et al. [58] presented a study deep convolutional network (CNN) to diagnose epilepsy in patients by analyzing EEG signals. A data set of five patients was used in the study; 100 EEG signals were analyzed for each data set. It was categorized into three parts: normal, preictal, and seizure. A normalization of signals was performed using Z-score, standard deviation, and zero means. Deep learning and artificial neural network (ANN) with 10-fold cross validation method were used to detect a seizure. The diagnostic accuracy, specificity, and sensitivity showed by the study were 88.7%, 90%, and 95%, respectively.

Benyoussef et al. [59] carried out a diagnostic study of Alzheimer's patients. The authors employed three distinct data mining techniques, such as decision tree, discriminant analysis, and logistic regression, and found that the classification results obtained using discriminant analysis were better than the outcomes of other two approaches. The highest rate of predictive accuracy achieved using the discriminant analysis was 66%.

Lama et al. [60] diagnosed Alzheimer's disease using three important data mining techniques, i.e., SVM, import vector machine (IVM), and rough extreme learning machine (RELM). The authors used an MRI images data set of 214 instances collected from the Alzheimer's disease neuroimaging initiative (ADNI) database. The author found a better diagnostic rate with RELM. The highest rate of prediction achieved using RELM was 76.61%.

S. R. Bhagya Shree et al. [61] conducted a study on Alzheimer's disease diagnosis using a naïve Bayes algorithm with the Waikato Environment for Knowledge Analysis (WEKA) tool. A data set of 466 patients was collected for the study by conducting the neuropsychological test. A number of steps were performed to preprocess the data such as attribute selection, imbalance reduction, and randomization. After data preprocessing, feature selection is performed using the wrapper method with synthetic minority over-sampling technique (SMOTE) filter. Data classification was performed using naive Bayes with cross-validation method. The model was evaluated on different platforms, viz. explorer, knowledge flow, and Java application programming interface (API).

Doyle et al. [62] proposed a study to diagnose Alzheimer's disorder in patients using brain images with the regression method. The study used a data set with 1023 instances and 57 attributes. Accuracy, specificity, and sensitivity were 74%, 72%, 77%, respectively.

Johnson et al. [63] used GA and logistic regression to diagnose Alzheimer's disease in patients. GA was used as feature selection and selects 8 features out of 11. Logistic regression was used as a classification technique applied with five folds.

Koikkalainen et al. [64] proposed to diagnose Alzheimer's disorder in patients using regression techniques. Authors have used 786 instances from the ADNI database. The predictive rate of accuracy given by the study was 87%.

J. Maroco et al. [65] proposed a model to predict dementia disorder in patients. Authors have compared the sensitivity, specificity, and accuracy of different data mining techniques, viz. neural networks, SVM, and random forests, linear discriminant analysis, quadratic discriminant analysis, and logistic regression. The study reveals that the random forest and linear discriminant analysis have shown the highest sensitivity, specificity, and accuracy, whereas neural networks and SVM have shown the lowest sensitivity, specificity, and accuracy.

The comparison of the performance of the studies is done by comparing the predictive rate of accuracy, sensitivity, and specificity achieved by different authors in their studies using different soft computing and machine learning techniques for several neurological disorder diagnoses. All these performance parameters are achieved by introducing confusion matrix, i.e., the concept of correct and incorrect actual and predictive events.

Accuracy is the ratio of correct prediction to the total prediction. Figure 12.8 shows the comparison of the predictive rate of accuracies achieved by different authors in their study.

Sensitivity is the ratio of true positives to the sum of true positives and false negatives. Figure 12.9 shows the comparison of sensitivity achieved by different authors in their study.

Specificity is the ratio of true negatives to the sum of true negatives and false positives. Figure 12.10 shows the comparison of specificity achieved by different authors in their study.

It is observed from Figure 12.8–12.10 that the range of accuracies, sensitivity, and specificity lies between 60–99.7%, 54–100%, and 62–98.2%, respectively. Parkinson and seizures have shown the highest performance among other disorders. Also, SVM techniques have shown better predictive results. The publication summary for diagnosis of different neurological disorders of the last 10 years has been shown in Figure 12.11.

In the last 10 years, maximum work was done with dementia, epilepsy, and stroke diagnosis using soft computing techniques. And in 2018, work done on multiple sclerosis is increasing consistently after epilepsy. Table 12.2 shows the detailed publication trends along with citations for some of the neurological disorder diagnosis articles.

From Table 12.2, it is observed that authors from different countries, viz. India, the United States, Korea, Malaysia, UK, Australia, Brazil, and Singapore, etc., have successfully employed different computing techniques for the diagnosis of different neurological disorders. The citation rate of these articles lies between 1 and 148. Also, the wide range of publishers, viz. Elsevier, Springer, BMJ, PMC, Hindawi, PubMed, and Public Library of Science, etc., have published articles on neurological disorder diagnosis using soft computing and machine learning techniques.

12.5 The Need for Neurological Disorders Diagnostic System

It is observed that different soft computing and machine learning techniques, such as naive Bayes, random forest, SVM, ANN, genetic algorithm, and PSO, etc., have been extensively used to diagnose different neurological disorders, viz. epilepsy, stroke, Alzheimer's, meningitis, etc. However, insignificant attention has been given to design and analyze the diagnostic methods for neurological disorders such as epilepsy, stroke, Alzheimer's, meningitis, Parkinson's, brain cancer, etc., by receiving data from different IoT devices. Therefore, an intelligent system is required for the early and precise diagnosis of different neurological disorders with IoT and big data concepts.

12.5.1 Design of Smart and Intelligent Neurological Disorders Diagnostic System

The intelligent diagnostic framework is proposed for the early diagnosis of different neurological disorders. The framework consists of five layers, viz. data source, big data storage, data preprocessing, big data analytics, and disease diagnosis, shown in Figure 12.12.

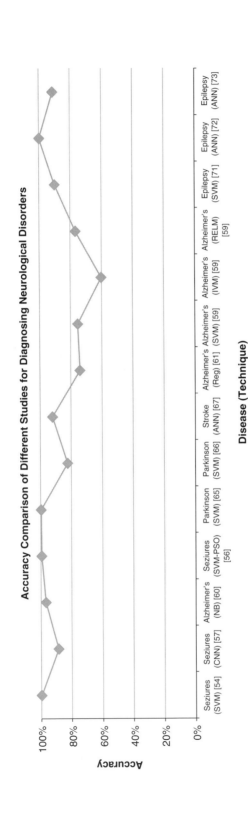

Figure 12.8 The accuracy achieved by different studies for neurological disorder diagnosis.

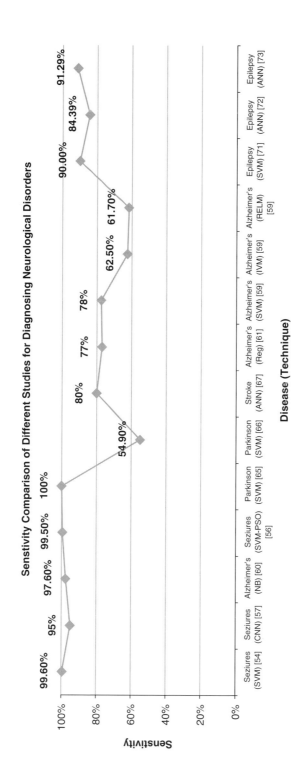

Figure 12.9 Sensitivity achieve by different studies for neurological disorder diagnosis.

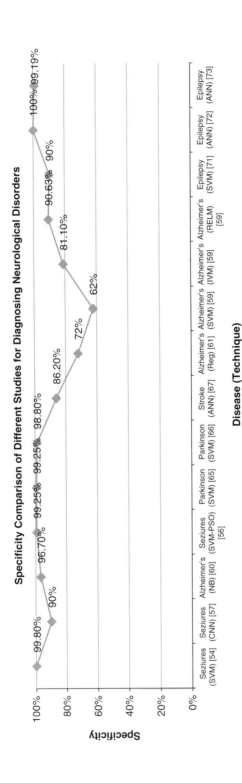

Figure 12.10 Specificity achieve by different studies for neurological disorder diagnosis.

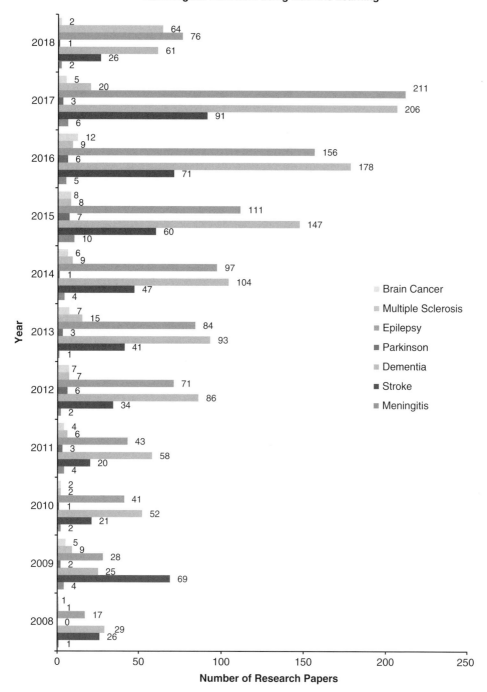

Figure 12.11 Publication trend from 2008 to 2018 for neurological disorder diagnosis using emerging techniques.

Table 12.2 Publications details along with citations used in the study.

Author	Author's country	University	Journal	Publisher	Citation
Alickovic et al. [55]	Sweden	Linkoping University	Biomedical Signal Processing and Control	Elsevier	11
Filho et al. [56]	Brazil	Instituto de Federal de Educação, Ciência e Tecnologia do Ceará (IFCE)	Computer Methods and Programs in Biomedicine	Pub Med	9
Acharya et al. [58]	Singapore	Ngee Ann Polytechnic	Computers in Biology and Medicine	Elsevier	92
Doyle et al. [62]	United Kingdom	King's College London	Plos One Journal	Public Library of Science	23
Johnson et al. [63]	Australia	Commonwealth Scientific and Industrial Research Organization	BMC Bioinformatics	BioMed Central	24
Koikkalainen et al. [64]	Finland	Kuopio University	Plos One Journal	Public Library of Science	25
Lama et al. [60]	Republic of Korea	National Research Center for Dementia	Journal of Healthcare Engineering	Hindawi	5
Maroco et al. [65]	Portugal	ISPA Instituto University	BMC Research Notes	BioMed Central	148
S. R. Bhagya et al. [57]	India	PES College of Engineering	Neural Computing and Applications	Springer	5
AbdulhamitSubasi et al. [57]	Saudi Arabia	Effat University	Neural Computing and Applications	Springer	11

Reference	Country	Journal	Publisher	Count
RajamanickamYuvaraj et al. [68]	India	Neural Computing and Applications	Springer	5
Vida Abedi et al. [69]	USA	Stroke	AHA Journals	7
Christopher P Kellner et al. [70]	USA	Journal of Neuro-interventional Surgery	BMJ Journals	1
MehrbakhshNilashi et al. [71]	Malaysia	Biocybernetics and Biomedical Engineering	Elsevier	10
Eun-Jae Lee et al. [72]	Korea	Journal of Stroke	PMC	12
Nanthini et al. [73]	India	Journal of Applied Sciences	Asian Network for Scientific Information	8
Nesibe et al. [74]	Turkey	Turkish Journal of Electrical Engineering & Computer Sciences	Tubitak	22
Patnaik [75]	India	Computer methods and programs in biomedicine	Elsevier	104

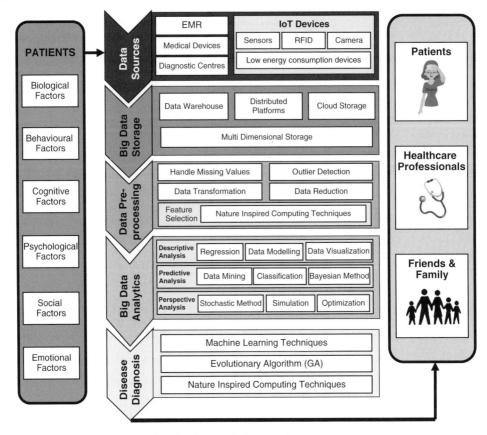

Figure 12.12 Neurological disorder diagnostic framework.

Initially, data of patients, viz. biological, behavioral, cognitive, psychological, social and emotional factors, have to be collected from different sources such as medical devices, diagnostic, and research centers, and electronic medical records (EMRs). Moreover, live data of patients can also be collected using IoT devices like sensors and wearable devices. IoT has been used to optimize energy as a large amount of energy is wasted in the health care sector by using different medical devices. This problem can be solved by introducing low power consumption IoT devices to collect neurological disordered patients' data sets. After collecting data from different sources, it is important to store this big data over different platforms for the effective utilization of data. Data warehouse, distributed platforms, multi-dimensional storage, and cloud storage are some important big data storage methods. Different big data analytics techniques like descriptive, prescriptive, and predictive can be used for data analysis [66]. Different preprocessing techniques will be used to handle noise, missing values, and to remove outliers from the data. Three different feature selection techniques called filters, wrapper, and embedded, and their hybridization can be used to select more suitable attributes. Additionally, emerging nature-inspired computing (ACO, PSO, ABC, FA, ALO, GWO, etc.) techniques may be used for better feature selection. The basic objective of features selection is to remove the unwanted attributes so that the data mining process can be expedited. The use of nature-inspired computing techniques will improve the predictive rate of accuracy and reduce the error rate as well as time to process data of the diagnostic system. Different big data analytics techniques, viz. descriptive, predictive, and perspective analytics, are used to perform effective analysis of the collected data set. Finally, the diagnostic analysis is performed by using machine learning, evolutionary, and nature-inspired computing techniques for early disease diagnosis of different neurological disorders. Finally, the rate of classification can be further improved by incorporating the concept of granular computing [67].

Features of Neurological Disorder Diagnostic Framework

- The system will be able to get data from different sources and in different formats.
- The IoT devices assist in collecting and monitoring live data of neurological disordered patients. For energy optimization, low power IoT devices will be used. Additionally, special types of software or applications will be used to further optimize the energy consumption.
- The proposed system will be able to store and manage big data.
- Different preprocessing and feature selection techniques will be employed to make data consistent and optimal.
- Nature-inspired computing along with several big data analytics techniques, viz. descriptive, predictive, and perspective, will be employed for early and precise diagnosis of neurological disorders [76, 77].
- In brief, the proposed framework is supposed to improve the overall health care of neurological disordered patients.

Challenges

- Data collection is difficult, as the data may be in unstructured and semi-structured formats.
- Energy optimization devices are required to improve the efficiency and power consumption used by IoT devices to collect live data.
- Data security is a critical issue to protect patients' data from unauthorized access.

12.6 Conclusion

Neurological disorders cover several nerve and brain-related issues and have cognitive, biological, and behavioral repercussions. Here, the focus was on the types of neurological disorders, and how they are different from psychological disorders. It is observed that dementia, stroke, and epilepsy havea high prevalence. However, fortunately, the death rate is zero. And unfortunately, the death rate of some disorders such as tetanus, brain cancer, motor neuron disease, and stroke is high, i.e., 27%, 19%, 17%, and 15%, respectively. Furthermore, it is observed from the statistics that China and India have shown a high prevalence of neurological disorders in their population. The role and relation of IoT and big data are presented in the study. Different authors have implemented various machine learning and soft computing techniques, viz. naive Bayes, random forest, ANN, SVM, GA, and PSO, etc., in order to diagnose different neurological disorders. And, the performance of each study was measured on three parameters, viz. accuracy, sensitivity, and specificity, whose range lies between 60–99.7%, 54–100%, and 62–100%, respectively. Parkinson's and seizures have shown the highest performance among other disorders. Also, SVM has shown better predictive results. Finally, to fulfill future requirements, an intelligent neurological disorder diagnostic framework is proposed.

References

1 WHO Library Cataloguing-in-Publication Data. Neurological Disorders public health challenges 2006.

2 Gautam, R. and Sharma, M. (2020). Prevalence and Diagnosis of Neurological Disorders Using Different Deep Learning Techniques: A Meta-Analysis. *Journal of Medical Systems* 44 (2): 49.

3 Meyer, M.A. (2016). *Neurologic Disease: A Modern Pathophysiologic Approach to Diagnosis and Treatment*. Heidelberg: Springer.

4 Roth, G. and Dicke, U. Evolution of nervoussystems and brains. In: *Neurosciences - from Molecule to Behavior: A University Textbook* (eds. C.G. Galizia and P.-M. Lledo). Berlin, Heidelberg: Springer Spektrum.

5 Sokoloff, L. (1996). Circulation in the central nervoussystem. In: *Comprehensive Human Physiology* (eds. R. Greger and U. Windhorst), 561–578. Berlin, Heidelberg: Springer.

6 Pleasure, D.E. (1999). Diseases affecting both the peripheral and central nervous systems. In: *Basic Neurochemistry: Molecular, Cellular and Medical Aspects*, 6e (eds. G.J. Siegel, B.W. Agranoff, R.W. Albers, et al.). Philadelphia: Lippincott-Raven.

7 https://www.ucsfhealth.org/conditions/neurological_disorders [accessed on 20 October 2018]

8 Aarli, J.A. (2005). Neurology and psychiatry: "OhEast is East and West is West ...". *Neuropsychiatric Disease and Treatment* 1 (4): 285–286.

9 Brandt, T., Caplan, L.R., and Kennard, C. (2003). *Neurological Disorders Course and Treatment*, 2e.

10 Charles, S. and Walinga, J. (2014). Definingpsychological disorders. In: *Introduction to Psychology, 1st Canadian Edition*. Vancouver, BC: BCcampus.

11 Sharifi, M.S. (2013). Treatment ofneurological and psychiatric disorders with deep brain stimulation; raising hopes and future challenges. *Basic and Clinical Neuroscience* 4 (3): 266–270.

12 https://medlineplus.gov/neurologicdiseases.html [Accessed on 20 October, 2018]

13 Galetta, S.L., Liu, G.T., and Volpe, N.J. (1996). Diagnostic tests in neuro-ophthalmology. *Neurologic Clinics* 14 (1): 201–222.

14 Tegueu, C.K., Nguefack, S., Doumbe, J. et al. (2013). The spectrum of neurological disorders presenting at a neurology clinic in Yaoundé, Cameroon. *The Pan African Medical Journal* 14: 148.

15 GBD 2015 Neurological Disorders Collaborator Group (2017). Global, regional, and national burden of neurological disorders during 1990–2015: a systematic analysis for the Global Burden of Disease Study 2015. *Global Health Metrics* 16 (11): 877–897.

16 Tsiatsis, V., Karnouskos, S., Holler, J. et al. *Internet of Things Technologies and Applications for a New Age of Intelligence*, 2. New York: Academic Press.

17 Elhoseny, M., Abdelaziz, A., Salama, A.S. et al. (2018). A hybrid model of the Internet of Things and cloud computing to manage big data in health services applications. *Future Generation Computer Systems* 86: 1383–1394.

18 Farahani, B., Firouzi, F., Chang, V. et al. (2018). Towards fog-driven IoTeHealth: promises and challenges of IoT in medicine and healthcare. *Future Generation Computer Systems* 78 (2): 659–676.

19 Firouzia, F., Rahmani, A.M., Mankodiy, K. et al. (2018). Internet-of-Things and big data for smarter healthcare: from device to architecture, applications and analytics. *Future Generation Computer Systems* 78 (2): 583–586.

20 Bhatt, C., Dey, N., and Ashour, A.S. (2017). *Internet of Things and Big Data Technologies for Next Generation Healthcare*. Cham: Springer.

21 Picciano, A.G. (2012). The evolution of big data and learning analytics in American highereducation. *Journal of Asynchronous Learning Networks* 16 (3): 9–20.

22 McAfee A. and Brynjolfsson E. (2012) Big data: the management revolution. Cambridge, MA: Harvard Business Review.

23 Big Data is the Future of Healthcare (2012). Cognizant 20–20 insights.

24 Gandomi, A. and Haider, M. (2015). Beyond the hype: big data concepts, methods, and analytics. *International Journal of Information Management* 35 (2): 137–144.

25 Kruse, C.S., Goswami, R., Raval, Y., and Marawi, S. (2016). Challenges and opportunities of big data in healthcare: asystematic review. *JMIR Medical Informatics* 4 (4).

26 Agarwal, P. and Mehta, S. (2014). Nature-inspiredalgorithms: state-of-art, problems and prospects. *International Journal of Computers and Applications* 100 (14): 14–21.

27 Mitchell, M. (1996). *An Introduction to GeneticAlgorithms*. Cambridge, MA: MIT Press.

28 Kaur, P. and Sharma, M. (2017). A survey on using nature inspired computing for fatal diseasediagnosis. *International Journal of Information System Modeling and Design* 8 (2): 70–91.

29 Blum, C. and Li, X. (2008). Swarm intelligence in optimization. In: *Swarm Intelligence*, Natural Computing Series (eds. C. Blum and D. Merkle). Berlin, Heidelberg: Springer.

30 Marco Dorigo and Giaani Di Caro (1992). Ant Colony Optimization Meta-Heuristic, PhD Thesis, Chapter 2.

31 Kennedy', J. and Eberhart, R. (1995). *Particle Swarm Optimization*. IEEE.

32 Karaboga, D. (2005). Artificial bee colony algorithm. *Scholarpedia* 5 (3): 6915.

33 Yang, X.-S. and He, X. (2013). Firefly algorithm: recentadvances and applications. *International Journal of Swarm Intelligence* 1 (1): 36–50.

34 Mirjalili, S. (2015). The ant lion optimizer, advances in engineering software. *Advances in Engineering Software* 83: 80–90.

35 Mirjalili, S., Mirjalili, S.M., and Lewis, A. (2014). Grey wolf optimizer. *Advances in Engineering Software* 69: 46–61.

36 Askarzadeh, A. (2016). A novel metaheuristic method for solving constrained engineering optimization problems: crow search algorithm. *Computers and Structures* 169: 1–12.

37 Mirjalili, S. and Lewis, A. (2016). The whale optimization algorithm. *Advances in Engineering Software* 95: 51–67.

38 Dhinesh Babu, L.D. and Krishna, P.V. (2013). Honey bee behaviour inspired the load balancing of tasks in a cloud computing environment. *Applied Soft Computing* 13 (5): 2292–2303.

39 Rajabioun, R. (2011). Cuckoo optimizationalgorithm. *Applied Soft Computing* 11 (8): 5508–5518.

40 Zheng, Y.-J. (2015). Water wave optimization: a new nature-inspired metaheuristic. *Computers & Operations Research* 55: 1–11.

41 Kaveh, A. and Khayatazad, M. (2012). A new meta-heuristic method: ray optimization. *Computers & Structures* 112: 283–294.

42 Chakraborty, A. and Kar, A.K. Swarm intelligence: A review of algorithms. In: *Nature-Inspired Computing and Optimization* (eds. S. Patnaik, X.S. Yang and K. Nakamatsu), 475–494. Cham: Springer.

43 Sharma, M. et al. (2013). Stochastic analysis of DSS queries for a distributed databasedesign. *International Journal of Computer Applications* 83 (5): 36–42.

44 Sharma, M., Singh, G., and Singh, R. (2018). A review of different cost-based distributed query optimizers. In: *Progress in Artificial Intelligence*, vol. 8.1, 45–62.

45 Long, N.C., Meesad, P., and Unger, H. (2015). A highly accurate firefly based algorithm for heart disease prediction. *Expert Systems with Applications* 42 (21): 8221–8231.

46 Fu, Q., Wang, Z., and Jiang, Q. (2010). Delineating soil nutrient management zones based on fuzzy clustering optimized by PSO. *Mathematical and Computer Modelling* 51 (11–12), 2010: 1299–1305.

47 Fürnkranz, J. et al. (2012). *Foundations of Rule Learning, Cognitive Technologies*. Springer-Verlag Berlin Heidelberg.

48 Royal Society (Great Britain). Machine learning: the power and promise of computers that learn by example, The Royal Society.

49 Jiawei, H., Micheline, K., and Jian, P. (2011). *Data Mining: Concepts and Techniques*, 3e. Hoboken, NJ: Elsevier.

50 Sharma, M., Sharma, S., and Singh, G. (2018). Performance analysis of statistical and supervised learning techniques instock data mining. *Data* 3 (4): 54.

51 Sharma, M., Singh, G., and Singh, R. (2017). Stark assessment of lifestyle based human disorders using data mining based learning techniques. *IRBM* 36 (6): 305–324.

52 Ian, W. and Eibe, F. (2005). *Data Mining: Practical Machine Learning Tools and Techniques*, 2e. Hoboken, NJ: Elsevier.

53 Kaur, P. and Sharma, M. (2018). Analysis of data mining and soft computing techniques in prospecting diabetes disorder in human beings: areview. *International Journal of Pharmaceutical Sciences and Research* 9 (7): 2700–2719.

54 Xu, D. and Tian, Y. (2015). A comprehensive survey of clustering algorithms. *Annals of Data Science* 2 (2): 165–193.

55 Alickovic, E., Kevric, J., and Subasi, A. (2018). Performance evaluation of empirical mode decomposition, discrete wavelet transform, and wavelet packed decomposition for automated epileptic seizure detection and prediction. *Biomedical Signal Processing and Control* 39: 94–102.

56 RebouçasFilhoa, P.P., Sarmentoa, R.M., Holandaa, G.B., and de Alencar Lima, D. (2017). A new approach to detect and classify stroke in skull CT images via analysis of brain tissue densities. *Computer Methods and Programs in Biomedicine* 148: 27–43.

57 Subasi, A., Kevric, J., and Abdullah Canbaz, M. (2019). Epileptic seizure detection using hybrid machine learning methods. *Neural Computing and Applications* 31.1: 317–325.

58 Rajendra Acharya, U., Oh, S.L., Hagiwara, Y. et al. (2017). Deep convolutional neural network for the automated detection and diagnosis of a seizure using EEG signals. *Computers in Biology and Medicine* 100: 270–278.

59 Benyoussef, E.M., Elbyed, A., and El Hadiri, H. (2017). Data mining approaches for Alzheimer's disease diagnosis. In: *Ubiquitous Networking* (eds. E. Sabir, A. García Armada, M. Ghogho and M. Debbah), 619–631. Cham: Springer.

60 Lama, R.K., Gwak, J., Park, J.S., and Lee, S.W. (2017). Diagnosis of Alzheimer's disease based on structural MRI images using a regularized extreme learning machine and PCA features. *Journal of Healthcare Engineering* 2017: 1–11.

61 Bhagya Shree, S.R. and Sheshadri, H.S. (2018). Diagnosis of Alzheimer's disease using naive Bayesian classifier. *Neural Computing and Applications* 29 (1): 123–132.

62 Doyle, O.M., Westman, E., Marquand, A.F. et al. (2014). Predicting progression of Alzheimer's disease using ordinalregression. *PLoSOne* 9 (8): 1–10.

63 Johnson, P., Vandewater, L., Wilson, W. et al. (2014). Genetic algorithm with logistic regression for prediction of progression to Alzheimer's disease. *BMC Bioinformatics* 15: 1–14.

64 Koikkalainen, J., Pölönen, H., Mattila, J. et al. (2012). Improved classification of Alzheimer's disease data via removal of nuisance variability. *PLoS One* 7 (2): e31112.

65 Maroco, J., Silva, D., Rodrigues, A. et al. (2011). Data mining methods in the prediction of dementia: a real-data comparison of the accuracy, sensitivity and specificity of linear discriminant analysis, logistic regression, neural networks, support vector machines, classification trees and random forests. *BMC Research Notes* 4 (299): 1–14.

66 Sharma, M., Singh, G., and Singh, R. (2018). Accurate prediction of life style based disorders by smart healthcare using machine learning andprescriptive big data analytics. *Data Intensive Computing Applications for Big Data* 29: 428.

67 Sharma, M., Singh, G., and Singh, R. (2018). An advanced conceptual diagnostic healthcareframework for diabetes and cardiovascular disorders. *EAI Endorsed Transactions on Scalable Information Systems* 5: 1–11.

68 Rajamanickam, Y., Rajendra Acharya, U., and Hagiwara, Y. (2018). A novel Parkinson's disease diagnosisindex using higher-order spectra features in EEG signals. *Neural Computing and Applications* 30 (4): 1225–1235.

69 Abedi, V., Goyal, N., Tsivgoulis, G., and Hosseinichimeh, N. (2017). Novel screening tool for stroke using artificialneural network. *Stroke* 48 (6): 1678–1681.

70 Kellner, C.P., Sauvageau, E., Snyder, K.V. et al. (2018). The VITAL study and overall pooled analysis with the VIPS non-invasive stroke detection device. *Journal of NeuroInterventional Surgery*: 1–7.

71 Nilashi, M., Ibrahim, O., Ahmadi, H. et al. (2018). A hybrid intelligent system for the prediction of Parkinson's disease progression using machine learning techniques. *Biocybernetics and Biomedical Engineering* 38 (8): 1–15.

72 Lee, E.-J., Kim, Y.-H., Kim, N., and Kanga, D.-W. (2017). Deep into the brain: artificial-intelligence in stroke imaging. *Journal of Stroke* 19 (3): 277–285.

73 SugunaNanthini, B. and Santhi, B. (2014). Seizure detection using SVM classifier on EEGsignal. *Journal of Applied Sciences* 14 (14): 1658–1661.

74 Nesibe, Y., Gülay, T., and Cihan, K. (2015). Epilepsy diagnosis using artificial neural network learned by PSO. *Turkish Journal of Electrical Engineering and Computer Sciences* 23: 421–432.

75 Patnaik, L.M. and Manyam, O.K. (2008). Epileptic EEG detection using neural networks and post-classification. *Computer Methods and Programs in Biomedicine* 91: 100–109.

76 Awad, M. and Khanna, R. Bioinspired computing: swarm intelligence. In: *Efficient Learning Machines*, 105–125. Berkeley, CA: Apress.

77 Bansal, J.C., Singh, P.K., and Pal, N.R. (eds.). *Evolutionary and Swarm Intelligence Algorithms*. Springer, Berlin, Heidelberg.

13

Comments-Based Analysis of a Bug Report Collection System and Its Applications

Arvinder Kaur and Shubhra Goyal

University School of Information and Communication Technology, Guru Gobind Singh Indraprastha University, Delhi, India

13.1 Introduction

In the new regime of the world wide web, plenty of structured and unstructured data is available. Structured data is associated with a database while unstructured data can be textual or nontextual. The analysis of unstructured data to discern patterns and trends that are relevant to the users is known as text mining or text analytics. Text mining emerged in the late 1990s and discovers hidden relationships and complex patterns from large textual sources. It uses several techniques such as classification, decision trees, clustering, link analysis, and so on. These techniques can be applied to high-dimensional data using dimensionality reduction statistical techniques such as singular value decomposition and support vector machine. Text analytics is evolving rapidly and has become an asset for researchers as it is capable of addressing diverse challenges in all fields.

In this research work, text analytics is used to uncover various patterns and trends from software bug reports reported in issue tracking systems. The issue tracking system is used to track and record all kinds of issues such as bug, new features, enhancements, tasks, and sub-tasks, or any other complaint in the system. There exists different issue tracking systems such as Bugzilla, Jira, Trac, Mantis, and many others. The Jira issue tracking system was developed by the Atlassian company and is used for bug tracking, issue tracking, and project management. Bug reports of 20 distinct projects of Apache software foundation (ASF) projects under the Jira repository over last nine years (2010–2018) are extracted. Several bug attributes such as Bug Id, Priority name, status, resolution, one-line description, developer assigned to, fix version, component they belong to, number of comments made for each bug, long description of bugs, and comments made among various contributors are extracted using the tool Bug Report Collection System (BRCS) [1]. Bug reports contain useful hidden information that can be beneficial for software developers, test leads, and project managers. Based on severity, bugs are broadly classified into two categories: severe

Intelligent Data Analysis: From Data Gathering to Data Comprehension,
First Edition. Edited by Deepak Gupta, Siddhartha Bhattacharyya, Ashish Khanna, and Kalpna Sagar.
© 2020 John Wiley & Sons Ltd. Published 2020 by John Wiley & Sons Ltd.

and nonsevere. Severe bugs are critical in nature and can cause system crash or can degrade the performance of software. Therefore, enormous open bugs of a high severity level are undesirable. To discover more valuable and profitable information from long descriptions and comments of bug reports, text analytics is used. Textual data is preprocessed using tokenization, stemming, and stop word removal. The most frequently occurring words are mined and correlated, and then these words are extracted to categorize bugs into various types of errors such as logical code error, input/output (I/O), network, and resource allocation errors, or to predict the severity of newly reported bugs. These aspects are analyzed to assist test leads, project managers, and developers to resolve critical open bugs, to classify into various categories of error and predict severity using keywords, and to predict duplicate bug reports using clustering, all without reading an entire bug report. This will aid in quick bug resolution to improve a system's performance and also saves time.

To the best of our knowledge, this is the first work that analyzes various trends from software bug reports to benefit test leads in maintenance processes of software projects. This analysis is performed in the form of research questions.

Research Question 1: Is the performance of software affected by open bugs that are critical in nature?

Research Question 2: How can test leads improve the performance of software systems?

Research Question 3: Which is the most error-prone areas that cause system failure?

Research Question 4: Which are the most frequent words and keywords to predict critical bugs?

Research Questions 5: What is the importance of frequent words mined from bug reports?

The main contributions of this research work are:

- Open bugs of various projects are analyzed and results exhibited that 24% of open bugs are of high severity level (blocker, critical, and major), which are critical and need to be resolved as a high priority to prevent failure of software systems.
- Most contributing developers of various projects are based on the maximum number of bugs resolved and comments made are extricated, which will assist test leads to assign a critical open bug to them.
- Co-occurring and co-related words are extracted and bugs are categorized into various types of errors, such as logical code error, network error, I/O error, and resource allocation error.
- It is established that the severity of a newly reported bug report can be predicted using frequently occurring words and most associated words.
- Frequent words of various projects are clustered using k-means and hierarchical clustering, which will help developers to detect duplicate bug reports.

The organization of the paper is as follows: Section 13.2 gives a brief description of the issue tracking system and bug report statistics studied in this work, followed by related work on data extraction process and various application of comments of bug reports in Section 13.3. Section 13.4 describes the data collection process and analysis of bug reports in done in Section 13.5. Threats to validity are discussed in Section 13.6, followed by a conclusion.

13.2   Background

13.2.1   Issue Tracking System

An issue tracking system is a software repository used by organizations for software maintenance and evolution activities. It provides a shared platform where team members of an
organization can plan, track, and report various issues and releases of software. It gives a
single view of all elements regardless of whether it is a bug, or a task, subtasks related to,
or a new feature request to a software team. This single view of information helps team
to prioritize their goals and to assign them to the right team member at the right time. It
also records the progress of every issue until it is resolved. There are several issue tracking
systems such as Bugzilla,[1] Codebeamer,[2] Trac,[3] and Jira.[4] The Jira issue tracking system is
developed by the Atlassian software company and provides an easy-to-use user interface. Its
prime advantage is that it provides easy integration with other Atlassian tools. It is devised
as multi-project request tracking system, which allows projects to be sorted in different
categories, where each project can have its own settings with respect to filters, workflows,
reports, issue types, etc. Jira is selected over other issue tracking repositories due to the
following reasons:

- Jira provides easy installation and does not requires any environment preparation.
- It provides intuitive and easy to use user interface.
- It supports a multi-project environment.
- It supports a variety of plugins to increase its functionality.

Among several types of issues maintained under the Jira repository, bug reports are an
essential software artifact. Thus, bug reports of various projects of ASF based on Jira repository are extracted. A bug report is characterized by useful standard fields of information
such as title of bug, its resolution, severity, developer assigned to, date, component to which
they belong, and so on. It also consists long description of bug reports and long comments
made by several developers to discuss and share experience about the bug resolution. In this
work, various fields of bug reports along with its long description and threaded comments
made among various contributors are extracted and analyzed.

13.2.2   Bug Report Statistics

In this section, bug reports in the Jira issue tracking system have been studied and analyzed.
Jira consists of more than a hundred projects of ASF. Twenty projects of ASF are studied
as the Apache project is a large open-source system and is prominently used in empirical
research. The bug reports are extracted from 1 January 2010 to 1 January 2018. For each
project, extracted data is recorded: (i) total number of issue reports (ii) total number of bug
reports, and (iii) total number of comments made as shown in Figure 13.1. Bug reports
of various projects are categorized on the basis of resolution and severity assigned. The
statistics are depicted in Figure 13.2a,b.

---

1 www.bugzilla.org.
2 www. Intland.com/products/codebeamer.html.
3 http://trac.edgewall.org.
4 www.atlassian.com/software/jira.

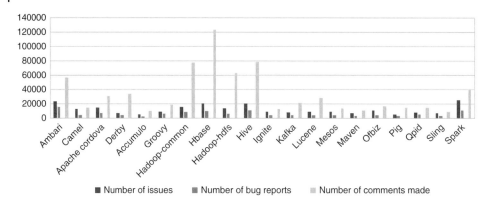

Figure 13.1 Statistics of bug reports of 20 projects of the Apache Software foundation.

13.3 Related Work

13.3.1 Data Extraction Process

The issue tracking system contains enormous data of various software artifacts such as bug reports, source code, change log, and other artifacts. Substantial research has been done on these artifacts in previous years. Mining these artifacts manually from issue tracking repositories is a tedious and exhausting process. To overcome this manual extraction of data, researchers have worked to improve the automated extraction of data. N. Bettenburg et al. [2] developed a tool named InfoZilla to extract structural elements, i.e., stack traces, source code, and patches from bug reports. It was believed that training machine learners on these elements gives a better performance. Various filters, namely, patch filter, stack traces filter, and enumeration filter, were implemented and the performance of tool is evaluated on eclipse bug reports. M. Nayrolles et al. proposed a tool named BUMPER (Bug Metarepository for developers and researchers) to extract information from bug report repositories and version control systems. It is an open-source web-based tool that extracts data based on queries entered by user. It accepts queries in two modes: basic and advanced. It supports five bug reports systems: Gnome, eclipse, github, netbeans, and ASF. The major limitation is that the end user should have a sound knowledge of query language to extract data using this tool [3]. Y. yuk et al propose two ways of extracting data by parsing bug report pages and then by refining them. In this, web pages containing information of bug attributes such as BugId, version, status, priority, and environment are crawled and then are transformed to data traces. These traces are then evaluated using Xmlwrench [4]. In contrast to these tools, which extracts text data in bug reports, a tool named configuration management system (CMS) was developed by R. Malhotra et al. This tool obtains log files from concurrent versioning systems (CVSs) to collect a number of defects from each class. It also computes the changes made during each version of the software [5]. A. Kaur et al. developed a tool named BRCS. This tool extracts bug reports of projects of ASF based on the Jira repository. The bug attributes such as bugId, resolution, status, priority, one-line description, and developer assigned to, are extracted along with long descriptions of each bug report. The main advantage is that each end user just has to mention the name of the project and range

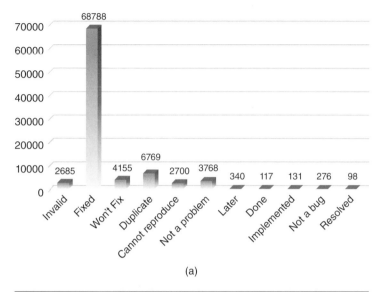

(a)

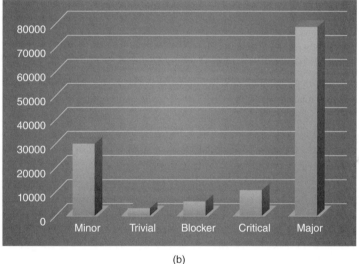

(b)

Figure 13.2 (a) Number of bug reports based on resolution. (b) Number of bug reports based on severity.

of bugIds, between which all bug reports will be extracted and saved in the form of Microsoft Excel files [1]. In this research work, an extension to tool BRCS is made. Along with several bug attributes, the number of comments reported for each bug and detailed conversations among various contributors in the form comments, along with the contributor's name and date of comment made are recorded. Stack traces, source code, and other structural information is also extracted. This extraction of information in the form of natural language is used in several applications, which are discussed in Section 13.3.2. A comparison of previous work done for the extraction of software artifacts is done in Table 13.1.

Table 13.1 Comparison of previous studies of data extraction.

Author, year	Software artifacts extracted	Proposed tool	Data sets used	Data extraction process	Limitations
N. Bettenburg et al. [2]	Bug Reports (stack traces, source code, patches)	InfoZilla	Eclipse	Several filters are implemented to extract these elements	Several filters based on elements needs to be implemented
Y. Yuk et al. [4]	Bug reports	—	Mozilla	Pages of bug reports are crawled and then transformed to data traces	Web pages are crawled through parsing method, no tool is used.
R. Malhotra [5]	Log files	CMS	JTreeView	Logs are obtained from CVS repository using "diff" command	—
M. Nayrolles et al. [3]	Bug reports	BUMPER	Gnome, eclipse, github, netbeans, apache software foundation	Web-based tool that extracts data based on queries in two modes: basic and advanced	Sound knowledge of query language is required to use the tool.
Our work	Bug Reports (Metadata, log description, comments)	BRCS	20 projects of Apache Software Foundation	Bug reports are accessed through REST APIs.	—

13.3.2 Applications of Bug Report Comments

Comments of bug reports are rich in content and valuable artifacts, which are used in several applications, such as bug summarization, emotion mining, duplicate bug report detection, and types of bug reports. This section briefly reviews the various research work done in these areas.

13.3.2.1 Bug Summarization

There are two methods of document summarization: abstractive and extractive. Abstractive summarization rephrases the words to generate a human-like summary while extractive summarization selects sentences from bug reports to generate a useful summary. Lots of research has been performed by researchers for text summarization. Various techniques have been used such as latent semantic analysis [6, 7], frame semantics [8, 9], computing centrality using similarity graph [10, 11], sentence regression [12], using inter-intra event relevance [13], sentence clustering for multidocument summarization [14], and automated highlighting. Apart from these, several machine learning algorithms are also used for summarization of the text document, such as genetic algorithm [15], neural network

[16], and conditional random fields [17]. Extractive document summarization has been done using query-focused techniques in which sentences are extracted based on queries entered [18–20]. Along with text summarization, summarization of bug reports is an emerging field that has motivated researchers to generate useful summaries. Summarization of bug reports is an effective technique to reduce the length of bug reports, keeping intact all important and useful description needed to resolve and fix bugs. S. Rastkar et al. used an e-mail summarizer developed by Murray et al. [21] to discover the similarity between e-mail threads and software bug reports. It was found that an e-mail summarizer can be used for bug summarization [22]. To improve the performance of this finding, Bug Report Corpus (BRC) was built during their successive study. BRC comprises 36 bug reports from four open-source projects; Mozilla, Eclipse, Gnome, and Kde. Manual summary of each bug report is created by three annotators and golden standard summary (GSS) is formed. GSS consists of those sentences that were elected by all three annotators, and a logistic regression classifier was trained on it. For automatic generation of summary of bug reports, four features, such as structural, participant, lexical, and length features were extracted and the probability of each sentence was computed. Sentences with high probability value form a summary of bug reports [23]. He. Jiang et al. modified the BRC corpus by adding duplicate bug reports of each master bug report. It was proposed that training a classifier on a corpus of master and duplicate bug reports could generate a more accurate summary of a master bug report. PageRank algorithm was applied to compute textual similarity between the master and duplicate bug reports and sentences were ranked based on similarity. Also, the BRC classifier proposed by Rastkar et al. [23] was used to extract features for each sentence and the probability of each sentence is calculated. Sentences extracted from PageRank and BRC classifier are merged using a ranking merger and finally, the summary of master bug reports is created [24]. I. Ferriera et al. ranks comments of bug reports using various ranking techniques and the approach is evaluated on 50 bug reports of open-source projects [25].

In contrast to supervised extractive summarization of bug reports, a couple of research works focus on unsupervised extractive summarization. Mani et al. proposed a centroid-based summarization approach and removes noise from sentences on the basis of questions, investigative, code, and other formats. The approach was evaluated by SDS and DB2 corpora [26]. This research work was extended by Nithya et al., which focuses on duplicate bug report detection from a noise reducer [27]. Lotufo et al. generates a summary of bug reports based on hypothesis; four summarizers were implemented, one for each hypothesis and one for all three hypotheses. The sentences were ranked based on relevance and the Markov chain method was used to generate a summary [28].

13.3.2.2 Emotion Mining

A. Murgia et al. analyzes whether issue comments convey emotional information and whether human raters agree on the presence of emotions in issue comments. To address these concerns, a pilot study is performed on 800 issue comments of various projects of ASF. Six primary parrot's emotions, i.e., love, joy, surprise, anger, sadness, and fear are identified. A pilot study confirmed that humans agree that issue comments represent emotions. A feasibility study was conducted to investigate these emotions. Machine learning classifiers were trained on a data set and models are built to identify three emotions: love, joy, and sadness. Emotions such as surprise, fear, and anger are not considered due to low occurrence in issue comments. Five variants of classifier based on each emotion were constructed based on support vector machine, naïve Bayes, single-layer perception, k-nearest neighbor, and random forest. The performance of each classifier is evaluated using a bootstrap validation approach. The results confirmed that issue comments do convey emotional information and it is possible to identify emotions through emotion-driven keywords [29]. Q. Umer et al. propose an emotion-based automatic approach for priority prediction of bug reports. For this approach, bug reports of four open-source projects from Bugzilla are extracted and summary of each bug report is preprocessed. After preprocessing, emotion-analysis is performed on each bug to identify emotion-words using emotion-based corpus and an emotion value is assigned. A machine learning classifier support vector machine is then trained on the data set based on emotion-value and the priority of bug reports is predicted. The approach is evaluated using various performance metrics such as precision, recall, and F-score. It was concluded that an emotion-based priority prediction helps developers to assign appropriate priority to bug reports [30]. G. Yang et al. proposed an approach to predict the severity of bugs using emotion similarity. For this approach, an emotion word-based dictionary (EWD) is created and similar emotions bug reports are identified using a smoothed unigram model (UM) based KL-divergence method. An emotion-score is assigned to each bug report and an emotion similarity multinomial (ES-multinomial) machine learning technique is used. The approach is evaluated on five open-source projects, namely, Eclipse, Gnu, Jboss, Mozilla, and Wireshark. Results are compared with other machine learning techniques, i.e., naïve Bayes multinomial and EWD multinomial. It was concluded that the proposed approach outperforms other techniques [31]. In contrast to emotions, the sentiment-based analysis is performed by J. Ding et al. and the authors developed a sentiment analysis tool SentiSW, which classifies issue comments into three categories: negative, positive, and neutral, and < sentiment, entity > tuple is generated. Six machine learning classifiers such as random forest, bootstrap aggregating, gradient boosting tree, naïve Bayes, ridge regression, and support vector machine are implemented in the Scikit-learn tool. The approach is evaluated using 10-fold cross-validation technique on 3000 issue comments. The tool achieves a mean precision of 68.71%, recall of 63.98%, and 77.19% accuracy, which outperforms other existing tools [32].

13.4 Data Collection Process

Bug reports and their related information of various Apache projects maintained under the Jira repository is extracted by the tool BRCS [1]. The tool BRCS extracts various attributes of bug reports, such as BugId, Priority name, severity, status, resolution, one-line description, developer assigned to, fix-version, bug creation date, component they belong to and a long description of each bug. These attributes are important in various fields such as a severity prediction of bug reports, duplicate bug report detection, bug fix time prediction, and others. With new and emerging fields of research, natural language processing has gained a lot of attention and researchers have focused on comments of bug reports for various applications as discussed in Section 13.3.2. To fulfill the need, BRCS has been extended, in which the total number of comments made by developers for each bug report

Figure 13.3 Example of bug report of Accumulo project.

is also extracted along with long threaded discussions in the form of comments among various contributors are also extracted. Along with comments, the name of the contributor, and the time and date of comments are also fetched. Bug reports of 20 projects of ASF are extracted, namely, Ambari, Camel, Apache Cordova, Derby, Accumulo, Hadoop-Common, Hadoop-Hdfs, Hbase, Hive, Ignite, Kafka, Lucene, Mesos, Maven, Ofbiz, Pig, Qpid, Sling, Spark, and Groovy. The information is retrieved in the form of various reports, which are discussed in Section 13.4.3. Figure 13.3 represents a snapshot of various attributes of bug reports that have been extracted.

13.4.1 Steps of Data Extraction

To carry out the extraction process, a sequence of steps is performed.

- To interact with the Jira server applications remotely, Jira REST APIs are used, which is standard interface of the Jira repository.

- BRCS access the Jira's REST APIs that provide access to data entities via URI paths. An HTTP request is generated and in response, bug reports data is collected that is parsed into objects.
- The communication format used by Jira REST APIs is JSON and some standard HTTP methods like GET, PUT, POST, and DELETE. The general structure of URI is as: http://host:port/context/rest/api-name/api-version/resource-name. Api-name used by our application is api, which is used for everything else in contrast to auth used for authentication-related operations. Current api-version is 2.
- In next step, the name of the Apache projects from which data needs to be extracted is entered. Along with the name, StartIssue Id and EndIssue Id of bug is entered. The process will then fetch all the bugs in between the entered IssueIds. The field IssueTypeId is checked, if response from the server is 1, it indicates that it is a bug, then all relevant information is extracted and if IssueTypeId is other than 1, then it denotes other types of issues and are generated in the form of error reports.

13.4.2 Block Diagram for Data Extraction

Figure 13.4 represents the various steps for extracting bug reports of various Apache projects from the Jira repository.

13.4.3 Reports Generated

The results of BRCS are generated in the form of various reports. They are described as follows:

13.4.3.1 Bug Attribute Report

The pertinent details of various bug attributes are generated in the form of this report. The distinct characteristics of each bug, such as bug Identifier, priority name, status of bug, resolution of bug, one-line description, bug assigned to, date of bug creation, bug fix version, components they belong to and number of comments made to resolve a bug are extracted. These characteristics are very significant as they can help an end user analyze the most prominent developer, name of a project, and many others that are described later in Section 13.5.

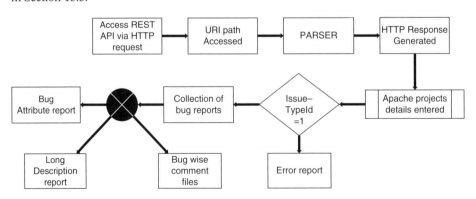

Figure 13.4 Data extraction process.

13.4.3.2 Long Description Report

This report contains detailed description of each bug. It includes a reason for the failure of bug, the possible solution to resolve the bug, and also tells information of other bug Ids with similar issues. This information is very useful in the emerging fields of text summarization, bug resolution, and other features through text data. The applications are discussed in detail in Section 13.5.

13.4.3.3 Bug Comments Reports

In this, the comments and conversation accomplished in between various developers in order to resolve a bug is reported. The comments start by describing the issue related to a bug, communication done among various developers describing various steps and actions taken to resolve the bug, whether successful and unsuccessful in resolving a bug, and the end describing the solution drawn to resolve each bug. The comments are generated in the form of separate text files of individual bug. Various features extracted are:

- Author name
- Author display name
- Author e-mail
- Date of comment creation
- Date of comment update
- Comments including stack traces, code segments, and enumerations

13.4.3.4 Error Report

This report contains information of those Issue Ids that were not a bug. Issues with Issue-TypeId other than 1, i.e., If IssueTypeId = 2 indicates a new feature, and if IssueTypeId = 5 indicates a subtask, then those bugIds are reported in this report.

13.5 Analysis of Bug Reports

This section analyzes bug reports of various software projects of ASF based on textual descriptions of bug reports and comments of various contributors. The results are analyzed in the form of various research questions.

13.5.1 Research Question 1: Is the Performance of Software Affected by Open Bugs that are Critical in Nature?

In software project, bugs are categorized based on severity, which defines how severe the bug is and its impact of a system's performance. Bugs are categorized into five severity levels: blocker, critical, major, minor, and trivial. Blocker bugs are those that need to be fixed instantly as it will cause the failure of a product while critical bugs require a quick fix as it may affect the functionality of a particular application of a product. Major bugs do not impact the functionality of a software but need to resolve for better performance. These three types of bugs are classified as severe bugs while minor and trivial bugs are nonsevere bugs. Nonsevere bugs are used for further enhancement in software projects. As severe bugs

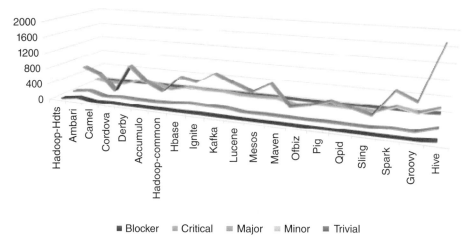

Figure 13.5 Number of open bugs of distinct severity level.

can alter the functionality of software projects and may cause system failure, a large number of open bugs of high severity are undesirable.

To evaluate this aspect, the number of open bugs of each severity level are obtained. It was established that 3.3% of open bugs are blockers, which can cause system failure, 7% are critical bugs and 12.6% are major bugs and are open in nature, which can cause failure of any functionality and reduce the performance of the software system. The results will assist test leads to assign these critical open bugs to developers in order to prevent system crash and to improve the performance of software systems. Figure 13.5 represents the number of open bugs per severity level for 20 distinct projects of ASF. It shows that each project has a greater number of open bugs with "major" severity level, which can degrade system's performance. Figure 13.6 depicts the total percentage of number of open bugs as per severity level, which needs to be resolved on a priority basis. It depicts that 9% of open bugs are blocker bugs, which can cause system failure and 18% and 33% of open bugs are critical in nature due to which system's performance can be declined.

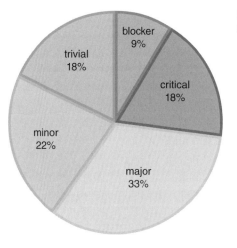

Figure 13.6 Percentage of open bugs as per severity level.

13.5.2 Research Question 2: How Can Test Leads Improve the Performance of Software Systems?

Open-source projects can be modified and shared by any end user as it is accessible publicly. In open-source software projects, many developers contribute in various projects in several ways such a bug fixing, bug resolution, adding new enhancements, requesting new features, providing feedback, and many others. As software evolves, the number of developers contributing to the software project changes. Some developers complete their assigned tasks and leave in the middle of the project. Only a few developers remain intact during the entire life cycle of software. In this scenario, it is challenging for test leads to find developers who can resolve high priority and high severity bugs at the earliest. To help test leads and project managers improve the system's performance, the top five developers of each project are retrieved based on the number of bugs resolved and their contribution in the form of the number of comments made for resolving bugs. The maximum number of comments made by developers exhibits those who are most involved in the bug resolution process. Therefore, test leads can assign bugs with high severity level to these developers in order for their quick resolution. Figure 13.7a–d represents the name of the developers for each project with maximum contribution in the bug resolution process. The results are retrieved from comments of bug reports of various projects of ASF.

13.5.3 Research Question 3: Which Are the Most Error-Prone Areas that Can Cause System Failure?

To explore most error-prone areas, textual attributes of bug reports, i.e., one-line description, long description, and several comments are analyzed. It was discovered that there are four common type of errors that are occurring frequently. These errors are logical errors, I/O errors, network errors and resource allocation errors. Certain keywords are determined and to help developers and test lead to uncover these errors, several associated words are identified. The presence of such keywords can indicate the presence of a particular type of error. This will help developers to locate the type of error without reading an entire bug report and can result in the quick resolution of a bug.

To accomplish this, data has been preprocessed using standard preprocessing steps in R language. It includes tokenization, stemming and stop word removal. For preprocessing, "tm" and "nlp" packages are installed. Tokenization is segmentation of a string of text into words called as tokens. Stemming eliminates the prefixes and suffixes from a word and convert it into a stem word and lastly, stop words are removed. Stop words are the most commonly used words in a language and do not contribute much to the meaning. These preprocessing steps are illustrated below:

Example: The boy was running after the bus.
Tokenization: "the," "boy," "was," "running," "after," "the," "bus"
Stemming: Running → run
Stop word removal: "the" is removed.

After preprocessing of text data, document term matrix is constructed and sparsity of a matrix is reduced by a factor of 0.98 using the removeSpaseTerms() function. The frequency of each word is computed using the findfreqterms() function. For each type of

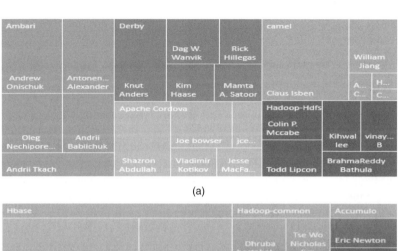

(a)

(b)

(c)

(d)

Figure 13.7 (a)–(d) Most contributing developer for 20 projects of Apache Software foundation.

Table 13.2 Categories of error and its significant keywords.

Categorization of error	Significant keywords
Logical code error	Throws, exception, divide, unhandled, pointer, throw, uncaught, null pointer, raises, system out of memory exception, trigger, dividezero, rethrow
Input/output error	Logging, build, classpath, inputformat, api-jarfilemissing, log, loggers, imports, initialized, requestchannels, map output location get file, displayed, console log
Resource allocation error	Task_allocation, buffers, memory, synchronized, configuration, memory failure, runtime, dynamic, bucketcache
Network related error	Datanode, localhost, address, port, domain, security, process, https, global, interfaces, binding, virtual, bindexcept, limits

```
associations <- findAssocs(dtm, "configuration", corlimit=0.1) # specifying a correlation limit of 0.95
associations_df <- list_vect2df(associations)[, 2:3]
head(associations_df)
ggplot(associations_df, aes(y = associations_df[, 1])) +
  geom_point(aes(x = associations_df[, 2]),data = associations_df, size = 3) +
  theme_gdocs()
```

Figure 13.8 Code for finding corelated words (a) Association graph for logical code error for Kafka project, (b) Association graph for input/output error for Kafka graph, (c) Association graph for network error for kafka project, (d) Association graph for resource allocation error of Kafka project.

error, correlated words around particular keywords are determined. Correlation indicates the cooccurrence of words in a document. FindAssocs() function is used, which always returns cooccurring words with the searched term. To classify a bug report into a particular type of error, a list of keywords is established. Table 13.2 represents frequently used keywords to identify a type of error. For illustration, the snapshot of implemented code is shown in Figure 13.8. The association graphs of various types of error are depicted for the Kafka project in Figure 13.9a–d.

13.5.4 Research Question 4: Which Are the Most Frequent Words and Keywords to Predict Most Critical Bugs?

To prevent failure of software systems and to improve the performance, it is necessary to predict the severity of a newly reported bug. The severity of a bug can be predicted using keywords that are extracted from severe bugs reported in the history of software systems. To analyze this aspect, most frequently occurring words of various projects are extracted. To predict the severity, most frequently cooccurring words with searched keyword such as "block" and "fail" are determined. This will help test leads and developers to predict the severity of a newly reported bug by matching words from a set of keywords. The word frequency graphs of words occurring more than 100 times in a project and associated graphs of the Apache Cordova project are depicted in Figure 13.10a–b. Table 13.3 lists some severe words of various projects.

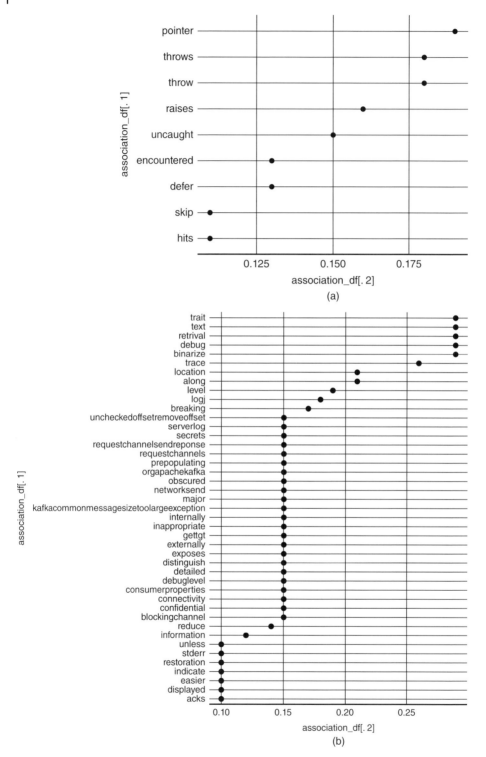

Figure 13.9 (a)–(d) Association graphs of various errors for Kafka project (a) Word-frequency plot of Apache Cordova project, (b) Association plot for high severe associated words.

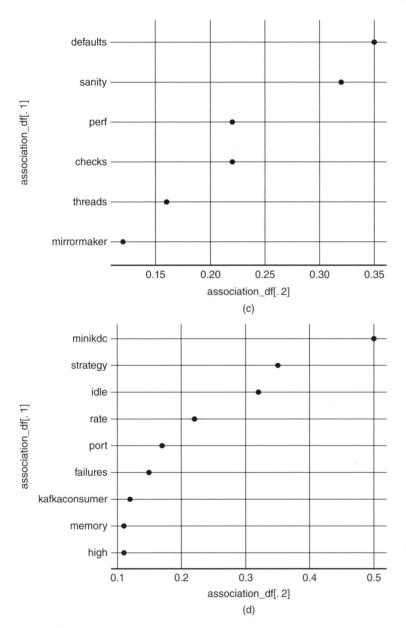

Figure 13.9 (*Continued*)

13.5.5 Research Questions 5: What Is the Importance of Frequent Words Mined from Bug Reports?

Frequent words mined from bug reports of various projects are of utmost importance as it forms an input to cluster analysis. Cluster analysis is an unsupervised machine learning method that groups set of similar objects in one cluster and are dissimilar than other clusters. Each cluster is a collection of useful information that can be used for

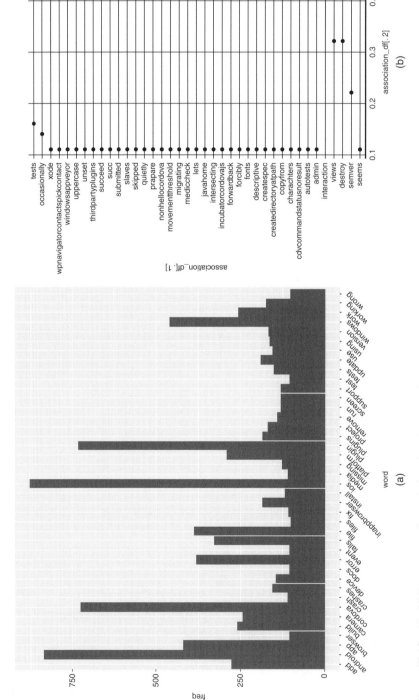

Figure 13.10 (a)–(b) Frequency and association plots for severe bugs.

Table 13.3 Frequent words for severe and nonsevere bugs.

Project	Nonsevere	Severe
Hadoop-hdfs	Fix, error, commands, message, typo, log, tests, report	Blocks, remove, failure, use, incorrect, storage, webhdfs
Apache cordova	Work, run, file, error, incorrect, remove, plugins, platform	Crashes, version, configxml, warning, xcode, event, destroy, forcibly
Hadoop-common	Message, default, log, error, typo, options, root	Unable, written, upgrade, stuck, timeout, tokens, zero, retry. Consequence, shutdown, close, returned
Hive	Error, exception, null, partition, throws	Join, columns, query, execution, dynamic, views, vectorize
Ambari	Error, default, message, check, button, page,	Upgrade, failed, python, install, hosts, clusters
Groovy	Access, breaks, default, causes, delegate, throws	Allow, assert, deadlock, closure, exception, fail
Hbase	Typo, warning, incorrect, missing, log, remove, wrong	Fails, abort, block, exception, split, deadlock, inconsistent, authfailedexception, datablock, compress, preemptive, fast
Lucene	Can, field, fix, incorrect, query, test	Bug, broken, causes, failure, exception, nullpointerexception
Mesos	Broken, add, log, link. Type, wrong, warning	Bad, crashes, fail, invalid, shutdown, implement, update
Maven	Error, plugin, pom, version, warning, fix	Causes, deploy, configuration, exception, failure, invalid, nullpointerexception
Qpid	Console, deleted, error, exchange, test, broker	Crash, deadlock, faie, exception, throws, timeout, unable
Sling	Tests, missing, due, integration, error	Null, volatile, exception, fails, content, npe
Spark	Fix, incorrect, error, return, wrong, missing, default	Block, fail, parquet, exception, executor, invalid, thrift
Accumulo	Error, fix missing, checking, test	Lock, blocks, deletion, misconfigured, nameerror, shell server it delete throws
Kafka	Unrelated, irregular, relative, ahead, commitsync	Indefinitely, unclosed, abstract fetcher thread shutdown, test broker failure

improved decision making. Clustering is of two types: k-means and hierarchical clustering. K-means clustering clusters the data based on their similarity and splits the data set into k-groups. Clusters are formed by computing the distance between each pair of observations. Euclidean distance is used to find similarity between two elements (a, b) defined in Eq. (13.1).

$$d(a, b) = \sqrt{\sum_{i=1}^{n} (a_i - b_i)^2} \tag{13.1}$$

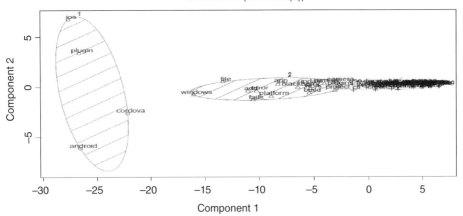

These two components explain 74.75% of the point variability.

Figure 13.11   K-means cluster group similar words.

Hierarchical clustering does not require a predefined number of clusters and represents data into a tree-based representation called as dendrogram. Cluster analysis had been widely used in various fields such as duplicate bug report detection [33], clustering defect reports [34], and categorization and labeling of bug reports [35]. To evaluate the importance of frequent words of bug reports, K-means and hierarchical clustering is performed on 20 projects of ASF and similar words are grouped into clusters. The results are illustrated for the Apache Cordova project. The K-means cluster in Figure 13.11 forms three clusters, showing dissimilarity of 74.75% in cluster 1 and 2, whereas cluster 3 contains immense similar words, thus depicting overlapping. Hierarchical clustering in Figure 13.12 depicts several clusters in the form of a dendrogram. Thus, clusters of similar words are used in many applications and is significant.

13.6   Threats to Validity

The main threat to internal validity is to correctly identify bug reports in several types of issues. To reduce this threat, various issues are manually inspected and identified that bug reports have the value of TypeId as 1. The tool BRCS is implemented to extract only bug reports from various issue types. One of the primary threats to external validity is that the tool does not work for multiple repositories. To reduce this threat, bug reports of 20 different projects of ASF based on the Jira repository are extracted. The tool will be implemented to extract the bug reports from other repositories to overcome this threat.

13.7   Conclusion

Software bug reports reported in the issue tracking system contains valuable and informative data that is useful to several applications, such as emotion mining, bug summarization,

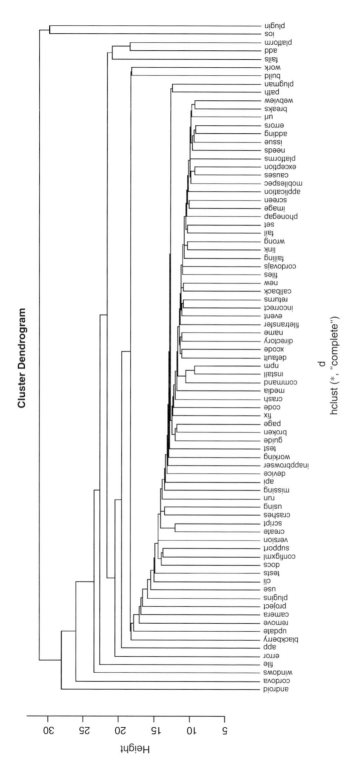

Figure 13.12 Dendogram of most similar words.

bug report duplicate detection, and many others. Analysis of this useful data is highly desirable. In this research work, quantitative analysis of data of bug reports is performed to discern various patterns and trends. The analysis is conducted on bug reports of 20 projects of ASF under the Jira repository. It was found that 24% of open bugs are critical in nature, which can cause system failure and are on high priority for resolution. Test leads can assign critical open bugs to the most contributing developers for quick resolution. Frequent words and most correlated words are extricated from long description and comments of bug reports, which can be used to categorize bugs into various types of errors and can predict the severity of a newly reported bug in the software. Clusters of frequent occurring keywords are formed using unsupervised clustering techniques to help developers detect duplicate bug report detection. This effective and practical analysis of bug reports of the issue tracking system will help software developers, test leads, and project managers in the software maintenance process.

References

1 Kaur, Arvinder. 2017. "Bug Report Collection System (BRCS)," 697–701. doi:https://doi .org/10.1007/1-84628-262-4_10.

2 Bettenburg, Nicolas, Thomas Zimmermann, and Sunghun Kim. 2008. "Extracting Structural Information from Bug Reports," 1–4.

3 Nayrolles, Mathieu. 2016. "BUMPER: A Tool for Coping with Natural Language Searches of Millions of Bugs and Fixes." doi:https://doi.org/10.1109/SANER.2016.71.

4 Yuk, Yongsoo, and Woosung Jung. 2013. "Comparison of Extraction Methods for Bug Tracking System Analysis," 2–3.

5 Malhotra, R. (2014). CMS tool : calculating defect and change data from software project repositories. *ACM Sigsoft Software Engineering Notes* 39 (1): 1–5. https://doi .org/10.1145/2557833.2557849.

6 Gong, Y. and Liu, X. (2001). Generic text summarization using relevance measure and latent semantic analysis. In: *Proceedings of the 24th Annual International ACM SIGIR Conference on Research and Development in Information Retrieval - SIGIR 01*, 19–25. ACM https://doi.org/10.1145/383952.383955.

7 Wang, D., Li, T., Zhu, S., and Ding, C. (2008). Multi-document summarization via sentence-level semantic analysis and symmetric matrix factorization. In *Proceedings of the 31st annual international ACM SIGIR conference on Research and development in information retrieval*, 307–314.

8 Hua, W., Wang, Z., Wang, H. et al. (2016). Understand short texts by harvesting and analyzing semantic knowledge. IEEE transactions on Knowledge and data Engineering 29 (3): 499–512.

9 Jha, N. and Mahmoud, A. (2018). *Using frame semantics for classifying and summarizing application store reviews. Empirical Software Engineering* https://doi.org/10.1007/s10664-018-9605-x.

10 Erkan, G. and Radev, D.R. (2004). LexRank: graph-based lexical centrality as salience in text summarization. *Journal of Artificial Intelligence Research* 22: 457–479. https://doi .org/10.1613/jair.1523.

11 Leskovec, Jure, and N Milic-Frayling. 2005. "Extracting Summary Sentences Based on the Document Semantic Graph." *Microsoft Research, ...* http://www.cs.cmu.edu/~jure/pubs/nlpspo-msrtr05.pdf.

12 Zopf, M., Mencía, E.L., and Fürnkranz, J. (2018). Which scores to predict in sentence regression for text summarization? In: *Proceedings of the 2018 Conference of the North American Chapter of the Association for Computational Linguistics: Human Language Technologies, Volume 1 (Long Papers)*, 1782–1791. https://doi.org/10.18653/v1/N18-1161.

13 Li, W., Wu, M., Lu, Q. et al. (2006). Extractive summarization using inter- and intra-event relevance. In: *Proceedings of the 21st International Conference on Computational Linguistics and the 44th Annual Meeting of the ACL - ACL 06*, 369–376. https://doi.org/10.3115/1220175.1220222.

14 Gupta, V.K., Operation, S., and Siddiqui, T.J. (2012). *Multi-Document Summarization Using Sentence Clustering*, 1–5. IEEE.

15 Simón, J.R., Ledeneva, Y., and García-Hernández, R.A. (2018). Calculating the significance of automatic extractive text summarization using a genetic algorithm. *Journal of Intelligent Fuzzy Systems, IOS Press* 35 (1): 293–304. https://doi.org/10.3233/JIFS-169588.

16 Sinha, Aakash, Abhishek Yadav, and Akshay Gahlot. 2018. "Extractive Text Summarization Using Neural Networks." http://arxiv.org/abs/1802.10137.

17 Shen, D., Sun, J.-t., Li, H. et al. (2004). Document summarization using conditional random fields. *Science* 7: 2862–2867. http://scholar.google.com/scholar?hl=en&btnG=Search&q=intitle:Document+Summarization+using+Conditional+Random+Fields#0.

18 Afsharizadeh, M., Ebrahimpour-Komleh, H., and Bagheri, A. (2018). Query-oriented text summarization using sentence extraction technique. In: *2018 4th International Conference on Web Research, ICWR 2018*, 128–132. IEEE https://doi.org/10.1109/ICWR.2018.8387248.

19 Ouyang, Y., Li, W., Li, S., and Lu, Q. (2011). Applying regression models to query-focused multi-document summarization. *Information Processing and Management* 47 (2). Elsevier Ltd): 227–237. https://doi.org/10.1016/j.ipm.2010.03.005.

20 Singh, M., Arunav Mishra, B., Oualil, Y., and Berberich, K. (2018). Advances in information retrieval. In: *40th European Conference on IR Research, ECIR 201*, vol. 10772, 657–664. https://doi.org/10.1007/978-3-319-76941-7.

21 Murray, G. and Carenini, G. (2008). Summarizing spoken and written conversations. In: *EMNLP 08: Proceedings of the Conference on Empirical Methods in Natural Language Processing*, 773–782. https://doi.org/10.3115/1613715.1613813.

22 Rastkar, S., Murphy, G.C., and Murray, G. (2010). Summarizing software artifacts. In: *Proceedings of the 32nd ACM/IEEE International Conference on Software Engineering – ICSE 10*, 505. https://doi.org/10.1145/1806799.1806872.

23 Rastkar, S., Murphy, G.C., and Murray, G. (2014). Automatic summarization of bug reports. *IEEE Transactions on Software Engineering* 40 (4): 366–380. https://doi.org/10.1109/TSE.2013.2297712.

24 Jiang, H., Nazar, N., Zhang, J. et al. (2017). PRST: a PageRank-based summarization technique for summarizing bug reports with duplicates. *International Journal* *of Software Engineering and Knowledge Engineering, World Scientific* 27 (06): 869–896. https://doi.org/10.1142/S0218194017500322.

25 Ferreira, I., Cirilo, E., Vieira, V., and Mourão, F. (2016). Bug report summarization: an evaluation of ranking techniques. In: *Proceedings – 2016 10th Brazilian Symposium on Components, Architectures and Reuse Software, SBCARS 2016*, 101–110. https://doi.org/10.1109/SBCARS.2016.17.

26 Mani, Senthil, Rose Catherine, Vibha Singhal Sinha, and Avinava Dubey. 2012. "AUSUM : Approach for Unsupervised Bug Report Summarization AUSUM : Approach for Unsupervised Bug Report SUMmarization Categories and Subject Descriptors," ACM Sigsoft, no. November. doi:https://doi.org/10.1145/2393596.2393607.

27 Nithya, R. and Arunkumar, A. (2016). Summarization of bug reports using feature extraction. *International Journal of Computer Science and Mobile Computing* 52 (2): 268–273. http://ijcsmc.com/docs/papers/February2016/V5I2201637.pdf.

28 Lotufo, R., Malik, Z., and Czarnecki, K. (2015). Modelling the 'hurried' bug report reading process to summarize bug reports. *Empirical Software Engineering* 20 (2): 516–548. https://doi.org/10.1007/s10664-014-9311-2.

29 Murgia, A., Ortu, M., Tourani, P. et al. (2018). An exploratory qualitative and quantitative analysis of emotions in issue report comments of open source systems. *Empirical Software Engineering* 23: 521–564. https://doi.org/10.1007/s10664-017-9526-0.

30 Umer, Q., Liu, H., and Sultan, Y. (2018). Emotion based automated priority prediction for bug reports. *IEEE Access* 6. IEEE: 35743–35752. https://doi.org/10.1109/ACCESS.2018.2850910.

31 Yang, G., Zhang, T., and Lee, B. (2018). An emotion similarity based severity prediction of software bugs: a case study of open source projects. *IEICE Transactions on Information and Systems* E101D (8): 2015–2026. https://doi.org/10.1587/transinf.2017EDP7406.

32 Ding, Jin, Hailong Sun, Xu Wang, and Xudong Liu. 2018. "Entity-Level Sentiment Analysis of Issue Comments," 7–13. doi:https://doi.org/10.1145/3194932.3194935.

33 Gopalan, R.P. and Krishna, A. (2014). Duplicate bug report detection using clustering. In: *Proceedings of the Australian Software Engineering Conference, ASWEC*, 104–109. https://doi.org/10.1109/ASWEC.2014.31.

34 Rus, V., Nan, X., Shiva, S., and Chen, Y. (2008). Clustering of defect reports using graph partitioning algorithms. In: *Proceedings of the 20th International Conference on Software and Knowledge Engineering*, 291–297. IEEE.

35 Limsettho, N., Hata, H., Monden, A., and Matsumoto, K. (2014). Automatic unsupervised bug report categorization. In: *Proceedings – 2014 6th International Workshop on Empirical Software Engineering in Practice, IWESEP 2014*, 7–12. https://doi.org/10.1109/IWESEP.2014.8.

14

Sarcasm Detection Algorithms Based on Sentiment Strength

Pragya Katyayan and Nisheeth Joshi

Department of Computer Science, Banasthali Vidyapith, Rajasthan, India

14.1   Introduction

Sentiments have become a puzzle to be solved these days. Verbal or written expressions of sentiments are tough to comprehend because of the innovative ways people have been adapting in order to express them. Where sentiments used to be a binary value earlier with just positive and negative values to look for, the advent of sarcasm has made the idea a little more explicit. Sarcasm is when someone decides to use words of opposite meaning to what he/she is actually feeling. Sarcasm is the new trendsetter and is so widely used and appreciated that those who do not know it have started to learn it. So, the text we come across in our day-to-day lives, be it on Amazon reviews or Twitter feeds or maybe the daily news headlines are a carrier of sarcasm, in some way or the other. If we wish to detect the sentiment values accurately, we need an algorithm that detects the types of sarcastic expressions along with the positive and negative emotions. According to linguistic Camp E. [1], broadly, sarcasm is of four types: propositional, embedded, "like"-prefixed, and illocutionary. Hyperbole, a type of embedded sarcasm is considered a strong marker of sarcasm. It is recognized when a sentence has both positive and negative polarity. The presence of these contradicting sentiment values is a pointer of hyperbolic sarcasm. Table 14.1 gives few examples for hyperbolic sarcasm.

The examples in Table 14.1, show hyperbolic sarcasm. As observed, each of the sentences have positive and negative polarity words in them. For instance, the sentence "A friend in need is a pest indeed" the word friend is a positive word and pest is negative word. Both the words are equally positive and negative in polarity. Hence, the sentence is a hyperbolic sarcasm. In all the other sentences of Table 14.1, we can find the bold words have positive as well as negative words in them, which are of similar weightage on the sentiment polarity scale. These sentences are few from the human annotated corpus of hyperbolic sarcasm.

Intelligent Data Analysis: From Data Gathering to Data Comprehension,
First Edition. Edited by Deepak Gupta, Siddhartha Bhattacharyya, Ashish Khanna, and Kalpna Sagar.
© 2020 John Wiley & Sons Ltd. Published 2020 by John Wiley & Sons Ltd.

Table 14.1 Examples for hyperbolic sarcasm.

The **trouble** of being punctual is that no one is there to **appreciate** it.

A **friend** in need is a **pest** indeed.

Everything I **like** is either **expensive, illegal**, or won't text me back.

Sweetie, leave the **sarcasm** and **insults** to the pros, you're gonna **hurt** yourself, go play in the traffic.

Think I'm **sarcastic**? Watch me **pretend** to **care**!

If you show me you don't give a **damn**, I'll show you I'm **better** at it.

I admit that my level of **weirdness** is above the national average, but I'm **comfortable** with that.

Don't **mistake** silence for **weakness**, smart people don't plan big moves out loud.

Success in life comes when you simply **refuse** to give up, with goals so strong that **obstacles, failure** and loss only act as motivation.

Love is a **weird** thing, **like** you just pick a human and you're like "yes, I **like** this one... I'll let this one **ruin** my life forever."

Not to **brag**, but I haven't had a mood swing in, **like**, 7 min.

Be selective in your **battles** for sometimes peace is **better** than being right.

Some of the **greatest hurt** will come from people you helped.

I'm not sure if I **attract crazy** or if I make them that way.

I'm **sorry** for whatever I did to make you feel I **care** about your feelings.

I would **like** to apologize to anyone I have not yet **offended**. Please be **patient**, I will get to you shortly.

Dear Friday, I'm so **glad** we are back together. I'm **sorry** you had to see me with Mon-Thursday, but I **swear** I was thinking of you the whole time.

People need to start **appreciating** the effort I put in to not be a serial **killer**.

Sarcasm: The **witty** will have **fun** but the **stupid** won't get it.

I don't **treat** people **badly**. I **treat** them accordingly.

I was **hoping** for a **battle** of wits, but you appear to be unarmed.

Bad at being **good, good** at being **bad**.

Drunk people, children, and leggings always tell the **truth**.

Sarchotic: when you're so **sarcastic**, people aren't sure if you're **joking** or if you're **joking** or if you're just **psychotic**.

This chapter works on a rule-based approach to detect at least one type of sarcasm (hyperbole) along with general sarcasm and non-sarcasm class. It uses SentiStrength[1] implemented in Python to find out the sentiment strengths. The algorithm proposed in this paper detects sarcastic, hyperbole, and non-sarcastic sentences. The sarcastic data set used for the experiment is taken from the social media pages dedicated to sarcastic posts on Facebook, Instagram, and Pinterest. The posts are collected manually from social media pages dedicated to sarcastic quotes and so are human annotated. The corpus has 1500 sentences with 500 sarcastic, 500 positive, and 500 negative sentences. Positive and

---

1 http://sentistrength.wlv.ac.uk.

Table 14.2 Examples for general sarcasm, positive sentences, and negative sentences.

Sarcasm:

*The trouble of being **punctual** is that **no one is there** to appreciate it.*

***I correct** the auto-correct **more than** the auto-correct **corrects me**.*

***Marriage** is one of the **chief causes** of **divorce**.*

Positive:

*The **best** preparation for tomorrow is **doing your best** today.*

***Learn** from yesterday, **live** for today, **hope** for tomorrow.*

***Positive action** combined with **positive thinking** results in **success**.*

Negative:

***Tears** come from the heart and not from the brain.*

*The person who tries to keep everyone happy often ends up **feeling lonely**.*

*The **worst** kind of **sad** is not being able to explain why.*

---

negative sentences are also taken from an already labeled data set of Amazon reviews from the University of California–Irvine (UCI) machine learning repository.[2]

The remaining chapter is arranged in the following manner: the literature survey of relevant papers elaborating various approaches of sarcasm detection, experimental setup, results, and evaluation of results.

Examples of sarcastic, positive, and negative sentences are given in Table 14.2.

14.2 Literature Survey

Dave and Desai [2] identified the challenges with sarcasm detection. Since the sentences are short and of ambiguous structure, i.e., they do not follow a correct grammar rule and hence are difficult to identify in textual data. These sentences also have URLs and emojis. In their study, the authors have detected various approaches of sarcasm detection including lexical analysis, likes and dislikes prediction, fact negation and temporal knowledge extraction.

Poria et al. [3] have introduced deep convolutional neural networks (CNNs) to analyze sarcasm deeply. The work recognizes the problems faced in computational sarcasm detection because of lack of contextual knowledge, general common sense, anaphora resolution, and logical reasoning. The researchers believed and proved that along with all the other factors user-profiling is also of importance. The probability of a tweet being sarcastic also depends upon the frequency of sarcastic tweets given by the user in the past along with several other factors like sentiment and emotion features. The model was trained using CNN as it uses the hierarchy of local features important for context learning and it gave good results for CNN-support vector machine (SVM) classification.

Joshi et al. [4] gave an initial attempt to use the concept of word-embeddings and the similarity score for detecting sarcasm. It follows two approaches: unweighted similarity

---

2 https://archive.ics.uci.edu/ml/datasets/Sentiment+Labelled+Sentences.

features and distance weighted similarity features to detect four types of word-embeddings: GloVe, LSA, Word2Vec, and dependency weights. It concludes experimentally that considering only word embedding features is not enough and sarcasm detection is better when it is used along with other features. Word embedding suffers from several issues like incorrect senses, contextual incompleteness, and metaphorical nonsarcastic sentences.

Joshi et al. [5] in their survey paper, have covered all the aspects of automatic sarcasm detection, including statistical approaches. They have identified learning algorithms such as SVM, logistic regression, and naïve Bayes as the most commonly used algorithms. They have also pointed out the use of the Winnow algorithm (balanced) to get high ranking features. The irony, political humor, and education labeled sentences area easily identified by naïve Bayes with the sequence nature of output labels is detected easily by SVM-HMM.

Joshi et al. [6] used the concept of "sentence completion" for detecting sarcasm. For this approach, the researchers have adopted two approaches: first; they have considered all the words that are occurring in a sentence as incongruous candidate words and then take the similarity measures; in the second approach, they have cut down the number of words to half by eliminating redundant comparisons for incongruous words. The researchers have used Word2Vec and WordNet as the similarity metric for both the approaches. The authors have observed that if the exact incongruous words are known, then the model gives much better results than other approaches. The model suffers errors due to the absence of Word-Net senses and inaccurate sentence completion.

Muresan et al. [7] have reported on a method for building a repository of tweets that are sarcastic in nature and are explicitly labeled by their authors. They used TwitterAPI to extract tweets of sarcastic nature detecting by their hashtags #sarcasm or #sarcastic and other sentiment tweets (exact positive or exact negative) by their hashtags for happy, lucky, joy, and sadness, angry, and frustrated. The authors have compared the annotated corpus with the positive and negative sentiment utterances and have investigated the effect of lexical features like n-grams or dictionary-based as well as pragmatic factors like emoticons, punctuation, and contexts on the machine learning approach used for sarcastic sentences. Their final corpus had 900 tweets falling under the category of sarcastic, positive, and negative in nature. They have used and compared the effects of three popular classification algorithms naïve Bayes, SVM, and logistic regression. They have conducted four classifications: Sarcasm_Positive_Negative, Sarcasm_NonSarcasm, Sarcasm_Positive, and Sarcasm_Negative.

Reganti et al. [8] have shown results for three classifiers: random forest, SVM, decision trees, and logistic regression. They have taken n-grams and sentiment lexica as bag-of-words features for detecting satire in a Twitter data set, newswire articles, and Amazon product reviews. Their results show that a combined approach with all features considered together gives better recognition rate than the features used alone.

Bharti et al. [9] recognized that verbal sarcasm is composed of heavy tonal stress on certain words or specific gestures like the rolling of eyes. However, these markers are missing in case of textual sarcasm. This paper has discussed sarcasm detection based on parsing and interjection words. They have used NLTK and TEXTBLOB packages of Python to find the POS and parse of tweets. The PBLGA algorithm, proposed by the paper takes tweets as input and detects sarcasm based on sentences with positive situation-negative sentiment and negative situation-positive sentiment. The second algorithm proposed by the paper detects sarcasm based on interjection words present in the sentence. These approaches give good recall and precision.

Rilof et al. [10] attempted to find sarcasm as a dividing line between a positive sentiment expressed in a negative context. For this bootstrap, learning is used where rich phrases of positive sentiments and negative situations are learned by the proposed model. Any tweet encountered with positive sentiment or a negative situation marks the source of sarcasm. Then those sentences are tested for contrasts in polarity. The sentences with a contrast between the two features are categorized as sarcastic.

Rajadesingan et al. [11] try to look into the psychological and behavioral aspect of sarcasm utterance and detection. This paper has tried to capture different types of sarcasm depending on the user's current and past tweets. Through this paper, the researcher has proposed SCUBA approach, i.e., sarcasm classification using a behavioral modeling approach. This approach basically proposes that given an unlabeled tweet from the user along with his past tweets then it can be automatically detected if the current tweet is sarcastic or not.

Kunneman et al. [12] identify linguistic markers of hyperbole as an essential feature to identify sarcasm. Since hyperbolic words are intensity carriers, their presence makes sarcasm detection much more relaxed than otherwise. For example: "the weather is good" is less likely to be sarcastic but "fantastic weather" is easily detected as hyperbolic sarcasm because of the word "fantastic." Intensifiers, thus, strengthen an utterance and play an essential role in hyperbole identification. Authors have used the balanced Winnow classification. Tweets were tokenized, less frequent words were removed, and the punctuations, emoticons, and capitalizations were preserved as possible sarcasm markers. N-grams (n = 1,2,3) were used as features for the data set.

Felbo et al. [13] used emoji occurrences with tweets as sarcasm indicators. The researchers show that the millions of texts with emojis available on social media can be used to train models and make them capable of representing emotional content in text. They have developed a pretrained DeepMoji model and observed that the diversity of emojis is crucial for its better performance. Their methodology includes pretraining on only English tweets without URLs having emoticons, which makes the model learn the emotional content in texts in a richer way. The DeepMoji model has used 256 dimension-embedding layers to pitch every word in the vector space.

Maynard et al. [14] have focused on hashtag tokenization because it contains crucial sentiment information. While tokenizing tweets, these hashtags get tokenized as one unit and effects the sentiment polarity of the sentence. Authors have used the Viterbi algorithm to find the best matches in the process. They have also used hashtags for detecting scope of sentiment in a tweet. In this paper, the authors have also considered the impact of different values of sarcastic modifiers on the meaning of tweets and have observed the change in polarity of sentiment expressed.

Joshi et al. [15] have discovered thematic structures in a big corpus. The paper attempts to find out the occurrence of sarcasm dominant topics and difference in the distribution of sentiment in sarcastic and nonsarcastic content. It also focuses on hyperbolic sarcasm having positive words with a negative implication according to the context. The probability of a word falling into sentiment or topic, sentiment distribution in tweets of a particular label and the distribution of topic over a label are the features taken up by the researchers. It was an initial but promising work where sarcasm detection was based on various sarcasm-prevalent topics. They detected topics based on two approaches, first with only sarcastic tweets and the second one with all kinds of tweets to capture the prevalence of sarcasm.

14.3 Experiment

This chapter attempts to find sarcasm along with positive and negative sentiments from sentences using the proposed algorithms and evaluate its performance using the confusion matrix. The experiment consists of the following steps: data collection, finding SentiStrength, classification, and evaluation.

14.3.1 Data Collection

The data for this experimental work includes 500 sarcastic sentences collected from various posts on pages from social media like Facebook and Instagram. These one- to two-sentence posts were explicitly taken from pages dedicated for sarcastic posts, so the sentences are human annotated and authentic sarcastic sentences. Out of these 500, 100 are specific Hyperbolic sentences, manually annotated by the authors. A separate set of 500 sentences each were taken for positive, and negative data set from a data set available online of sentiment labeled sentences at the UCI machine learning repository [16]. The total data set is of 1500 sentences, 100 hyperbolic, 400 general sarcasm, 500 positive, and 500 negative sentences.

14.3.2 Finding SentiStrengths

The Windows version of online web application of SentiStrength has been used for the experiment, and the sentiment strengths are recorded for the corpus of 1500 sentences. SentiStrength estimates the sentiment weight of a sentence by giving weights to all the positive and negative words used in a sentence and based on that gives a value of positive and negative sentiment strength to the sentence. These positive and negative strengths are recorded and used for further analyzation of the corpus. The strength value range varies from 1 to 5 for positive sentiment, 1 being neutral and 5 for strongly positive. Similarly, the range for negative sentiment strength ranges from -1 to -5 where (-1) is for neutral and (-5) for strongly negative (Figure 14.1).

Figure 14.1 Sentiment strengths and their elaboration as given by SentiStrength windows application.

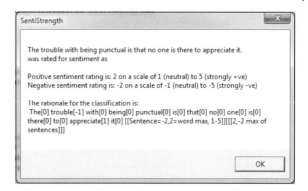

14.3.3 Proposed Algorithm

The main aim of this work is to propose an algorithm that not only detects sarcasm but tries to dig a bit deeper and detect at least one type of sarcasm, i.e., hyperbole in this case. The algorithm takes in two inputs and analyzes them to predict hyperbole, sarcastic, and nonsarcastic sentences. This algorithm was implemented in Python and the results are discussed later in the document. The basic algorithm and its extended version are as follows:

Algorithm 14.1 Sarcasm_Detector_with_SentiStrength

Input: Positive Strength (P), Negative Strength (N), i = iteration

Sarc_Detector:

```
score_i = P_i + N_i
IF (score_i <= 0):
        IF (score_i = 0):
                IF (P_i != 1 AND N_i != -1):
                        print ("Hyperbole")
                ELSE:
                        print ("Neutral")
        ELSE:
                print ("Sarcasm")
ELSE:
        print ("Non-Sarcasm")
```

Flowchart 14.1   Sarcasm_Detector_with_SentiStrength

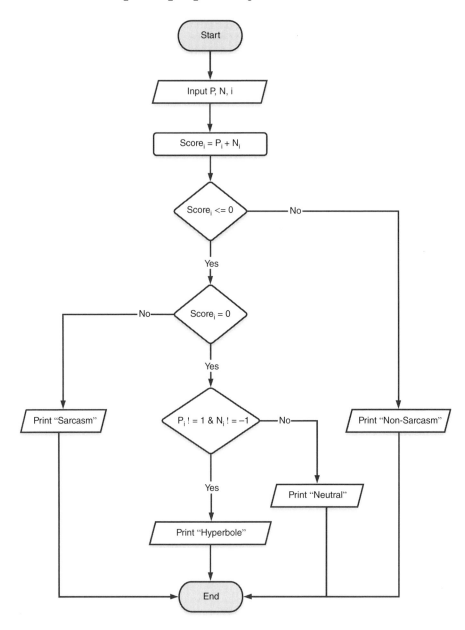

Algorithm 14.2 Sarcasm_Detector_with_SentiStrength_Extended

```
Input: Positive Strength (P), Negative Strength (N), i =
iteration

Sarc_Detector:

score_i = P_i + N_i
IF (score_i <= 0):
        IF (score_i = 0):
                IF (P_i != 1 AND N_i != -1):
                        print ("Hyperbole")
                ELSE:
                        print ("Neutral")
        ELSE:
                IF (P_i = 1 AND N_i < -1):
                        print ("Negative")
                ELSE:
                        print ("Sarcasm")
ELSE:
        IF (P_i > 1 AND N_i = -1):
                print ("Positive")
        ELSE:
                print ("Hyperbole*")
```

The Algorithm 14.2. attempts to detect hyperbole, along with positive and negative sentences. However, hyperbole* needs human annotation to be sure. The algorithms are discussed with details in the next subsection.

14.3.4 Explanation of the Algorithms

The first algorithm (Algorithm 14.1) takes the positive and negative sentiment strength and gets a score by adding them up. Further, it checks the score for a zero or negative value. If the overall score is zero and it is because both the positive and negative strengths are having similar strengths (nonneutral), then the sentences are marked hyperbolic.

If the score is not zero but negative, then it is marked sarcastic, and if none of the above conditions qualify, it is marked nonsarcastic sentence.

The extended algorithm (Algorithm 14.2) taps in a little more depth than Algorithm 14.1 by looking for pure negative and pure positive sentences along with sarcastic and hyperbolic sentences. Like the first algorithm, it reads the positive and negative strength value of a sentence and then looks for nonneutral strengths causing a zero score and marks them

Flowchart 14.2   Sarcasm_Detector_with_SentiStrength_Extended

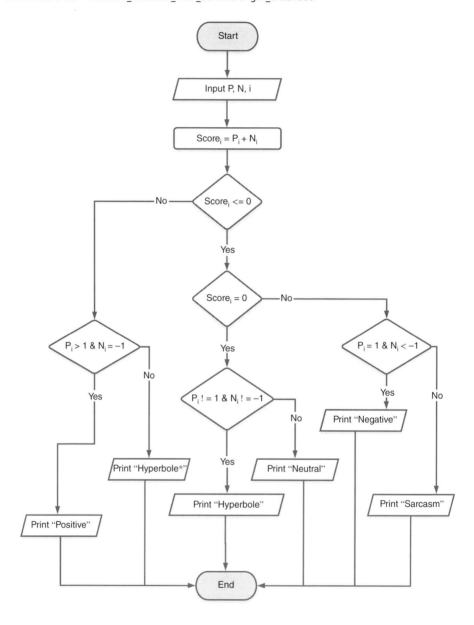

as hyperbolic in a similar way as above. Further, it looks for nonzero negative scores and finds if the negative score is a result of neutral-positive strength. In such case, the algorithm marks the sentence as pure negative as its score is a result of the only negative strength of the sentence.

In case of a negative score if the positive component is not neutral but less than the negative component, then the sentence is likely to be sarcastic. Moreover, if the overall score is positive, then the components are checked for a neutral negative and nonneutral positive value. In such case, the sentence is a pure positive sentence. If the positive score of a sentence is a result of significant positive strength value but a nonneutral negative value then again, the sentence can be a possible "hyperbole*." This hyperbole needs to be annotated manually to get good results. Such cases can be hyperbolic sarcasm as well as impure positive sentences.

The formulae used in the Algorithm 14.1 and 14.2 are as follows:

For positive:

$$Pi + Ni > 0, \tag{14.1}$$

where,

$Pi \neq 1$ and
$Ni = -1.$

The sentences with only positive sentiment words and no negative words are classified as purely positive sentences.

For negative:

$$Pi + Ni < 0, \tag{14.2}$$

where,

$Pi = 1$ and
$Ni \neq -1.$

The sentences spotted with high negative strengths and neutral positive strength are categorized as purely negative sentences.

For sarcasm:

$$Pi + Ni <= 0, \tag{14.3}$$

where,

$Pi \neq 1$
$Ni \neq -1$ and
$|Ni| > Pi.$

Sarcastic sentences are observed to be usually carrying negative words, i.e., general practice is to express a positive feeling wrapped up in negative words. So, sentences having both polarities in them but the negativity supersedes the positivity, that has good chances of being sarcastic in nature.

For hyperbole:

$$Pi + Ni = 0, \tag{14.4}$$

where,

$P_i = |N_i|$ and
$P_i, |N_i| \neq 1$

That means, there can be two cases for Eq. (14.4):

Case 1: $P_i = N_i = 0$; i.e. $P_i = 1$ and $N_i = -1$ (Neutral scores for both sentiments)

Example: *Life is a book and there are thousand pages I haven't yet read.*

Explanation: The above sentence doesn't hold any sentiment word and so SentiStrength gives it neutral score on both positive and negative sentiment. That renders the score of the complete sentence as neutral.

Case 2: $N_i = -P_i$, where, P_i and $N_i \neq 0$; (Nonneutral scores for both sentiments)

Example: *The **trouble** of being punctual is that no one is there to **appreciate** it.*

Explanation: The sentence has nonneutral sentiment scores for both positive ($P_i = 2$) and negative sentiments ($N_i = -2$) but the scores are leveled up in weight so they cancel out their overall effect on the complete sentence and render the sentence with a neutral SentiStrength.

In a nutshell, we can summarize all above equations in Eq. (14.5):

$$\text{Sentiment Analysis Conditions} = \begin{cases} \textbf{Pi + Ni > 0,} \text{ where Ni} = -1, \text{Positive} \\ \textbf{Pi + Ni < 0,} \text{ where Pi} = 1, \text{Negative} \\ \textbf{Pi + Ni = 0,} \text{ where Pi} = |\text{Ni}| \neq 1, \text{Hyperbole} \\ \textbf{Pi + Ni < = 0,} \text{ where Pi,} |\text{Ni}| \neq 1 \, |\text{Ni}| > \text{Pi, Positive} \end{cases} \quad (14.5)$$

14.3.5 Classification

The above algorithms were derived on the following observation:

a) The sentences having all positive words and no negative words can be classified as positive sentences.

b) The sentences having all negative words and no positive words can be classified as negative sentences.

c) Hyperbolic sentences are known to have both positive and negative sentiments expressed in them in equal manner.

d) General sarcasm has varied proportions of negative and positive sentiment strengths.

Table 14.3 describes the cases of sentiment prediction considered by the algorithms and Table 14.4 gives a deeper understanding with suitable examples.

14.3.5.1 Explanation

In the Table 14.4, which shows example cases for the logic discussed in Table 14.3, the first, second, and third sentences are all neutral as per their scores but in a little different way. The first sentence doesn't have any sentiment word contributing in its formation and hence both of its positive and negative scores are neutral (1 and −1) summing up to give a neutral overall score to the sentence (case1). Similar case is with the second sentence where there is no sentiment word involved but the sentence, as per human annotator, is a sarcastic sentence. Hence it is wrongly classified by the algorithm as neutral (case1). Moreover, the third sentence has two sentiment words "trouble" and "appreciate" both having equal negative and positive strengths (i.e., −2 and 2) respectively. So, their overall sentiment score again sums up to be neutral but the algorithm rightly classifies it as hyperbole (case2).

Table 14.3   Shows the patterns used by extended Algorithm 14.2 to detect the positive, negative, hyperbole, and sarcasm.

Cases	Score (PS + NS)	PS	NS	Result
Case1	Neutral	Neutral	Neutral	Neutral
Case2	Neutral	Nonneutral	Nonneutral	Hyperbole
Case3	Positive	Nonneutral	Neutral	Positive
Case4	Positive	Nonneutral	Nonneutral	Hyperbole*
Case5	Negative	Neutral	Nonneutral	Negative
Case6	Negative	Nonneutral	Nonneutral	Sarcasm

Table 14.4   Shows example cases for Table 14.3.

Sl. No.	Sentences	P_i	N_i	$Score_i$	Result
1.	*Life is a book and there are thousand pages I haven't yet read.*	1	−1	0	Case1
2.	*I correct the auto-correct more than the auto-correct corrects me.*	1	−1	0	Case1[a)]
3.	*The **trouble** of being punctual is that no one is there to **appreciate** it.*	2	−2	0	Case2
4.	*The **best** preparation for tomorrow is doing your **best** today.*	2	−1	1	Case3
5.	*Learn from yesterday, live for today, **hope** for tomorrow.*	3	−1	2	Case3
6.	***Sorry** for being late, I got caught up **enjoying** the last few minutes of not being here.*	3	−2	1	Case4
7.	***Difficult** roads lead to **beautiful** destinations.*	3	−2	1	Case4
8.	*Marriage is one of the chief causes of **divorce**.*	1	−2	−1	Case5[a)]
9.	***Tears** come from the heart and not from the brain.*	1	−4	−3	Case5
10.	*I **smile** all day so that nobody knows how **sad** I am.*	3	−4	−1	Case6
11.	*Hard work beats **talent** when talent **doesn't** work hard.*	2	−3	−1	Case6

a)   Wrongly classified, when the sentence is actually sarcastic.

The fourth and fifth sentences are both good examples of case3 sentences. They both have positive sentiment words, i.e., "best" and "hope," which have good positive scores (i.e., 2 and 3) and these sentences lack any negative sentiment words so their overall score sums up to be positive rendering the sentences purely positive.

The sixth and seventh sentences are the examples for the case4 where there's equal chances of the sentence being a hyperbolic sarcasm or a positive sentence. The fifth sentence has a negative token "sorry" and a positive word "enjoying," but the weightage of positive word is more than the negative one (i.e., 3 and −2), which leaves the score of the sentence as positive. The sentence, however, is an example of hyperbolic sarcasm. The seventh sentence of Table 14.4 also has the similar sentiment strengths like the sixth sentence. Due to the presence of a positive and negative word in the same sentence "difficult" and "beautiful" with positive sentiment strength more than the negative one, i.e., 3 and −2, the sentence qualifies as a positive sentence, which is indeed right. That's why the case4 requires human annotation to be sure because the case where both sentiment words are present but higher positive sentiment strength beats a low negative strength and gives a positive score, can be both a hyperbolic sarcasm as well as a positive sentence.

The eighth and ninth sentences are case5 examples. In the eighth sentence "Marriage is one of the chief causes of divorce," the word divorce has a negative sentiment strength of −2 but ironically, its opposite word "marriage" doesn't has a positive score so the sentence is rendered negative in the absence of positive sentiment score. Such anomalies are exceptions and rare and the algorithm fails to recognize them correctly. The sentence is, as a matter of fact, a sarcastic sentence but the algorithm classifies as negative. The ninth sentence as a lot of negative words, i.e., "dark," "sad," and "broken" and no positive ones so the sentence turns out to be neutral on positivity and get a −4 on negative sentiment strength. This sentence is correctly classified as a case5 negative sentence.

Last but not the least, sentences like the tenth and eleventh ones in Table 14.4, which have both positive and negative sentiment words like "smile" (weight: 3) and "sad" (weight: −4) in tenth sentence and "talent" (weight: 2), and "doesn't" (weight: −3) in the eleventh sentence but the negative weight is more than the positive one and the overall score of the sentences comes negative. Such sentences containing both positive and negative sentiment weights but giving overall negative score are classified as sarcastic that is the case6 of the algorithm.

14.3.6 Evaluation

The algorithms proposed by the paper classify the sentences as positive, negative, sarcastic, and hyperbolic. The classification results then undergo evaluation on the following criteria:

Accuracy. It is the balance of true answers in the sample,

$$Accuracy = \frac{No.of\ TrueP + No.of\ TrueN}{No.of\ TrueP + FalseP + FalseN + TrueN}$$

Precision. It is a balance of TrueP against all positive answers,

$$Precision\ (Prec.) = \frac{No.of\ TrueP}{No.of\ TrueP + FalseP}$$

Recall. It is the balance of the true positives versus all the true results,

$$Recall\ (Rec.) = \frac{No.of\ TrueP}{No.of\ TrueP + FalseN}$$

F-Score. It is the accuracy measure of the test. It is calculated on the values of precision and recall,

$$F - Score\ (F) = 2 * \frac{Prec. * Rec.}{Prec. + Rec.}$$

14.4 Results and Evaluation

The results obtained after implementing the extended algorithm are shown in Tables 14.5 and 14.6. Table 14.5 shows the total number of instances given to algorithms for testing and how many were correctly identified by the algorithm. It also presents the total number of sentences that were not correctly identified by the extended algorithm. The extended algorithm was able to detect 68 hyperboles out of 100 hyperbolic sentences, whereas it detected 270 out of 400 general sarcastic sentences correctly. For positive and negative sentences, the extended algorithm was able to detect 440 out of 500 and 350 out of 500 correctly respectively. The classification results were evaluated on the confusion matrix on the criteria of accuracy, recall, precision, and F-measure. Table 14.6 displays the evaluation results of the extended algorithm. The table shows the accuracy, precision, recall, and F-measure values for all four types of sentiments. Accuracy values for all four sentiments (in bold) show the performance of the extended algorithm. The extended algorithm was 88% accurate in case of hyperbole, 84% in case of identifying overall sarcasm and gave 87% and 77% accurate results for positive and negative sentiments respectively. Apart from accuracy values, the algorithm shows 68% and 88% recall for hyperbole and positive sentiments respectively, and 68% and 75% precision values for sarcasm and negative sentiments. These results are depicted graphically for analysis purposes (Figures 14.2 and 14.3). In the first graph (Figure 14.2) the

Table 14.5 True positive and True negative values of the classification result.

	Total instances	True positive	True negative
Hyperbole	100	68	32
Sarcasm	400	270	130
Positive	500	440	60
Negative	500	350	150

Table 14.6 Evaluation results of the classification done by the extended algorithm.

	Hyperbole	Sarcasm	Positive	Negative
Accuracy	**0.88**	**0.84**	**0.87**	**0.77**
Precision	0.34	**0.68**	0.77	**0.75**
Recall	**0.68**	0.54	**0.88**	0.70
F-measure	0.45	0.59	0.81	0.72

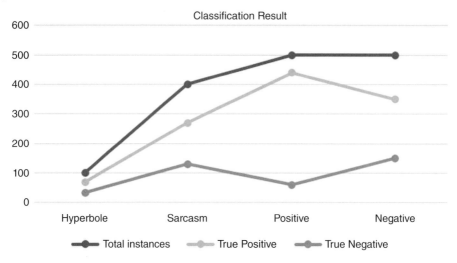

Figure 14.2 Chart showing classification results of all four sentiments.

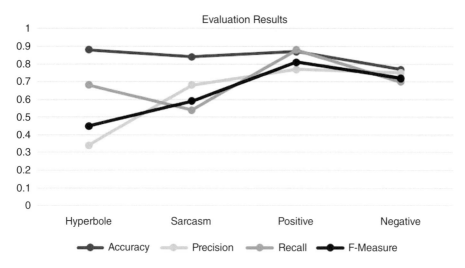

Figure 14.3 Chart showing evaluation results.

topmost line shows total number of instances, and the next line is for correctly identified instances, which is visibly close to the total number of instances. The third line is for the sentences that were not correctly classified by the algorithm, which is visibly close to the 0 line. This shows that the performance of algorithm is very good since the errors are very low in count. The next graph (Figure 14.3) shows evaluation results graphically. Observation of the graph tells us that the algorithm gave best results for positive sentiment as all four evaluation criterion lie above the 70% mark. Apart from that, the accuracy line for all for all four sentiments tops the chart, which shows that the proposed extended algorithm was able to give great results at identifying the sentiments accurately.

14.5 Conclusion

Sarcastic sentences are known to be of a different behavior than the regular sentiment sentences. Due to this ambiguous nature of sarcasm, it is complicated to identify. This paper has tried to adopt a logical yet straightforward approach of catching the scope of sarcasm over a sentence via its sentiment strength. We calculate the sentiment strength of sentences. These strengths are in two parts, positive strength and negative strengths. The devised algorithms then calculate these strengths. The algorithms are formulated on the basis of human observations of positive, negative, sarcasm, and hyperbole. It takes in the positive strength, negative strength, and their sum to decide in a unique way if the sentence falls into one of the four sentiment categories or is neutral. This experiment was performed on a data set of 1500 sentences. The results show that this may prove to be a right approach to rule-based sarcasm detection. This approach saves the time and complexity of data cleaning. The algorithms can be extended further and can be made to detect all types of sarcasm along with all modes of positive and negative sentiments. Future works may aim toward developing a system or algorithm that can detect all sentiments altogether.

References

1 Camp, E. (2012). Sarcasm, pretense, and the semantics/pragmatics distinction*. *Noûs* 46 (4): 587–634.

2 Dave, A.D. and Desai, N.P. (2016). A comprehensive study of classification techniques for sarcasm detection on textual data. In: *International Conference on Electrical, Electronics, and Optimization Techniques (ICEEOT)*, 1985–1991. IEEE.

3 Poria, S., Cambria, E., Hazarika, D., & Vij, P. (2016). A deeper look into sarcastic tweets using deep convolutional neural networks. *arXiv preprint arXiv:1610.08815*.

4 Joshi, A., Tripathi, V., Patel, K., Bhattacharyya, P., & Carman, M. (2016). Are Word Embedding-based Features Useful for Sarcasm Detection?. *arXiv preprint arXiv:1610.00883*.

5 Joshi, A., Bhattacharyya, P., and Carman, M.J. (2017). Automatic sarcasm detection: a survey. *ACM Computing Surveys (CSUR)* 50 (5): 73.

6 Joshi, A., Agrawal, S., Bhattacharyya, P., & Carman, M. (2017). Expect the unexpected: Harnessing Sentence Completion for Sarcasm Detection. *arXiv preprint arXiv:1707.06151*.

7 Muresan, S., Gonzalez-Ibanez, R., Ghosh, D., and Wacholder, N. (2016). Identification of nonliteral language in social media: a case study on sarcasm. *Journal of the Association for Information Science and Technology* 67 (11): 2725–2737.

8 Reganti, A., Maheshwari, T., Das, A., and Cambria, E. (2017). Open secrets and wrong rights: automatic satire detection in english text. In: *Companion of the 2017 ACM Conference on Computer Supported Cooperative Work and Social Computing*, 291–294. ACM.

9 Bharti, S.K., Babu, K.S., and Jena, S.K. (2015). Parsing-based sarcasm sentiment recognition in twitter data. In: *2015 IEEE/ACM International Conference on Advances in Social Networks Analysis and Mining (ASONAM)*, 1373–1380. IEEE.

10 Riloff, E., Qadir, A., Surve, P. et al. (2013). Sarcasm as contrast between a positive sentiment and negative situation. In: *EMNLP, Association of Computational Linguistics*, vol. 13, 704–714.

11 Rajadesingan, A., Zafarani, R., and Liu, H. (2015). Sarcasm detection on twitter: a behavioral modeling approach. In: *Proceedings of the Eighth ACM International Conference on Web Search and Data Mining*, 97–106. ACM.

12 Kunneman, F., Liebrecht, C., Van Mulken, M., and Van den Bosch, A. (2015). Signaling sarcasm: from hyperbole to hashtag. *Information Processing & Management* 51 (4): 500–509.

13 Felbo, B., Mislove, A., Søgaard, A., Rahwan, I., & Lehmann, S. (2017). Using millions of emoji occurrences to learn any-domain representations for detecting sentiment, emotion and sarcasm. *arXiv preprint arXiv:1708.00524*.

14 Maynard, D. and Greenwood, M.A. (2014). Who cares about sarcastic tweets? Investigating the impact of sarcasm on sentiment analysis. In: *LREC*, 4238–4243.

15 Joshi, A., Jain, P., Bhattacharyya, P., & Carman, M. (2016). 'Who would have thought of that!': A Hierarchical Topic Model for Extraction of Sarcasm-prevalent Topics and Sarcasm Detection. *arXiv preprint arXiv:1611.04326*.

16 Kotzias, D., Denil, M., De Freitas, N., and Smyth, P. (2015). From group to individual labels using deep features. In: *Proceedings of the 21th ACM SIGKDD International Conference on Knowledge Discovery and Data Mining*, 597–606. ACM.

15

SNAP: Social Network Analysis Using Predictive Modeling

Samridhi Seth and Rahul Johari

USIC&T, GGSIP University, New Delhi, India

15.1 Introduction

Predictive analytics uses archive data, machine learning, and artificial intelligence for prediction. Predictive analytics has two steps: first, archive data is fed into algorithms and patterns are created, and second, current data is used on same algorithms and patterns for predictions. Digitization of every resource has aided the predictive analytics. Currently, hundreds of tools have been invented that can be used to predict probabilities in an automated way and decrease human labor.

Predictive analytics involves identifying "what is" needed from archive data, then to study whether the archive data used meets our needs. Then the algorithm is modified to learn from a data set to predict appropriate outcomes. It is important to use an accurate and authentic data set and update the algorithm regularly.

To the Best of Our Knowledge and Wisdom, Predictive Analytics Process Can be as Detailed as: [1]

1) *Define project*: All the minute details like outcomes, scope, deliverables, objectives, and data sets to be used are defined.
2) *Data collection*: It involves data mining with complete customer interaction.
3) *Data analysis*: Extracting important information via inspection, cleaning, and modeling.
4) *Statistics.* Validation and testing.
5) *Modeling*: Accurate predictive models are prepared.
6) *Deployment*: Models prepared are deployed in everyday decision-making process to get results and output.
7) *Model monitoring*: Maintenance of the models.

15.1.1 Types of Predictive Analytics Models: [1]

1) *Predictive models*: Relation between samples of models are found.
2) *Descriptive models*: These quantify relationships in data in view to classify into different clusters.
3) *Decision models*: Finds relation between known data, decision, and forecast results.

Intelligent Data Analysis: From Data Gathering to Data Comprehension,
First Edition. Edited by Deepak Gupta, Siddhartha Bhattacharyya, Ashish Khanna, and Kalpna Sagar.
© 2020 John Wiley & Sons Ltd. Published 2020 by John Wiley & Sons Ltd.

15.1.2 Predictive Analytics Techniques: [2]

15.1.2.1 Regression Techniques

Mathematical equations within the relations are found. Different models applied on this technique are:

a) *Linear regression model*: Relation between rasp (or dependent) predictor (or independent) variables is found.
b) *Discrete choice model*: Used where response variable are continuous and has unbounded range.
c) *Logistic regression*: Binary dependent variables are transformed into unbounded continuous variables.
d) *Multinomial logistic regression*: Used where dependent variables have more than two unordered categories.
e) *Prohibit regression*: It is also used for modeling categorical-dependent variables but the criterion used for categorization is different from multinomial logistic regression.
f) *Time series model*: Used to predict future behavior of variables.
g) *Survival or duration analysis (or time to event analysis):* Used to study characteristics of survival data, such as, censoring, and nonnormality, since they are difficult to analyze using conventional statistical models.
h) *Classification and regression trees*: These are nonparametric decision tree learning technique produces classification or regression trees based on dependent variables, that is, whether it's classification or regression respectively.
i) *Multivariate adaptive regression splines*: A nonparametric technique used to build flexible models by fitting piecewise linear regression.

15.1.2.2 Machine Learning Techniques

Machine learning techniques imitate human acts and learn from training examples to forecast the future. It is used in applications where a future prediction is directly possible by carefully studying relations, or in cases where relations are very complex or formulating mathematical equations is not possible. Different models applied on this technique are:

a) *Neural network*: Nonlinear techniques to model complex functions.
b) *Multilayer perception*: It has an input layer, output layer, and one or more hidden layers of nonlinearly activating nodes. A weight vector is assigned to each node.
c) *Radial basis functions*: Functions are created to build distance criterion with respect to center used for interpolation and smoothening of data.
d) *Support vector machines*: Detection and exploitation of complex patterns in data using clustering, classifying, and ranking in data.
e) *Naive Bayes*: Based on Bayes' conditional probability rule to perform classification tasks.
f) *k-nearest neighbors*: A pattern recognition statistical methods in which no pre-assumptions are made but a training set is used.
g) *Geospatial predictive modeling*: Based on principle that event occurrences are united in distribution, which is based on spatial environment factors.

Some case studies where predictive analytics can play a major role are as follows:

- While deciding cutoffs for college admissions, previous year's cutoffs are studied. Highest key benefit of predictive analytics has been in the field of industry and business.
- Losses incurred due to natural calamities like tsunami and earthquake can be reduced by adopting timely predictive analytics practices by analyzing seismic zones and evicting people from those zones.
- Predictive analytics has continuously been used in weather forecasting for many years.
- Predictive analytics can be used in the field of pharmacy to decide the production of medicine with the arrival of a particular season.
- Similarly, predictive analytics is used in the industry to decide the sales and production of goods in different regions.
- For prediction of jobs in the IT industry.
- To predict the agricultural produce for the season and requirements of the products such as pesticides, tools, etc.
- To predict rainfall for the season.
- Stock market.

Some pitfall areas where predictive analytics cannot be applied:

- Unforeseen scenarios or abrupt incidents.
- Successful implementation of government policies for urban and rural citizens.
- Accuracy and authenticity of archived data and the tools used.
- Whether the movie released will be successful or not.
- Sport matches outcome.

Some applications [2] of predictive analytics are customer relationship management, health care, collection analytics, cross sell, fraud detection, risk management, direct marketing. Predictive analytics can be used in statistical analysis and visualization, predictive modeling and data mining, decision management and deployment, as well as big data analytics.

The literature survey of different research papers undertaken as part of the current research work is summarized below.

15.2 Literature Survey

In [3], author(s) have discussed what is social network analysis and the need of social network analysis, that is, why is it important to study social networks and their impacts. Author(s) have taken into consideration multiple parameters and presented the results in graphical forms.

In [4], author(s) have described what is predictive analytics, its importance in IT industry, relation between big data and predictive analytics and why predictive analytics has gained so much hype in recent years. Importance of tools in this field is also explained. Management of big data and results of predictive analytics is another big challenge that needs to be taken care of seriously. Some solutions to the management challenges are recognizing the barrier, exploiting operational data, improving mining techniques, and increasing marketing effectiveness with the increasing of data at a faster pace; it becomes important to pick unique combination of competitive strengths and optimize analytics dimensions.

In [5], authors(s) have discussed what is big data, its history, emergence of big data, big data analytic tools, issues, and challenges of big data. Multiple big data applications have been presented. Comparison on data mining techniques has also been considered.

In [6], author(s) have discussed the importance of predictive analytics in the industry for a decision-making process. Technological cycle of predictive analytics is described. Cost is an important characteristic of big data epoch, thereby it becomes a duty to select the best technology for predictive analytics. Therefore, author(s) have proposed architectural solutions based on international standards of data mining, creating analytic platforms for industrial information analytics systems. Issues of mobility and computational efficiency in predictive analytics technology have also been discussed.

In [7], author(s) have described what is predictive analytics, its uses, applications, and technologies. Then, author(s) have applied predictive analytics at both micro- and macro-levels of granularity. Articles applied at the "macro" level entitled "Predicting Elections for Multiple Countries Using Twitter and Polls," "Financial Crisis Forecasting via Coupled Market State Analysis," and "Dynamic Business Network Analysis for Correlated Stock Price Movement Prediction" have been discussed.

In [8], author(s) have discussed the importance of predictive analytics in education field by addressing fields such as enrollment management and curriculum development. Predictive analytics help organizations grow, compete, enforce, improve, satisfy, learn, and act. Goal-directed practices are observed for organization success. Author(s) have presented a case study at Delhi Technological University, sponsored by All India Council for Technical Education to determine the effectiveness of predictive analytics in the field of education. The process used involves the following four steps:

1) Data collection
2) Build the predictive model
3) Model validation
4) Analytics enabled decision making

According to author(s) opinion, deployment of cloud computing techniques will facilitate cloud-based predictive analytics in a cost-effective and efficient manner.

In [9], author(s) have presented a predictive visual analytics system for predicting event patterns. Future events are predicted by combining contextually similar cases that had occurred in the past. Social media data and news media data has been used for analysis. Two cases have been discussed for evaluating the system: German winds clashed in the Alps on March 24, 2015 and a heavy snow storm on the east coast of the United States on January 26, 2015. Steps for predictive value analytics are: topic extraction, predictive analysis, and visual analytics.

Predictive modeling can be fully utilized for improvements where total possible outcomes are known. So, in [10], authors(s) have predicted the outcomes of a sports match. To have more accurate predictions, a model based on knowledge discovery in database has been used. The model is capable of providing suggestions for improvements in itself to provide better predictions and reconstruct itself. The model predicts on the basis of various factors, for example, performance of the individual, performance of the team, atmosphere, type of match, health problems, etc. This model is best suited for games in groups.

In [11], author(s) have applied big data analytics on agricultural practices to increase the profit by increasing the production and helping the farmers, further reducing suicide rates of farmers. The main factors considered in this work are crop yield and land area under cultivation. Five different data sets were integrated and a regression technique was used to find the relation between the two factors stated above.

Since Python programming engines are really few, in [12], author(s) have proposed a Python predictive analysis by observing neighboring calculations. Their aim is to detect bugs. A novel encoding scheme has been incorporated for handling of dynamic features. A prototype is prepared and evaluated on real-world Python programs. The prototype framework has two steps: first is to collect traces of execution of passing run, the second is encoding the traces and some unexecuted branches into symbolic constraints.

In [13], author(s) have discussed the importance of social network analysis with the help of a Twitter data set. Analysis is important for the well-being of the society. It becomes necessary to understand the social networking today in the world of technology and gain useful knowledge. Author(s) have analyzed Twitter user profiles on the basis of some characteristics. Analysis include clustering, classification, and detection of outliers among the user profiles. The incorporated approach is related to big data in terms of volume of data generated. Hence, big data analytics can also be considered. The presented approach involves two steps: extraction and analysis.

In [14], author(s) have discussed the application of social network analysis in the field of data fusion. Parallel computing based methodology has been proposed for data extraction. The methodology will also help in enhancing the fusion quality. Hop count weighted and path salience approaches were taken into consideration. Results are visualized as a cumulative associated data graph. Results also show sensitivity of betweenness, closeness, and degree centrality measure of social network.

Big data plays an important role in the field of education. With data comes predictive modeling. Analytics is important to deduce results and impacts. In [1], author(s) have used prediction- and classification-based algorithms on students' data to analyze the performance of first-year bachelor students in a computer application course. Different algorithms were used and then comparative analysis has also been done on the WEKA tool to find which one was the best. According to author(s) results, multilayer perception algorithm performed the best.

In [2], author(s) have discussed predictive modeling using machine learning techniques and tools in the field of health care. It is an emerging field since there's a need for reduction in price of medicines and other health care–related things, such as, health tests, health care monitoring devices, surgical instruments, etc. Author(s) have also discussed other machine learning applications and showcased its importance in the field of health care. Author(s) have discussed what is machine learning, its techniques, algorithms, and its tools.

In [15], analysis of management of big data in a Windows environment is being carried out on a cloud environment with the help of the Aneka tool. A remote electronic voting machine application has been analyzed using MongoDB – a big data tool. The paper presents a secured voting based application by using a login ID and password, which is generated randomly, hence, difficult to hack. Further, views are encrypted and saved in the database

Social network is a wide area of research these days. So, in [16], author(s) have done a comparative study of two different mathematical models based on mixed integer linear programming for an analysis of social network. Their aim is to determine the existing models that best suits the social network analysis criteria. First, the model works on the minimization of the maximum diameter of cluster diameters while the second model works on minimizing minimum distance between objects of same cluster. As per author(s) results, both models are adequate choices with respect to different parameters.

In [17], author(s) have presented predictive modeling approach on the Bombay stock exchange, the Dow Jones Industrial Average, the Hang Seng Index, the NIFTY 50, the NASDAQ, and the NIKKEI daily index prices. Equity markets have been analyzed in detail to check whether they follow pure random walk–based models. Dynamically changing stock market prices have been studied thoroughly. Predictive modeling uses different machine learning algorithms for building frameworks to predict future index prices. Some of the algorithms are: adaptive neuro-fuzzy inference system, support vector regression, dynamic evolving neuro-fuzzy inference system, Jordan neural network, and random forest. As per analysis conducted, prices can be forecasted beforehand using effective predictive modeling algorithms. Pure random walls are not followed. According to author(s), this research can provide business markets with huge profits. It can also help in developing different trading strategies.

Predictive analytics is used for better decision making and thereby having better results. Due to complexity and dynamic nature, these predictive analytics techniques cannot be used on cloud computing. Therefore, in [18], author(s) have devised methods for predictive analytics on cloud systems based on quality of service parameters. Along with the models, some simulation tools have been developed for graph transformation systems, models for self-adaptive systems, and adaptation mechanisms.

In [19], author(s) have proposed an error recovery solution for a micro electrode dot array using real-time sensing and droplet operations. These methodologies work on predictive analytics and adaptive droplet aliquot operations. Analytics has been used for determining minimum droplet aliquot size that can be obtained on a micro electrode dot array biochip. A predictive model has been proposed describing completion rate of mixing leading to mixed error recovery control flow.

In [20], author(s) have used a Hadoop platform for processing k nearest neighbors and classification. Advantage of using Hadoop is that it is scalable, easy to use, and robust to node failures. Aim of the analysis was to find fast processing techniques of classification for enhancing classification accuracy, utilizing logistic regression, and k nearest neighbors. Further, the analysis enhances the rate of accuracy, true positive, false positive, precision recall, sensitivity, and the specificity rate of classification. Prediction-based classification process involves the following four steps:

1. Selecting appropriate data set.
2. Temporal distribution of data set.

3. Applying k nearest neighbor algorithm.
4. Use logistic regression technique.

K nearest neighbor approach is used for accuracy to identify appropriate model and classification for classifying data perfectly.

In [21, 22], author(s) have done a statistical survey on data mining techniques, described the importance of big data, its techniques, and applications. To show the effectiveness of data mining techniques, simulations have been carried out of a real-time data set in MongoDB.

In [23], author(s) have used prediction system for energy consumption evaluation, which would help in making better energy utilization. Author(s) have proposed four models for the same, which are used at different stages. The four models are: energy profile model, energy correlation analysis model, new energy profile model, and new energy pprofile model. Data used for analysis was provided by the national energy data system.

In [24], trending patterns have been proved for the Twitter data set using big data technology – spark streaming. Tweets and popular hashtags have been analyzed. Hashtagging helps to focus on a particular item. The research was conducted to find the fake groups working with hashtagging to gain popularity.

15.3 Comparative Study

In Section 15.2, various applications have been discussed in the field of predictive modeling and social network analysis. Table 15.1 summarizes the application name and the area of application.

15.4 Simulation and Analysis

The data set [14] used is a real time data set from Cambridge University. In a three-day conference, nine people were given nodes and their movement was recorded for all the three days, that is, when these nodes came in contact with each other, duration of contact, etc. The data set describes the contacts that were recorded by all devices distributed during this experiment.

A few random rows of the data set are given below.

Source node	Destination node	Initial time of contact	End time of contact	No of time of contact	Time between two consecutive meeting
1	8	121	121	1	0
1	3	236	347	1	0
1	4	236	347	1	0
1	5	121	464	1	0
1	8	585	585	2	464

Table 15.1 Comparison table of literature work.

Paper reference number	Application domain	Application
3	Predictive modeling	Industry for decision-making process
4	Predictive modeling	Stock market prices, election, financial crisis
5	Predictive modeling	Education
6	Predictive modeling	Event patterns
7	Predictive modeling	Sports
8	Predictive modeling	Python
9	Social network analysis	Twitter data set
10	Social network analysis	Data fusion
11	Predictions and classification	Education
12	Machine learning and predictive modeling	Healthcare
13	Social network analysis	Mixed integer linear programming
14	Predictive modeling	Business development, stock market
15	Predictive analytics	Decision making
16	Predictive modeling	Error recovery solution for micro electrode dot array
17	Prediction-based classification	K nearest neighbor
20	Predictive modeling	Energy consumption evaluation

15.4.1 Few Analyses Made on the Data Set Are Given Below

15.4.1.1 Duration of Each Contact Was Found

Source node	Destination node	Initial time of contact	End time of contact	No of time of contact	Time between two consecutive meeting	Duration of each contact
1	8	121	121	1	0	0
1	3	236	347	1	0	111
1	4	236	347	1	0	111
1	5	121	464	1	0	343
1	8	585	585	2	464	0

15.4.1.2 Total Number of Contacts of Source Node with Destination Node Was Found for all Nodes

For example; analysis for node 1 is given below. Total number of times node 1 came in contact with each node was found.

Number of contacts with each node		
Source node	Destination node	Number of contacts
1	2	19
1	3	8
1	4	17
1	5	16
1	6	2
1	7	27
1	8	24
1	9	17
Total contacts of node 1		130

According to this analysis, the most active node was node 2 and the most passive node was node 5.

15.4.1.3 Total Duration of Contact of Source Node with Each Node Was Found
For example, duration of contact for first node with each other node is given below.

Total duration of contact with each node		
Source node	Destination node	Total duration of contact with each node
1	2	1241
1	3	19 708
1	4	38 963
1	5	6556
1	6	7
1	7	2921
1	8	1176
1	9	1212
Total duration of node 1		71 784

According to this analysis, the most active node was node 3 and the least active was node 5.

15.4.1.4 Mobility Pattern Describes Direction of Contact and Relation Between Number of Contacts and Duration of Contact
It can be inferred from the data whether the source node is coming in contact (for interaction) with the destination node in a single time or if the nodes are contacting again and again.

Only those nodes have been considered that had interaction in less than 50% of total number of contacts, that is, they might have come in contact with each other many times but the number of times they interacted is less than half of total number of contacts. For example, suppose node A contacted B 10, but, out of 10 times node A interacted with B only 4 times and the rest 6 times they only came close to each other but did not have any interaction. These nodes were considered for this analysis.

The analysis is given below.

	Mobility pattern			
Source node	Destination node	NOC	Total DOC	No of interactions
1	2	19	1241	6
1	6	2	7	1
1	7	27	2921	127
1	8	24	1176	7
1	9	17	1212	4
2	1	32	4803	13
2	5	11	595	3
2	8	10	1597	4
3	5	3	0	0
3	8	2	579	1
5	2	2	0	0
5	3	1	589	1
5	8	16	2269	7
5	9	2	4	1
6	1	3	0	0
7	1	26	2185	10
7	4	16	55 761	7
7	5	3	4	1
7	6	13	2728	5
7	8	1	0	0
7	9	8	6364	3
8	3	13	844	5
8	5	23	1210	6
8	7	7	1277	3
9	1	3	122	1
9	5	1	0	0
9	8	1	0	0

15.4.1.5 Unidirectional Contact, that is, Only 1 Node is Contacting Second Node but Vice Versa Is Not There

Such nodes are:

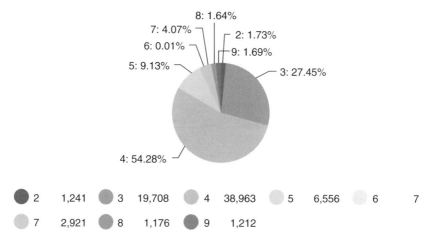

15.4.1.6 Graphical Representation for the Duration of Contacts with Each Node is Given below

Duration of contact of node 1 with each node:

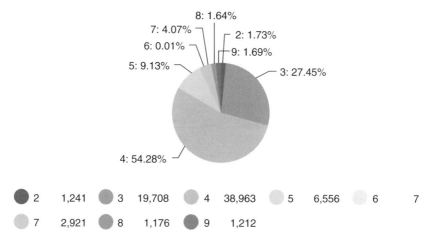

Duration of contact of node 2 with each node:

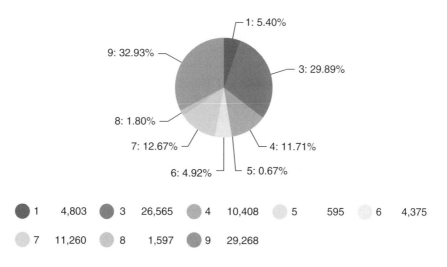

Duration of contact of node 3 with each node:

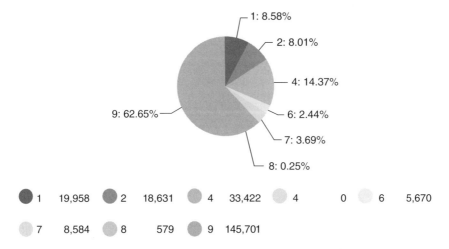

Duration of contact of node 4 with each node:

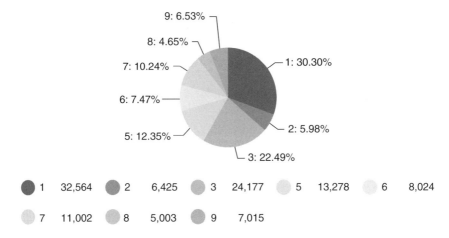

Duration of contact of node 5 with each node:

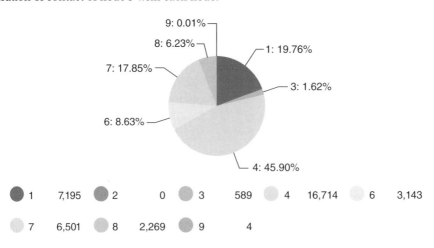

Duration of contact of node 6 with each node:

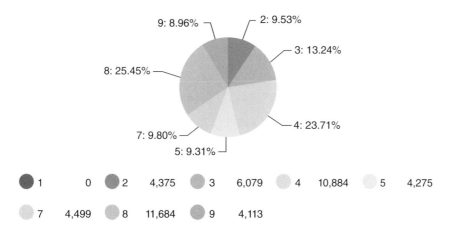

Duration of contact of node 7 with each node:

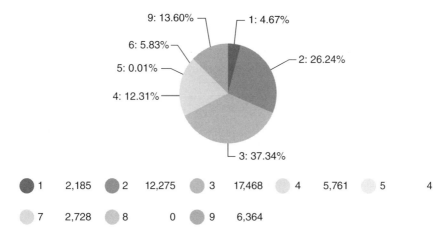

Duration of contact of node 8 with each node:

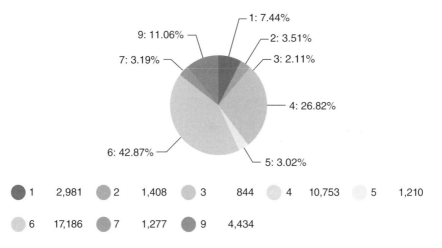

Duration of contact of node 9 with each node:

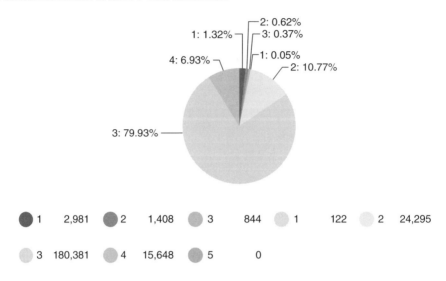

15.4.1.7 Rank and Percentile for Number of Contacts with Each Node

Rank defines relative position of an element among a list of objects [24]. Percentile defines the percentage below which the given value falls in the group of observations [25–29].

Rank and percentile has been calculated using inbuilt data analysis feature of Microsoft Excel.

Rank and percentile for number of contacts of node 1 with each node:

	Number of contacts with each node	
Source node	Destination node	Number of contacts
1	2	19
1	3	8
1	4	17
1	5	16
1	6	2
1	7	27
1	8	24
1	9	17
Total contacts of node 1		130

Point	Column 1	Rank	Percent (%)
6	27	1	100.00
7	24	2	85.70
1	19	3	71.40
3	17	4	42.80
8	17	4	42.80
4	16	6	28.50
2	8	7	14.20
5	2	8	0.00

Rank and percentile for number of contacts of node 2 with each node:

	Number of contacts with each node	
Source node	Destination node	Number of contacts
2	1	32
2	3	59
2	4	24
2	5	11
2	6	20
2	7	51
2	8	10
2	9	34
Total contacts of node 2		241

Point	Column 1	Rank	Percent (%)
2	59	1	100.00
6	51	2	85.70
8	34	3	71.40
1	32	4	57.10
3	24	5	42.80
5	20	6	28.50
4	11	7	14.20
7	10	8	0.00

Rank and percentile for number of contacts of node 3 with each node:

Number of contacts with each node		
Source node	**Destination node**	**NOC**
3	1	6
3	2	43
3	4	29
3	5	3
3	6	7
3	7	22
3	8	2
3	9	66
Total contacts of node 3		178

Point	**Column 1**	**Rank**	**Percent (%)**
8	66	1	100.00
2	43	2	85.70
3	29	3	71.40
6	22	4	57.10
5	7	5	42.80
1	6	6	28.50
4	3	7	14.20
7	2	8	0.00

Rank and percentile for number of contacts of node 4 with each node:

Number of contacts with each node		
Source node	**Destination node**	**NOC**
4	1	36
4	2	16
4	3	19
4	5	18
4	6	28
4	7	16
4	8	34
4	9	10
Total contacts of node 4		177

Point	Column 1	Rank	Percent (%)
1	36	1	100.00
7	34	2	85.70
5	28	3	71.40
3	19	4	57.10
4	18	5	42.80
2	16	6	14.20
6	16	6	14.20
8	10	8	0.00

Rank and percentile for number of contacts of node 5 with each node:

Number of contacts with each node		
Source node	Destination node	NOC
5	1	13
5	2	2
5	3	1
5	4	11
5	6	7
5	7	17
5	8	16
5	9	2
Total contacts of node 5		69

Point	Column 1	Rank	Percent (%)
6	17	1	100.00
7	16	2	85.70
1	13	3	71.40
4	11	4	57.10
5	7	5	42.80
2	2	6	14.20
8	2	6	14.20
3	1	8	0.00

Rank and percentile for number of contacts of node 6 with each node:

Number of contacts with each node		
Source node	Destination node	NOC
6	1	3
6	2	12
6	3	8
6	4	31
6	5	6
6	7	17
6	8	11
6	9	9
Total contacts of node 6		97

Point	Column 1	Rank	Percent (%)
4	31	1	100.00
6	17	2	85.70
2	12	3	71.40
7	11	4	57.10
8	9	5	42.80
3	8	6	28.50
5	6	7	14.20
1	3	8	0.00

Rank and percentile for number of contacts of node 7 with each node:

Number of contacts with each node		
Source node	Destination node	NOC
7	1	26
7	2	36
7	3	45
7	4	16
7	5	4
7	6	13
7	8	1
7	9	8
Total contacts of node 7		149

Point	Column 1	Rank	Percent (%)
3	45	1	100.00
2	36	2	85.70
1	26	3	71.40
4	16	4	57.10
6	13	5	42.80
8	8	6	28.50
5	4	7	14.20
7	1	8	0.00

Rank and percentile for number of contacts of node 8 with each node:

Number of contacts with each node		
Source node	**Destination node**	**NOC**
8	1	24
8	2	6
8	3	13
8	4	35
8	5	23
8	6	39
8	7	7
8	9	24
Total contacts of node 8		171

Point	Column 1	Rank	Percent (%)
6	39	1	100.00
4	35	2	85.70
1	24	3	57.10
8	24	3	57.10
5	23	5	42.80
3	13	6	28.50
7	7	7	14.20
2	6	8	0.00

Rank and percentile for number of contacts of node 9 with each node:

Number of contacts with each node		
Source node	Destination node	NOC
9	1	3
9	2	34
9	3	71
9	4	25
9	5	1
9	6	8
9	7	9
9	8	1
Total contacts of node 9		152

Point	Column 1	Rank	Percent (%)
3	71	1	100.00
2	34	2	85.70
4	25	3	71.40
7	9	4	57.10
6	8	5	42.80
1	3	6	28.50
5	1	7	0.00
8	1	7	0.00

15.4.1.8 Data Set Is Described for Three Days Where Time Is Calculated in Seconds. Data Set can be Divided Into Three Days. Some of the Analyses Conducted on the Data set Day Wise Are Given Below

i. Total Duration of Contact and Total Number of Contacts for Node 2 on Day 1 with all the other nodes.

Source	Destination	Duration	No of times of contact
2	1	358	6
2	3	115	3
2	4	237	3
2	5	7	2
2	6	241	4
2	7	1209	13
2	8	1438	3
2	9	0	0

Similar analyses can be conducted for other nodes for all three days.

ii. Total Duration of Contacts and Total Number of Contacts for three days individually for Node 1.

Node 1 duration of contact			
Destination	**Day 1**	**Day 2**	**Day 3**
2	21	1220	0
3	462	19 246	0
4	18 596	19 775	592
5	6556	0	0
6	7	0	0
7	2182	739	0
8	1176	0	0
9	0	111	1101

Node 1 total contacts			
Destination	**Day 1**	**Day 2**	**Day 3**
2	8	11	0
3	2	6	0
4	8	8	1
5	16	0	0
6	1	1	0
7	15	12	0
8	20	0	4
9	0	4	12

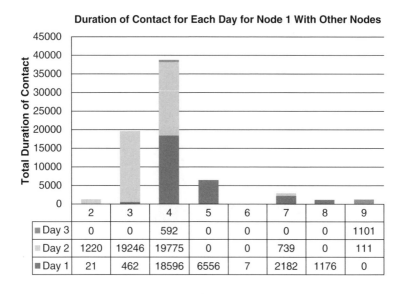

Duration of Contact for Each Day for Node 1 With Other Nodes

	2	3	4	5	6	7	8	9
■ Day 3	0	0	592	0	0	0	0	1101
Day 2	1220	19246	19775	0	0	739	0	111
■ Day 1	21	462	18596	6556	7	2182	1176	0

Total Number of Contacts Each Day for Node 1 With Other Nodes

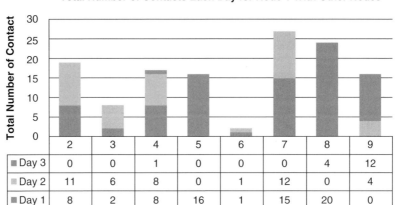

	2	3	4	5	6	7	8	9
■ Day 3	0	0	1	0	0	0	4	12
■ Day 2	11	6	8	0	1	12	0	4
■ Day 1	8	2	8	16	1	15	20	0

Similar analyses can be conducted for other nodes.

iii. Total Duration of Contacts and Total Number of Contacts for all three days for Node 1.

Source	Destination	Duration	No of times of contact
1	2	1241	19
1	3	19 708	8
1	4	38 963	17
1	5	6556	16
1	6	7	2
1	7	2921	27
1	8	1176	24
1	9	1212	16

Total number of times Node 1 contacted other nodes in 3 days

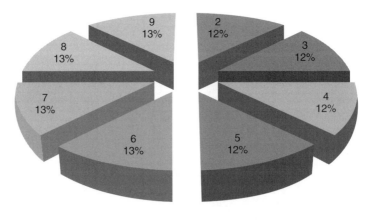

Total Duration of Contact of Node 1 for 3 days

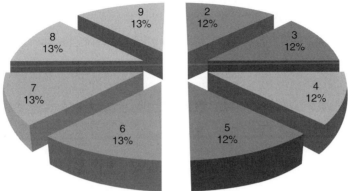

15.5 Conclusion and Future Work

Predictive analytics is a vast topic and can be used in many fields for future predictions and analytics. Predictive analytics has led to many developments and will continue to do so in the future.

In this paper, the data set chosen is from Cambridge University where social contact between people varies from the way people interact in India. Hence, in future a data set within India will be chosen for analysis to provide much better results in accordance with Indian society.

Secondly, the study of social network analysis can be extended to any field where there is some human interaction, like in a party, college, hospital, offices, etc.

In future, implementations in Python and predictive analysis will also be carried out.

References

1 Sivasakthi, M. "Classification and prediction based data mining algorithms to predict students' introductory programming performance." Inventive Computing and Informatics (ICICI), International Conference on. IEEE, 2017.

2 Nithya, B., and V. Ilango. "Predictive analytics in health care using machine learning tools and techniques." Intelligent Computing and Control Systems (ICICCS), 2017 International Conference on. IEEE, 2017.

3 Srivastava, Atul, and Dimple Juneja Gupta. "Social network analysis: Hardly easy." Optimization, Reliabilty, and Information Technology (ICROIT), 2014 International Conference on. IEEE, 2014.

4 Earley, S. (2014). Big data and predictive analytics: What's new? *IT Professional* 16 (1): 13–15.

5 Bhardwaj, Vibha, and Rahul Johari. "Big data analysis: Issues and challenges." Electrical, Electronics, Signals, Communication and Optimization (EESCO), 2015 International Conference on. IEEE, 2015.

6 Dorogov, A. Yu. "Technologies of predictive analytics for big data." Soft Computing and Measurements (SCM), 2015 XVIII International Conference on. IEEE, 2015.

7 Brown, D.E., Abbasi, A., and Lau, R.Y.K. (2015). Predictive analytics: predictive modeling at the micro level. *IEEE Intelligent Systems* 30 (3): 6–8.

8 Rajni, J. and Malaya, D.B. (2015). Predictive analytics in a higher education context. *IT Professional* 17 (4): 24–33.

9 Yeon, H. and Jang, Y. (2015). Predictive visual analytics using topic composition. In: *Proceedings of the 8th International Symposium on Visual Information Communication and Interaction*. ACM.

10 Grover, Purva, and Rahul Johari. "PAID: Predictive agriculture analysis of data integration in India." Computing for Sustainable Global Development (INDIACom), 2016 3rd International Conference on. IEEE, 2016.

11 Zhao, Baojin, and Lei Chen. "Prediction Model of Sports Results Base on Knowledge Discovery in Data-Base." Smart Grid and Electrical Automation (ICSGEA), International Conference on. IEEE, 2016.

12 Xu, Z. et al. (2016). Python predictive analysis for bug detection. In: *Proceedings of the 2016 24th ACM SIGSOFT International Symposium on Foundations of Software Engineering*. ACM.

13 Iglesias, J.A. et al. (2016). Social network analysis: Evolving Twitter mining. In: *2016 IEEE International Conference on Systems, Man, and Cybernetics (SMC)*. IEEE.

14 Farasat, A. et al. (2016). Social network analysis with data fusion. *IEEE Transactions on Computational Social Systems* 3 (2): 88–99.

15 Varshney, Karishma, Rahul Johari, and R. L. Ujjwal. "BiDuR: Big data analysis in UAF using R-voting system." Reliability, Infocom Technologies and Optimization (Trends and Future Directions)(ICRITO), 2017 6th International Conference on. IEEE, 2017.

16 Pirim, Harun. "Mathematical programming for social network analysis." Big Data (Big Data), 2017 IEEE International Conference on. IEEE, 2017.

17 Ghosh, I., Sanyal, M.K., and Jana, R.K. (2017). Fractal inspection and machine learning-based predictive Modelling framework for financial markets. *Arabian Journal for Science and Engineering*: 1–15.

18 De Oliveira, Patricia Araújo. "Predictive analysis of cloud systems." Software Engineering Companion (ICSE-C), 2017 IEEE/ACM 39th International Conference on. IEEE, 2017.

19 Zhong, Zhanwei, Zipeng Li, and Krishnendu Chakrabarty. "Adaptive error recovery in MEDA biochips based on droplet-aliquot operations and predictive analysis." Computer-Aided Design (ICCAD), 2017 IEEE/ACM International Conference on. IEEE, 2017.

20 Raj, J. Adarsh Samuel, J. Lenin Fernando, and Y. Sunil Raj. "Predictive Analytics on Political Data." Computing and Communication Technologies (WCCCT), 2017 World Congress on. IEEE, 2017.

21 Samridhi Seth, Rahul Johari. "Evolution of Data Mining Techniques: A case study using MongoDB." Proceedings of the 12th INDIA Com; INDIACom-2018; IEEE Conference ID: 42835 2018 5th International Conference on "Computing for Sustainable Global Development", 14th – 16th March, 2018 2018.

22 Samridhi Seth, Rahul Johari. "Statistical Survey Of Data Mining Techniques: A Walk Through Approach Using MongoDB." International Conference On Innovative Computing And Communication, 5th-6th May, 2018.

23 Sun, Z. et al. (2018). Energy evaluation and prediction system based on data mining. In: *2018 33rd Youth Academic Annual Conference of Chinese Association of Automation (YAC)*. IEEE.

24 Garg, P. et al. (2018). Trending pattern analysis of twitter using spark streaming. In: *International Conference on Application of Computing and Communication Technologies*. Singapore: Springer.

25 https://crawdad.org

26 https://en.wikipedia.org/wiki/Predictive_analytics

27 https://www.predictiveanalyticstoday.com/what-is-predictive-analytics/#content-anchor

28 https://en.wikipedia.org/wiki/Rank

29 https://en.wikipedia.org/wiki/Percentile

16

Intelligent Data Analysis for Medical Applications

Moolchand Sharma[1], Vikas Chaudhary[2], Prerna Sharma[1], and R. S. Bhatia[3]

[1]*Department of Computer Science & Engineering, MAIT, Delhi, India*
[2]*Department of Computer Science & Engineering, KIET, Ghaziabad, India*
[3]*Department of Electrical Engineering, NIT, Kurukshetra, India*

16.1   Introduction

Most developing countries are faced with increased patient mortality from several diseases primarily as a result of the inadequacy of medical specialists. This insufficiency is impossible to overcome in a brief span of time. However, institutions of higher learning are capable of taking immediate action by producing as many doctors as possible. Many lives will be impacted while waiting for the student's journey of becoming a doctor and then a specialist. Currently, for proper diagnosis and treatment, it is required that patients consult specialists. Certain medical practitioners do not possess enough experience or expertise to handle certain high-risk diseases. Nonetheless, the diseases may increase in severity by the time patients get access to specialists, which may take a few days, weeks, or even months. A patient may suffer for the rest of their life if the high-risk disease cannot be addressed at an early enough stage.

With the use of computer technology, the number of mortality and wait times to see specialist can be effectively reduced. Doctors can be assisted in making decisions without consultation with specialists by developing specialized software that emulates human intelligence. Software cannot replace a specialist, yet it can aid doctors and specialists in the process of predicting a patient's condition and diagnosis from certain rules/experiences. Patients can then be shortlisted for further treatment on the basis of high-risk factors or symptoms or if they are predicted to be highly affected with certain diseases or illness [1–3]. It can result in the reduction of time, cost, human expertise, and medical error by applying intelligent data analysis (IDA) techniques in medical applications.

A program was developed to make clinical decisions that help health professionals. It is called the medical decision support system (MDSS). The process involves dealing with medical data and knowledge domains in order to diagnose a patient's condition and recommends a suitable treatment for a particular patient. Another system developed for assisting the monitoring, managing, and interpreting of a patient's medical history as well as providing aid to the patient and the medical practitioner is a patient-centric health information

Intelligent Data Analysis: From Data Gathering to Data Comprehension,
First Edition. Edited by Deepak Gupta, Siddhartha Bhattacharyya, Ashish Khanna, and Kalpna Sagar.

system. This system would improve decision making in medical science, increase patient compliance, and iatrogenic diseases, and medical errors could be reduced.

IDA research in medicine is, just like any research, targeted at direct or indirect enhancements in provisions of health care. Techniques and methods can only be tested through test cases that form problems of the real world. Functional IDA proposals for medicine are followed by conditions that specify the range of real-life functions discussed by such propositions, a resulting system's in-depth interpretation thus constitutes a crucial feature.

The use of IDA systems in the clinical environment is another consideration of IDA. It is used as a well-informed assistant in order to gather data and comprehend it in order to assist physicians in performing their tasks with greater efficiency and effectiveness. The physician will diagnose the disease in a stipulated time and perform appropriate actions if they get correct information at the right time. The revolution in information has built a viable collection and storage of a large volume of data across various origins on information media. This data can be single-case (single patient) or multiple-case based (multiple patients). Utilization of raw data in problem solving is infeasible due to its noisiness or insufficient data. However, if important information is picked out by computationally intelligent ways, the conversion of data to mine is possible. After mining, relevant and operational knowledge/information, extracted at the correct level, is then ready to use and aid in the diagnosis of a patient by a physician. Rapidly emerging globality of information and data arises to support the following critical issues:

(i) The provision of standards in terminology, vocabularies, and formats to support multilinguality and sharing of data.
(ii) Standards for interfaces between different sources of data.
(iii) Integration of heterogeneous types of data, including images and signals.
(iv) Standards for the abstraction and visualization of data.
(v) Reusability of data, knowledge, and tools.
(vi) Standards for electronic patient records.

A panel discussion of artificial intelligence (AI) in (AIME 97) has identified these above-cited issues. The exercise of determining guidelines of any variety is a convoluted duty. Nonetheless, some guidelines are essential to perform intercommunication and thus integration between different origins of data. Correct patient management is the most important aim of health care, which is directly related to proper utilization of a very invaluable resource that is clinical data. Hence it is thoroughly justified that investment is done in the process of development of appropriate IDA methods, techniques, and tools that can be used for analysis of clinical data. A major focus needs to be given to research by relevant communities and groups.

In the 1990s, we witnessed enhancement in studies in intelligent systems for the effective utilization of systems based on current requirements. Various researches were done by combining two or more techniques and the function of the systems as utilized for ensuring the performance of the system. An intelligent referral system for primary child health care was established with a primary objective to reduce mortality in children, specifically in rural areas. The system succeeded in addressing common pediatric complaints after a thorough consideration of crucial risk factors, such as immunization, weight monitoring, development nutrition, and milestones. The intelligent referral system for primary child health care used an expert system for the purpose of collecting data of a patient's history. Various expert systems have been made, viz. HERMES, an expert system dealing with diagnosis of chronic liver diseases. SETH is an expert system for administration of curing acute drug poisoning. PROVANES is a hybrid expert system for critical patients under anesthesiology, and ISS for diagnosis of sexually transmitted diseases [11].

16.1.1 IDA (Intelligent Data Analysis)

Every human activity is witnessing a deepening gap between the process of data generation and data comprehension. It is extremely crucial that this gap is bridged, especially in the field of medical science as decision making in this field has to be augmented by arguments from medical practitioners along with information of trends and regularities derived from data. IDA is one of the most pressing issues pertaining to artificial intelligence and information that can overcome the gap of decision making. Potentially valuable information or knowledge, which is previously unknown has been revealed by IDA on huge amounts of data. IDA can be viewed as a kind of decision assistance technique. IDA systematically derives necessary information and interesting models from data present online with the use of machine learning, pattern recognition, artificial intelligence, statistics, database, and visualization technology; this information is then used for the process of aiding decision makers to achieve correct conclusions [4]. The stages involved in the process of IDA are:

(1) Data construction.
(2) Data mining or rule finding.
(3) Result explanation and validation.

Data construction constitutes selection of requisite data from relatable data source and incorporating this in a database is then used in the process of data mining. Data mining involves figuring rules enclosing data in a database by use of defined techniques and algorithms. Finally, the third stage of result validation mandates the examination of these rules and results in the explanations providing comprehensible, reasonable, and intuitive descriptions by logical reasoning [10].

Extracting useful knowledge is the major goal of IDA; the process demands a combination of analysis, classification, conversion, extraction, organization, reasoning, and other techniques. It is good and challenging figuring out appropriate methods to resolve the difficulties experienced in the process. Continued challenges are presented by IDA methods and tools, as well as the authenticity of obtained results.

AI techniques research and applications in data analysis provide the main framework of IDA issue examination. These applications include every part of data visualization, data preprocessing, i.e., editing, sampling, fusion, transformation, database mining techniques, data engineering, applications and tools, statistical pattern recognition, use of domain knowledge in data analysis, knowledge filtering, and machine learning, postprocessing, evolutionary algorithms, big data applications, neural nets, and fuzzy logic [5]. Specifically, papers where development of AI is discussed are preferred; this includes AI development–related data analysis techniques, architectures, methodologies, and applicability to numerous domains.

To provide methods that facilitate data understanding is the main aim of IDA, so, although its proximate goal is not classification or prediction by construction of data models, IDA must frame the relations and principles to provide assistance in making a decision in complex situations. It should be clear that IDA is neither data mining nor knowledge discovery in databases (KDD), although it may have a portion of data mining methods and some its process are also KDD. The union of background knowledge (BK) and data analysis is the main objective of IDA for providing the user with information.

Preparation of data and its mining: the problem that comes in preparation of data and its mining is the lack of resources, or could be said lack of true resources. The major sources of data preparation and data mining are:

a. Medical records taken for research purposes
b. Textbooks
c. Insurance claims
d. Hospital's/clinic's personal records

a. *The medical records taken for research purposes*: Such data is a secondary data and not primary data, i.e., it depends on the source of data collection, aka humans. Such data could then be age-specific, gender-specific, lifestyle-specific, and diet-specific, thus depending upon physical and chemical properties and aspects that thus could not be trusted solely. A very common aspect to consider this is the exposure to radiation, as most of the human beings and the animals are now exposed to radiation, and each individual's exposure level depends upon 1. residential and working locations from the radiation source 2. past diseases, as some medications could have eventually built immunity from certain radiation exposure, and 3. diet, physical health, and exercise/yoga; thus, for research purposes, data could prove to be an effective data set but can't be fully trusted.

b. *Textbooks*: Theoretical information may stand useful or else in practical applications; for example, not all fevers are the same, and not all heart surgeries are the same (some hospitals do require the patient's family's legal approval before operating on serious cases). The information could be from old sources, and since, in today's world most of the people are exposed to radiation, thus some or other changes do exist in them as a result of which some medical operation couldn't be done, but such people are a very crucial source of information for future purposes and deeper research in the medical field.

c. *Insurance claims*: Insurance claims could be falsely made for profit and business purposes; there have been way too many cases filed as a result of false medical claims. Data abstracted from such sources could lead to false information, which would not only decrease the efficiency of the IDE-based model, but would also increase the input of financial- and time-based resources, and could lead to the patient's death or some serious complication otherwise.

d. *Hospital's/clinic's personal records*: Hospital's/clinic's personal records are some of the most vital sources for data preparation and data mining, but most of the conversations that take place in such environments are casual, i.e., they take place in personal chats like text SMS, etc., or are face-to-face talks, thus no record exists of such crucial data. Sometimes it's important to ensure certain information doesn't come out on paper in order to prevent complexities for a medical professional's personal career and as well as the hospital's reputation. One such example might be that information like a sudden critical abnormality arising in the patient's body, such as internal bleeding, swelling, etc., and the doctor's immediate response to it, which comes from years of practice. No records of such experienced and immediate thoughts that come to a doctor's mind could ever be made, as they are like solutions to a problem that come to one's mind under pressure only.

One thing that is common in all of the above is that data that would be true would surely be present in all the records; hence IDA researchers seek for the information that is common in all.

And one of the major problem besides all the above is that the model once prepared would eventually fall back, i.e., become outdated within months, because the medical field is the most dynamic field, and situations vary from country to country, human to human, and also the research on a micro- and macro- level in a medical field grows exponentially each day. Old treatment patterns could suddenly become new with certain changes and the latest in treatments could fail due to not being in common use because of a change in medicine chemical composition and new machineries.

16.1.1.1 Elicitation of Background Knowledge

The elicitation of background data or an effective data set that could mark greater efficiency is one of the most important issues that needs to be resolved in order to make IDA and its related models efficient. IDA philosophy suggests one of Bayesian statistics: posterior knowledge is obtained by altering previous knowledge (background data). The main issue is that physicians require appropriate backgrounds, if not, then with at least effective backgrounds, but they are unable to provide prior experience or priors. This problem can be avoided by Bayesian statistics with the use of flat priors that are incapable to assert background data, hence restricting the application of Bayesian approach in the settings of an IDA system. The Bayesian community have provided some answers to this problem, priors being obtained through intervals/ranges updated on the basis of the collection of data. Other solutions exist but they refuse the Bayesian statements, and first are elicited followed by updating the background data using different approaches, similar to hierarchical structuring.

16.1.2 Medical Applications

IDA research aims directly or indirectly at the enhancement in provision of health care similar to any other research in the medical field. Benchmark tests for these techniques and processes can essentially be problems for the real world. IDA proposals that might prove to be viable for the medical field must have detailed needs that outline the range of problems from the real world. Critical aspects are formed by in-depth examination of resulting systems explained in such proposals. The clinical environment is another consideration of IDA systems. Reducing differences between data comprehension and data gathering is the main role of IDA, which enables a physician to achieve his job in efficient and effective manner. Armed with the right information at the right time, a physician becomes well equipped to arrive at the correct diagnosis or to carry out the right option in the stipulated time frames. The revolution in information has built a viable collection and the storage of this large volume of data across various origins on information media is sizeable. This data can be categorized as multiple-case based for multiple patients or single-case based for a single patient. Utilization of unanalyzed data in problem solving is infeasible due to its noisiness or insufficient data. However, if important information is picked out by computationally intelligent ways, a conversion of data to mine is possible. After mining, relevant and operational knowledge/information, extracted at the correct level, is then ready to use and aid in the diagnosis of a patient by a physician.

AI in the medical field was primarily related to the development of medical expert systems in the late seventies and eighties whose primary aim was aiding diagnostic decision making in medical domains that were specialized [6]. MYCIN by Shortliffe was a keystone research in this field that was accompanied by various attempts that led to a specific diagnostic and rigoristic expert system, for example: PUFF, PIP, VM, CASNET, HODGKINS, PIP, HEADMED, CENTAUR, ONCOCIN, MDX, ABEL, GALEN, and various alternates. Systems made to support diagnosis in internal medicine, which are the most generalized and comprehensive: INTERNIST-1, which supports expert-defined rules and its follower, CADUCEUS, which along with INTERNIST-1 constitutes a structure of pathophysiological states reflecting wide expertise about the issues [7]. Expert system research deals with main obstacles like knowledge acquisition, as well as its reasoning, representation, and explanation at this early stage.

From the early days of knowledge-based systems, rules were formulated, and particularly when using expert systems, was a prime conformity for expression of knowledge in a symbolic way. The advantage of rules is uniformity, simplicity, transparency, and inferential ease, which has resulted in them becoming the most commonly adopted representation of real-world information. Rules expressed at a level of abstraction are derived from domain experts that are correct from the expert's perspective and are certainly understandable to the expert since these formulations are their rule of thumb. However, rules defined by humans carry a huge risk that they would come laced with the individual prejudice of an expert, although it may appear that a clear modular chunk of information is formed by each rule. The entire knowledge of the whole would be replete with gaps, inconsistencies, and other defects primarily because of their flat organizations (which is the absence of a global, hierarchical, and comprehensive organization of the rules).

Furthermore, the expert would provide essential knowledge required for focusing and guiding the learning. Irrespective of the fact that these rules are directly received or learned from experts, the layout has to be intuitive, acceptably expressive, and simple in order to be applied in a particular application.

16.2 IDA Needs in Medical Applications

The combination of clinical, medical, and pharmaceutical innovations and technology is one of the serious issues in intelligent data analytics in the health care industry. The industry that can never face a recession is health care; even in a recession, it would grow further because of the increase in medical demands during that time period. Moreover, being one of the best business sectors, people tend to need more to earn more, and eventually this need turns out to be helpful for mankind, as new advancements not only increase the rate of successful operations but also help in preventing various epidemics from spreading beforehand (a combination of IDA in the medical field and the environment).

As an outcome of increasing complexities, demands, researches, and advancements in technicalities, health care businesses are facing a dilemma to control a large amount of data over hard or soft copy formats. Thus, IDA favors such organizations directly and mankind indirectly by improving services and managing adversities in the health care sector (record storage and data processing).

One of the most important reasons, or which could be interpreted as the only sole purpose, of integrating data analysis with health care is to increase the care, save the life of a patient, and reduce the cost of health facilities offered.

16.2.1 Public Health

With the help of analytics approach issues like the study on disease types and patterns and other factors related to the disease, a patient's health (physical and mental) can be improved. Huge amounts of data help decide the necessities, services, and also helps predict and prevent future problems, which benefits people. Thus problems faced while operating on a patient wouldn't ever be encountered in the operation of some other patient, or even if faced, could be solved before it becomes fatal by utilizing AI (implemented from IDA).

16.2.2 Electronic Health Record

An electronic health record (EHR), or electronic medical record (EMR), is the systematized collection of patient and population electronically stored health information in a digital format. These records can be shared across different health care settings. EHRs may include a range of data, including demographics, medical history, medication, and allergies, and immunization statistics–like information.

16.2.3 Patient Profile Analytics

16.2.3.1 Patient's Profile
Individuals who can benefit a from particular approach are identified by patient profile analytics. This may incorporate changes in lifestyle such as diet, exercise, etc. Big data analytics in health care is used to derive patterns and insights dynamically from patient data and this information can be used to determine improved decisions. Consider, for instance, patients diagnosed with chronic illnesses. As of now, data segments of patients accounts for 78% of all health care spending, 91% of prescription-related data, 81% of in-patient stays, 98% of home health care visits, and 76% of physical visits.

Despite availability of such a large number of resources dedicated for the care of chronic illnesses patients, there is immense data that isn't addressed. Such nonadherence brings up serious consequences. If such large data could be analyzed efficiently, then not only the health care industry would flourish but also the personal health of each individual would improve.

16.3 IDA Methods Classifications

16.3.1 Data Abstraction

In general, the first step in the database design architecture process is data abstraction. A full database is a way to provide a complex system that is to be generated without developing a simplified framework first. It is possible for developers due to data abstraction to start from the very first element, i.e., data abstractions, followed by the addition of information and details to create the final model [8]. Abstraction, in general, is the process of reducing the number of characteristics of database into a set of essential characteristics.

The most efficient way to understand the data abstraction is to think of the way the word "abstract" is used when one talks about long documents. The abstract is the shortened, and the simplified, form. We often read it to get a glimpse of the facts before reading the entire document.

The three formal abstraction layers we usually use are:

- User model: How the user describes the database, it is very important to ensure fixed writing patterns in it.
- Logical model: Much more formal and more detailed database – often rendered as an entity relationship (ER) model.
- Physical model: Further addition of details – indexing, data types, etc.

16.3.2 Data Mining Method

Data mining is the process in which data after abstraction is mined, i.e., further filtration for data for precise and detailed information.

Techniques used for data mining:

1. *Anomaly detection*: Anomaly detection technique is used to decide if an entity is considerably diverse from others. Director of the BYU Data Mining Lab, Christophe Giraud-Carrier has outlined the explanation using an instance of how anomaly detection is used to monitor gas turbines in order to ensure that the turbines are functioning efficiently or at least properly. "These turbines have physics associated with them that predicts how they are going to function and their speed and all that jazz," Giraud-Carrier said. Sensors are used to monitor changes like temperature and pressure, etc., to keep a check if anything exceptional is observed over time.

2. *Association learning*: Association learning or market-basket analysis, is used to predict which things occur either in groups or pairs during specific time zones or after a period of time. An example given by Giraud-Carrier: "How Walmart reviews purchases by people who buy bread and also purchase milk, or people who buy baby formula also purchase diapers, thus placing them nearby for better sales and consumer's comfort. It's all about trying to find this association among products and the idea being that, again, you can hold this information." A similar ideology could also be seen in casinos, by offering free services to the players to ensure they don't give up playing.

3. *Cluster detection*: Cluster detection is identifying diverse subcategories or groups within a database. Machine learning algorithms are efficient in detecting significantly different or varying subsets within a data set.

   Giraud-Carrier said "Humans generally differentiate things into groups for easier access to each; keeping related items together. We put people in buckets: people we like, people we don't like; people we find attractive, people we don't find attractive," he said. "In your head you have these notions of what makes people attractive or not, and it's based on who knows what; your mind would automatically categorize the next persona you meet in one of the above-mentioned groups, without even realizing, and this is how **AI** based models of IDE works."

4. *Classification*: Classification manages things that already have some well-defined labels, unlike cluster detection. This is known as data training – information that is used to train the model on, or rather easily classify it, using some algorithm. For instance, differences between content found in spam and legitimate messages is identified by spam filters. This has been achievable through identification of large sets of e-mails as spam, and is followed by training of the model.

5. *Regression*: This technique works by making predictions from the connection, which exist in the database or data sets. For instance, future engagements of a particular user on Facebook can be predicted from the data of the user's history – likes, photo tags, comments, hash tags, frequently contacted friends, and there similar data and communications with different users as well, friend requests received and sent, and also various activity(s) on the site. Other instances could be the use of data on the relationship between education level and income of a family to anticipate the best choice of neighborhood or society to shift their residency to. Thus, it is very clear that regression allows all relations within the data set/database to be studied for predicting the future behavior/future outcome.

16.3.3 Temporal Data Mining

Temporal data mining (TDM) is referred to as TDM and is a method of searching for the existing patterns or correlations in huge temporal data sets. TDM was and is being evolved from data mining and is heavily influenced by the applications of temporal reasoning and temporal databases, which are two major sets of TDM. Major TDM techniques act on the available temporal database and convert the temporal data into a static representational data set, which is used to exploit the existing "static" machine learning techniques (i.e., providing efficient data for machine training purposes). Recently, there is an exponential growth in the field of the development of TDM techniques and strategies. An extension of the known decision trees induction algorithm put forward by Console et al. is specifically for the temporal dimensions, i.e., the temporal database. One of the advantages of using temporal decision trees is that the output of the algorithm is a tree that can immediately be used for pattern recognition purposes, however, a disadvantage of this method is that it cannot be applied to time intervals but only to time points [9].

Requirement of Temporal Abstraction: Temporal abstractions is based on the selected removal of time-oriented, i.e., time-based raw data; it is done using temporal reasoning methods [18]. It has become a center of interest in research as an intermediate and fundamental reasoning process for intelligent data preparation, i.e., intelligent analysis of temporal data in fields of tasks such as diagnosis and patient's health monitoring, which are the most crucial in the medical sector [10, 11].

16.4 Intelligent Decision Support System in Medical Applications

It can't be denied that in health care, the field of intelligent support systems is expanding with a rapid pace, or to be more precise, it is growing at an exponential rate. This pace is driven forward because of the availability of enormous amounts of information in health data that are unprecedented, yet useful in various sectors. But because of such exponential growth, health care is becoming more incomprehensive and more complex with time compared with its situation at the beginning of the decade. It is undeniable that the amount of new information entering the medical sector will make the existing platforms more complex. Decision support is an important step for decision makers in almost all the industries and sectors of life as well. Decision support systems help the officials/person responsible for decision making to gather and interpret the available information from the provided data set, thus building a core foundation for decision making and situation analysis [12, 13]. These systems range from a simple software to a complex deep neural network, as well as an artificial, and complicated information-based intelligence. Decision support processes can be database-oriented, text-centered, or spreadsheet-oriented in nature.

In the field of health care, clinical decision support systems (CDSS) plays a pivotal task. Evidence-based rules and clinical guidance from medical science are used by health care service providers in making routine clinical decisions. Through IDSS (intelligent decision support systems), interpretative analysis of massive patient data with intelligent information-based processes, allows nurses and doctors to gather information instantly and assists them in diagnosis and treatment [8, 19]. IDSS may be used in health care and in many fields, such as examining real-time data from various monitoring devices, analysis of patient and family history to be used in prognosis, reviews of general features and trends in medical record data sets, and several different fields. As below (Figure 16.1) the conventional flow of process shows how decision and implementation is done for decision making in IDSS and also below, Figure 16.2 shows the new flow of process for selecting the intelligent system/expert system for IDSS.

16.4.1 Need for Intelligent Decision System (IDS)

As intelligent decision systems (IDSs) can handle big and complex data, by supervised or unsupervised learning, they offer a flexible support to the medical decision-making process, which also benefits from the immense computing power available currently [17]. The intelligent medical data analysis became an unexpectedly fruitful niche within the CAD field for the intensive exploitation of ML algorithms. Regarding privacy, legality, and ethics, the European Union (EU), for instance, developed mechanisms to monitor these important issues:

- Privacy concerns.
- Resistance/job fears of doctors – assisting clinicians instead of replacing them.
- Making recommendations (supporting decisions), not directly and finally diagnosing.
- Inclusion of users (doctors, hospital managers, IT department) in the design and testing processes.
- Responsibility and ethics during the "best accuracy race" between proposed IDSs

16.4.2 Understanding Intelligent Decision Support: Some Definitions

Understanding intelligent clinical decision supports requires defining some important details and related conditions. To assist decision support in the field of health care, various

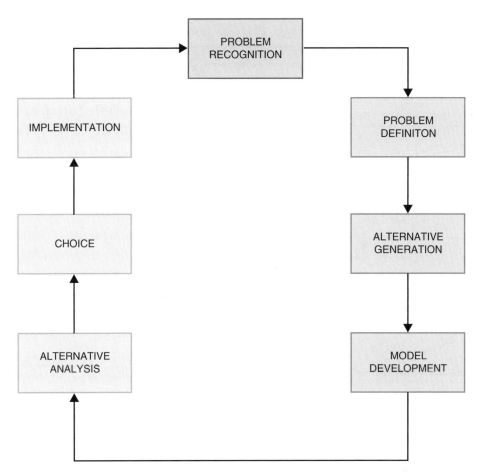

Figure 16.1 Conventional decision support system.

approaches are present while AI continues to be one of various methodological areas from which excellent and essential ideas are incorporated.

1. *Clinical decision support systems (CDSS)*: A computerized application that supplements and assists clinicians in informed decision making by giving them proof-based information of patient's data. CDSS involves three constituents: a language system, an information system, and a problem-processing system [14, 19].

2. *Intelligent decision support system (IDSS)*: It is determined by a system that assists decision making demonstrating intelligent behavior, which might involve reasoning as well as learning. Knowledge-based system, rule-based expert systems, or neural network systems are applied to achieve such learning and reasoning.

3. *Artificial intelligence (AI)*: AI refers to the art of arming computers with intelligence at par with humans. It can be attained by a combination of software and hardware systems that can execute jobs that are rule-based and involve decision making [15].

4. *Artificial neural network (ANN)*: Functional aspects and a structure of biological neural network is implied in a mathematical model. It mimics in a plain way, similarly as the

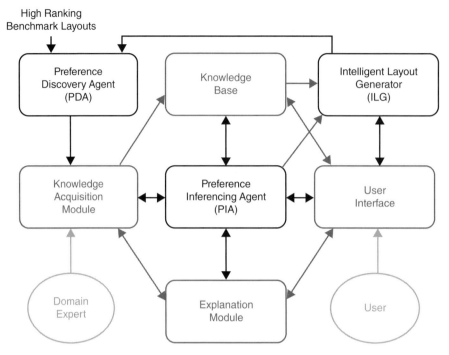

Figure 16.2 Intelligent system for decision support/expert analysis in layout design.

brain processes information. It consists of a number of highly connected processing elements (neurons/nodes). Artificial neural networks (ANNs) further improve to extract patterns/trends or recognize meaning from difficult data that are too complicated to be acknowledged by any other computer-based techniques [16]. After training, ANN can achieve expertise in the field of information it has been provided for examination. This expertise consequently predicts as well as answers questions in new situations.

16.4.3 Advantages/Disadvantages of IDS

- IDS can play an important role as "complementary medicine," in particular, due to the transparency of the data analysis.
- Various IDSs perform roughly the same. However, there are significant differences between them, mostly depending on the database and data type.
- Instead of using standalone algorithms, a better idea is to use a committee-machine approach, involving a group of algorithms working together to solve the problem.
- Last but not least, physicians usually do not trust automated diagnosis, and resent the idea that computers could replace them.
- When choosing between ML techniques, physicians prefer those with high explanation ability (e.g., Bayesian decision, classification tree).
- Clinical diagnosis still relies mostly on physicians' expertise and personal intuition.
- Medical diagnosis is "subjective" and depends both on available data and on the experience of the physician.

- Possible hospital regulations may prevent the extensive use of automated diagnosis.
- In enough cases, the ML algorithms are surpassed by classical medical methodologies (e.g., the biopsy) or human expertise.

16.5 Conclusion

Recent advancements of information systems in modern hospital and health care institutions has resulted ina surge in the volume of the medical data collected. This mandates that relevant systems are put in place that can be used to obtain relevant characteristics of medical data and how to address them. IDA is in its nascent research stages and was developed to overcome this requirement and primarily intends to reduce differences between data gathering and interpretation. The idea of a knowledge-driven data abstraction method was designed in the mid-eighties, however, active research in this field took place less than 10 years ago (thereby representing one of the early attempts), which was triggered by ongoing research in temporal reasoning in the medical field (thus the interest focused on temporal data abstractions).

To date, the two technologies are progressing exclusive of each other, even when both have a common aim of analyzing patient data in an intelligent manner, but for separate purposes: machine learning for extracting knowledge and data abstraction for formulation of a high degree of useful information on a single patient. As suggested above, the use of data abstraction in the context of machine learning is an idea worth exploring. The technology of data abstraction needs to become more mature. The main focus as of now was a derivation of temporal trends. This work needs to be continued but also be made applicable to deal with the different forms of abstraction, such as periodic abstractions. The huge raw data involved requires computational efficiency; and for systems to be operated in real time is a major concern.

The final stage of experimental design involves a meticulous choice of patients, their characteristics, and various hypotheses for confirmation. The introduction of data warehouses changes this selective approach to data collection and all data gathered is devoid of any pointed purpose. Still medical data collected and collated in warehouses is a very important resource for the probable unearthing of new knowledge. IDA researchers ought to provide methods for dual ends of this range, as the evidence of quality of such tools will be their applicability in medical institutions and their acceptability by experts of the medical field.

References

1 Yoshida, H., Jain, A., Ichalkaranje, A. et al. (eds.) (2007). *Advanced Computational Intelligence Paradigms in Healthcare*, vol. 1. Heidelberg: Springer.

2 Vaidya, S., Jain, L.C., and Yoshida, H. (eds.) (2008). *Advanced Computational Intelligence Paradigms in Healthcare*, vol. 2. Heidelberg: Springer.

3 Sardo, M., Vaidya, S., and Jain, L.C. (eds.) (2008). *Advanced Computational Intelligence Paradigms in Healthcare*, vol. 3. Heidelberg: Springer.

4 Intelligent Data Analysis in Medicine and Pharmacology: A Position Statement Riccardo Bellazzi 1 and Blaž Zupan 2 10.11.1.26.166.

5 S. Brahman & L.C. Jain (Eds.): Adv. Comput. Intell. Paradigms in Healthcare 5, SCI 326, pp. 3– http://10.springerlink.com

6 Patel, V.L., Shortliffe, E.H., Stefanelli, M. et al. (2009). The coming of age of artificial intelligence in medicine. *Artificial Intelligence in Medicine* 46 (1): 5–17.

7 Chapter Intelligent Decision Support Systems in Healthcare Sheryl Brahman and Lakshmi C. Jain.

8 Temporal Data Mining Based on Temporal Abstractions Robert Moskovitch and Yuval Shahar Medical Informatics Research Center Department of Information Systems Engineering Ben Gurion University, P.O.B. 653, Beer Sheva 84105, Israel.

9 Khare, V.R. and Chougule, R. (2012). Decision support for improved service effectiveness using domain aware text mining. *Knowledge-Based Systems* 33: 29–40.

10 Khademolqorani, S. and Hamadani, A.Z. (2013). An adjusted decision support system through data mining and multiple criteria decision making. *Procedia – Social and Behavioral Sciences* 73: 388–395.

11 Keravnou, E., Garbay, C., Baud, R., Wyatt, J. (Eds.) Artificial Intelligence in Medicine 6th Conference in Artificial Intelligence in Medicine, Europe, AIME '97, Grenoble, France, March 23–26, 1997, Proceedings.

12 D. Foster, C. McGregor, S. El-Masri, 2005,Asurveyofagent-based intelligent decision support systems to support clinical management and research.

13 Viademonte, S. and Burstein, F. (2006). From knowledge discovery to computational intelligence: a framework for intelligent decision support systems, chapter 4. In: *Intelligent Decision-Making Support Sytems* (eds. J.N.D. Gupta, G.A. Forgionne and M.T. Mora), 57–78. Springer-Verlag London Limited.

14 Basu, R., Fevrier-Thomas, U., and Sartipi, K. (2011). *Incorporating Hybrid CDSS in Primary Care Practice Management*. McMaster eBusiness Research Centre.

15 Kukar, M., Kononenko, I., and Silvester, T. (1996). Machine learning in prognosis of the femoralneck fracture recovery. *Artificial Intelligence in Medicine* 8: 431–451.

16 Lavrăc, N. and Mozetĭc, I. (1992). Second generation knowledge acquisition methods and their application to medicine. In: *Deep Models for Medical Knowledge en- Gineering* (ed. E.T. Ker-avnou), 177–199. Elsevier.

17 Kononenko, I., Bratko, I., and Kukar, M. Application of machine learning to medical diagnosis. In: *Machine Learing and Data Mining: Methods and Applications* (eds. R.S. Michalski, I. Bratko and M. Kubat). Willey in press.

18 Larizza, C., Bellazzi, R., and Riva, A. (1997). Temporal abstractions for diabetic patient's management. Proc. AIME-97. In: *Lecture Notes in Artificial Intelligence*, vol. 1211, 319–330. Springer.

19 El-Sappagh, S.H. and El-Masri, S. (2014). A distributed clinical decision support system architecture. *Journal of King Saud University-Computer and Information Sciences* 26 (1): 69–78.

17

Bruxism Detection Using Single-Channel C4-A1 on Human Sleep S2 Stage Recording

Md Belal Bin Heyat[1], Dakun Lai[1], Faijan Akhtar[2], Mohd Ammar Bin Hayat[3], Shafan Azad[4], Shadab Azad[5], and Shajan Azad[6]

[1]*Electronic Science and Engineering, University of Electronic Science and Technology of China, Chengdu, China*
[2]*Computer Science and Engineering, University of Electronic Science and Technology of China, Chengdu, China*
[3]*Science, Orbital Institute, UP, India*
[4]*Mechanical Engineering, Dr. A.PJ. Abdul Kalam Technical University, UP, India*
[5]*Library Science, U.P. Rajarshi Tandon Open University, UP, India*
[6]*Medical Nursing, Apollo Hospital, UP, India*

17.1 Introduction

Snooze bruxism is a verbal multipurpose disease caused by crushing or gritting of the fangs during sleep, i.e., related to an extreme snooze stimulation movement. The tautness and twisted reasons are glitches in the powers, muscles, and other assemblies near the jaw-bone, ear aches, annoyances, grazes to the fangs, and syndromes in the mouth joints. These indications as an entirety are typically labeled as temporomandibular joint (TMJ) pain. In certain persons, bruxism can be recurrent [1], as a result of annoyances, injured fangs, and other glitches. Since one can experience essential snooze bruxism, and also be ignorant of it, pending problems mature [2], and it is significant to distinguish the ciphers and indications of bruxism in order to pursue dental repair. The snooze bruxism throughout daylight or restlessness is usually a semi-voluntary "compressing" movement of the jowl (Figure 17.1). Throughout the darkness, the bruxism incidents are attended by a lurid involuntary crushing. This research will reference solitary to bruxism throughout snooze. Snooze bruxism is distinct in the global cataloging of snooze syndromes, as a labeled crusade syndrome considered by crushing or compressing of the teeth during snooze [3, 4].

The main symptoms of bruxism are augmented tooth compassion, a cloudy headache, depressions on your patois, fangs compressed or cracked, injury from mastication on your teeth, tenderness, revealing deeper coatings of your spine, weary jaw influences, etc. The main causes of bruxism are irregular arrangement of higher and lower fangs, snooze glitches, anxiety, aggressive, anger, stress, Parkinson's syndrome, etc. The snooze bruxism is mostly found in youngsters.

Intelligent Data Analysis: From Data Gathering to Data Comprehension,
First Edition. Edited by Deepak Gupta, Siddhartha Bhattacharyya, Ashish Khanna, and Kalpna Sagar.
© 2020 John Wiley & Sons Ltd. Published 2020 by John Wiley & Sons Ltd.

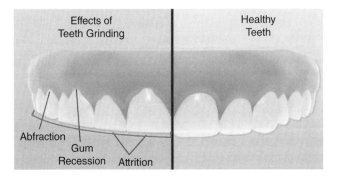

Figure 17.1 Differences of bruxism patient teeth and normal human teeth.

Bruxism snooze syndrome is a mental disturbance by the human and is very harmful to the teeth. Its syndrome is a generated parallel disease of the teeth, such as pain in the jaw etc. There is only 60 seconds of data used in this work. This work is very helpful in the immediate prognostic and has reduced the problem of headache, jaw pain, brain stroke, etc., because all of the above generates the side effect of the snooze bruxism or snooze syndrome. This work stops the generating of new diseases as a side effect of snooze bruxism. The cost of the prognostic and time using our new method is much less. In this work, a snooze bruxism syndrome patient's prognosis results in only one channel and is in an S2 snooze stage. These channels are a very good and accurate result. Therefore, it is very accurate to the other prognostic systems.

17.1.1 Side Effect of Poor Snooze

If snooze is not experienced in the proper way, it affects human behavior:

a) **Mental Health**
   The human snooze is disturbed, thus it increases the chance of depression and anxiety.
b) *Medical*
   The human snooze is disturbed, thus the risk of medical conditions increases, such as heart disease, diabetes, and Alzheimer's.
c) *Weight*
   The human snooze is disturbed, thus it increases the chance of putting on weight.
d) *Accidents*
   If snooze is not completed, thus a person more vulnerable to accident and injury, both on and off the job.
e) *Learning*
   A lack of snooze can negatively influence your learning abilities.
f) *Relationship*
   Poor snooze can negatively influence any relationship you have with others.
g) *Immune*
   A lack of snooze can disrupt your immune system.
h) *Productivity*
   Poor snooze contributes to decreased employee productivity.

i) *Economy*

According to the British economy, snooze deprivation cost 40 billion euro in one year.

j) *Mortality*

Poor snooze increases the chance of mortality risk.

17.2   History of Sleep Disorder

The prominence of sleep has been introduced before many centuries ago in ancient books such as the Bible and Babylonian Talmud. The renowned scientist Charles Dickens first time discovered the sleep disorder in 1836. After long time later in 1950 and 1960, the scientists William Dement and Nathaniel Kleitman identified the stages of sleep. After ten years later in 1970, the Dement introduced the diagnosis of sleep disorders. Additionally, the French scientist Henri Pieron has published the book in 1913. He introduced that the sleep is checked by the physiologically. This book is introduced the sleep research on modern techniques using physiologically [21–23].

The sleep disorders are enormously maladaptive and disturb many persons in negative ways. Sleep disorders are not usually present in a person's entire life. Some sleep disorders are short-term and originate about after the age of imitation. Besides, some sleep disorders are degraded by the stress that our intimates may not have proficient. Sleep disorders generate many harmful diseases of the human. Therefore, we say that it is the beginner symptoms of harmful diseases like heart, mental, and urinary diseases. It's covered the 30 percent of the current population [24].

17.2.1   Classification of Sleep Disorder

The Sleep disorders are insomnia, narcolepsy, nocturnal frontal lobe epilepsy (NFLE), sleep apnea, bruxism, and periodic limb movement disorder.

a) *Sleep Apnea*

Sleep apnea is an instinctive interruption of breathing that arises in the sleep. The main symptoms of sleep apnea are shallow breathing and snoring during sleep. The sleep apnea is seen more often men than women. Sleep apnea is the source of many diseases such as heart attack, stroke, obesity, and diabetes. It is divided into three types such as obstructive, central, and complex sleep apnea [25–27].

- *Obstructive Sleep Apnea (OSA):* In the obstructive sleep apnea, the breathing is disturbed by a blockage of airflow. The 5% men and 3% in women report it for the obstructive sleep apnea [28–31]. In the current time, only 10% of obstructive sleep is diagnosed.
- *Central Sleep Apnea:* In the central sleep apnea, the brain breathing control centers are extreme in the sleep. Additionally, the effects of sleep unaccompanied can eliminate the brain's directive for the body to respire [32–34].
- *Complex Sleep Apnea:* Complex sleep apnea is a combination of central sleep apnea and obstructive sleep apnea [35–36]. The prevalence ranges of this apnea are 0.56 to 18%. It is generally detected in the treatment of obstructive sleep apnea with continuous positive airway pressure (CPAP) [37].

b) *Nocturnal Frontal Lobe Epilepsy (NFLE)*

The NFLE is a disorder primarily characterized through seizures occurring entirely in the sleep. The semiology of which proposes a frontal lobe origin, the involvement of the orbitofrontal regions. It has usually deliberated as a moderately homogeneous epileptic syndrome. The NFLE is associated with temporobasal cortical dysplasia. The large numbers of patients do not appear to exist with gross mental disturbance. Also, the many NFLE patients criticize of chronically disturbed sleep and daytime sleepiness [38–40].

c) *Insomnia*

Insomnia is an indication, not a standalone identification. Since it is "difficulty initiating," it could be due to diminished capacity for sleep. Many people remain unaware of the behavioral and medical options available to treat insomnia [41–46]. There are three types of insomnia:

- *Short-term Insomnia:* The indications are 1 week to 3 weeks.
- *Transient Insomnia:* The indication lasts less than 1 week.
- *Chronic Insomnia:* The indication lasts longer than 3 weeks.

d) *Narcolepsy*

Narcolepsy is asleep and neural disorder produced by the mind's incapability to control sleep-wake phases. The core structures of narcolepsy are cataplexy and fatigue. The syndrome is also frequently related to unexpected sleep assaults. To appreciate the fundamentals of narcolepsy, it is essential to understand the first analysis of the structures of normal sleep. Slumber happens in succession. We primarily enter bright sleep stages and then grow into gradually deeper stages. A deep sleep stage is called non-rapid eye movement (NREM) slumber. Narcolepsy touches both genders similarly and usually first mature in youth and may continue unrecognized as they slowly grow. The example of a familial association with narcolepsy is rather small but a mixture of inherent ecological issues may be the source of this sleep syndrome [47–51].

e) *Bruxism*

Bruxism is a one type of sleep disorder, in which voluntarily grind and clinches the teeth. The symptoms of bruxism are fractured, flattered, and chipped teeth. The main causes of bruxism are disturbances in the sleep and asymmetrical arrangement of the tooth. The risks of bruxism are a side effect of psychiatric medicines, drinking of the alcohol, and smoking [52].

f) *Rapid Eye Movement Behavioral Disorder (RBD)*

The dreaming is purely a "mental" movement that is practiced in the mind while the body is at rest. However, people who suffer from RBD start acting out their dreams. They physically move their limbs or even get up and engage in activities associated with waking. Persons with RBD lack this muscle paralysis, which permits them to act out their dreams in dramatic and violent manner while they are in the REM sleep stage. Sometimes they start talking, twitching, and jerking during dreaming for years before they fully act out their REM dreams. The main symptoms of the RBD are leaping, punching, kicking, etc. People with neurodegenerative disorders like Parkinson's disease, with multiple system atrophy are also at higher risk to suffer from RBD [53–54].

17.2.2 Sleep Stages of the Human

Human sleep follows in periods ranging at length about 1 hour and 30 minutes. The sleep stages are divided into two parts such as NREM and REM [55].

a) *NREM Sleep Stage*

The human finishes this stage approximately 70% through the total sleep. The NREM sleep stage is dividing into many parts.

- **NREM-1:** In this stage, the human snooze is started, a somnolence, tired snooze in which the patient is separately awakened easily. The human eye is closed and opens slowly, so that the motion of the eye is gradually slow. The human body might evoke the bit graphic pictures when awakened from dream. The mind evolution from alpha-wave frequency is 8–13 Hz, whereas delta-wave frequency is 4–7 Hz. Five percent to 10% of the total snooze is completed in this stage.
- **NREM-2:** The eye movement is fully closed and brainwaves slow down. The theta-wave observed snoozer has developed a slower awakening stage. In this stage, 45–50% of the total snooze happens. Snooze axles range from 12–14 Hz.
- **NREM-3:** The human eye movement stops and the brain slows down in this stage.
- **NREM-4:** The human brain completely builds a delta wave. In this stage the human goes all the way to deep snooze. The all-human organs like brain, muscles, etc., are in relaxed mode. This totals 15–25% of the snooze completed by the human [8].
- **NREM-5:** The eye is closed but snooze will be a disrupted in this stage. Only 1% of the snooze is completed in this stage.

b) *REM Snooze Stage*

In this stage, the human breathing is fast, unequally deep, eye movement is dissimilar, and limb muscles are briefly paralyzed. The blood pressure and heart rate rises, and 20–25% of the total snooze is completed by the human [5].

17.3 Electroencephalogram Signal

The British physician Richard Caton discovered that animal brains generated electricity in the nineteenth century. In 1924, German psychiatrist, Hans Berger in Jena, recorded the electric field of the human brain for the first time. Human signals are converted into electrical form with the help of an EEG and record the data-using computer or display devices or recording devices. An EEG is a common tool used in the medical field for performing sleep research [56–57].

17.3.1 Electroencephalogram Generation

An EEG signal is a result of the flow of a synaptic current that flow in dendrites of the neurons in the cerebral cortex. These currents generate an electrical field, which is measured by the EEG system as an EEG signal. An EEG signal is a result of the flow of positive ions like sodium, calcium, and potassium, as well as negative ions like chlorine, across the cell membrane. This potential is due to action potential and it is recorded by scalp electrodes placed overhead for EEG measurement [10].

17.3.1.1 Classification of Electroencephalogram Signal

An EEG signal consists of a few important waves. They are named as gamma, beta, alpha, theta, and delta.

- *Alpha waves.* These waves have the frequency range of 8 to 12 Hz. This wave appears as a round or sinusoidal-shaped signal. These waves are related to relaxation and disengagement. These are slower waves. The intensity of alpha waves increases during peaceful thinking with eyes closed. These waves found behind the head near the occipital and in the frontal lobe of the human brain. These waves experience an increase after smoking. Alpha waves are shown in the posterior half of the head and are usually found over the occipital region of the brain. They can be detected in all parts of the posterior lobes of the brain.
- *Beta waves.* These waves have the frequency range of 12 to 30 Hz. These waves relate to focused concentration. These waves are small and fast in nature. They are detected both in the central and frontal areas of human brain. These waves become frequent when we are resting or suppressing movement, or doing mathematical calculations. It also found that intensity of beta waves increases with the intake of alcohol, leading to the hyperexcitable state. The amplitude of beta rhythm is generally within 30 μV.
- *Gamma waves.* These waves have a frequency range of 31 Hz and above. They reflect mechanisms of consciousness. They are low amplitude waves and occur rarely. The detection of these rhythms can be used to ascertain brain diseases.
- *Delta waves.* These waves range from 0.5 to 4 Hz. These waves are slowest in nature. This wave is observed while snoozing. These waves represent deep sleep.
- *Theta waves.* These waves have frequency range of 4 to 8 Hz. These waves are related to insufficiency and daydreaming. Small values of theta waves show very little difference between being awake or in sleep. We can say it is a transition phase from consciousness to drowsiness. These waves are the result of emotional stress like frustration, disappointment, etc.

17.4 EEG Data Measurement Technique

17.4.1 10–20 Electrode Positioning System

The 10–20 system is a standard method for positioning electrodes over the scalp. The system depicts interconnection between the locations of an electrode and the underlying area of the cerebral cortex. The 10 and 20 numbers represent the distance between the adjacent electrodes. Here 10/20 means 10% or 20% of the total front-back or right-left distance of the skull. The skull is divided into five different sites known as central, frontal, occipital, parietal, and temporal. For measurement purposes, we divide the brain into two parts. The first part is the right hemisphere and the second one is the left hemisphere. In the measurement, each lobe is identified using a letter. A number is used with the lobe to find the type of hemisphere. The number can be the odd number or even number. Even numbers indicate the location of electrodes in the left hemisphere, for example, two, four, six, and eight. Odd numbers indicate the location of electrodes in the right hemisphere, for examples: one, three, five, and seven. The midline position of electrodes shows that using a "z" subscript.

According to this system, we have three different sights that help in electrode placements. These are inion, nasion, and preauricular points anterior to the ear. A prominent bump is present at the back of the head. It is the smallest point of the skull from the backside of the head and is known as the inion. The nasion is defined as a point that lies in between forehead and nose.

17.4.1.1 Procedure of Electrode placement

- First, a measuring tape is used. Its centimeter side is considering for measurement. A middle line is a measure over the scalp and the measurement is made from nasion point to inion point. This total length is called the midline position. For example, consider the total length to be 38 cm. Here the preliminary marks are denoted as Nz and Iz.
- Consider 50% of the total length. This point is the preliminary mark denoted by Cz.
- Mark 10% up from nasion and 10% up from inion. These are the so-called preliminary marks denoted by Fpz and Oz.
- Measure 20% either from the preliminary mark Fpz or from Cz. These are the preliminary points Fz and Pz.
- Now consider measurement from one preauricular point to another preauricular point. These points are available anterior to the ear.
- Find 50% of the above measurement. It will be 50% of 40, which is 20 cm. Now mark the intersection point of this measurement and the one made at the beginning between nasion to inion. This intersection point is the true Cz point.
- Find a mark 10% up from the preauricular points. These points should be marked as T3 and T4.
- Start measuring from the first mark of T3 to Cz. Calculate the total length. After that, measure from the first mark of T4 to Cz. Calculate the total length.
- Now calculate the measure and mark a point at 50% of the total length calculated in the previous step. These points are nothing but the preliminary marks C3 and C4.
- Cross section marks on Fpz are drawn to get the true Fpz mark.
- Now at the back of head, start encircling from the 10% Fpz point to the 10% Oz point. This is the total circumference of the head. Start measuring 50% of the total circumference from Fpz to the back of the head. At the cross section with your preliminary Oz mark is your true Oz mark.
- Now start measuring 5% of the total circumference to the left and right side of Oz. These are the true marks named O1 and O2.
- Now mark 5% of the total circumference to the left and right side of Fpz. These are the true Fp1 and Fp2 marks.
- Start measuring 10% down from the Fp1 and Fp2 points marked earlier. These are the marks for F7 and F8.
- Now measure and mark from F7 to F8 point and measure the distance.
- Now start measuring the half of the distance between F7 and F8 point [11]. Note the point of the intersection with the preliminary mark Fz. This is the true mark for Fz.
- Now mark from the F7 point to the Fz point. Note the total distance. Similarly, measure from F8 point to Fz point and note the total distance.
- Start measuring half of the distance between F7-Fz and F8-Fz. These are preliminary marks named F3 and F4.

- Mark the distance that is 20% of the nasion to inion distance from FP1 to F3. At the intersection, point there will be the point named the true F3 mark. Similarly, mark 20% of the nasion to inion distance from FP2 to F4. At the intersection, point there is the true F4 mark.
- Now obtain the preliminary mark C3 by measuring from Fp1 to O1; similarly, obtain the preliminary mark C4 by measuring from Fp2 to O2.
- Find the intersection point of the first and second mark. This intersection point is the true C3. Similarly, mark half of the distance Fp2-O2. Also, find the intersection point of the first and second mark. This intersection point is the true C4.

17.5 Literature Review

This paper represents the meaning between nap bruxism, verbal badgering at school, and life enjoyment among Brazilian adolescents. The high school years are age with changes and conflicts. In previous research, bruxism is a wellness issue, which squashes the lesser and higher teeth. Deregibus et al. [69] studied that the repeatability in identifying sleep bruxism events by mutual external electromyography (EMG) and heart rate signals measured using the Bruxoff device. The ten subjects including five males and five females and the average are 30.2 years. Castroflorio et al. [70] compare the detection of bruxism by the combination of heart rate and EMG using the compact device. They used eleven healthy and fourteen bruxism subjects in the work. They find that compact device is good inaccuracy. Additionally, the researcher is used in the portable EMG and Electrocardiogram (ECG) for the detection of bruxism. The accuracy of the systems are 62.2 % [71]. Pawel S Kostka and Ewaryst J Tkacz [72] are used twelve patients for the primary ten hours sleep recording. They find that many sources of data analysis with sympathovagal estimation are helpful in the early detection of bruxism disorder. Nantawachara Jirakittayakorn and Yodchanan Wongsawat [73] are designed the detection system of bruxism on the masseter muscle using an EMG instrument. Our research teams are diagnosed by the bruxism using psychological signals such as ECG, EEG, and EMG. We have used around three hundred samples from bruxism and healthy subjects. We are also used the two stages of sleep such as REM and w, We applied the low pass filter, hamming window technique, and power spectral density method to achieved the normalized power. Our specificity of the research in different signals were 92.09, 94.48, 77.07, and 77.16 [7, 52, 74].

17.6 Subjects and Methodology

17.6.1 Data Collection

The proposed work is completed in some steps; it is mentioned in Figure 17.2. We used a single C4-A1 channel of the EEG signal for the S2 sleep stage. the single–channel were more reliable and accurate for the detection of patients [74–76]. Additionally, we used 18 subjects including bruxism and normal. The proposed methods are analysis the channel and estimate the normalized value of the power spectral density. The data was collected

Figure 17.2 Flow chart of the proposed work.

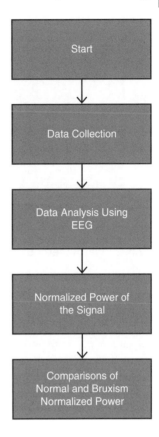

by the physionet website (http://physionet.org), this website is used in the research work [77–79].

17.6.2 Low Pass Filter

We use Low Pass Filter (LPF) for the cleaning of the data, the low pass filter passes the low frequency and block the high frequency (Figure 17.3). We worked in the 25 Hz cutoff frequency for the filtration of the data [7, 43, 52, 74]. The low pass filters exist in different forms including electronic circuits like audio, anti-aliasing, and analog to digital conversion.

17.6.3 Hanning Window

We used the Hanning window for the sampling of the data because this window is less noisy. The renowned Australian scientist Julius Von Hann designed this window. It's also known as different names such as hann filter, raised cosine window and von hann window [80–81].

$$W_0(x) \triangleq \left[\frac{1}{2} \left\langle 1 + \cos\left(\frac{2\pi x}{L}\right) \right\rangle = \cos^2\left(\frac{\pi x}{L}\right) \right], \quad |x| \leq \frac{L}{2} 0, \quad |x| \leq \frac{L}{2}$$

Where, $W_0(x)$ is the Hanning window, L is the length of the window.

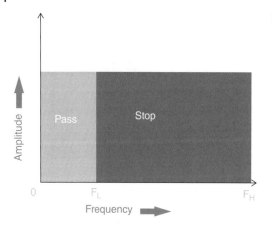

Figure 17.3   Low pass filter.

17.6.4   Welch Method

We used the estimation of power spectral density using the welch method. The welch method is designed by the renowned scientist PD Welch. It is used in different fields such as engineering, physics, and math for estimation of power in different frequencies of the signals. It is based on the concept of periodogram estimation. It is convert the time signal into frequency domain signal. This method is the updated version of the periodogram estimation and Bartlett method. The welch method is negligible in noise so it's useful for the improvement and accuracy of the system [82–83].

17.7   Data Analysis of the Bruxism and Normal Data Using EEG Signal

We analyzed the sleep recording of subjects such as bruxism and normal. The recorded channels of the bruxism data in the S2 sleep stage are C3-P3, C4-P4, C4-A1, DX1-DX2, EMG1-EMG2, ECG1-ECG2, FP1-F3, FP2-F4, F3-C3, F4-C4, F7-T3, F8-T4, P3-O1, P4-O2, ROC-LOC, SX1-SX2, T3-T5, and T4-T6 (Figure 17.4). Additionally, the normal data in the S2 sleep stage are ADDOME, C3-P3, C4-P4, C4-A1, Dx1-DX2, EMG1-EMG2, ECG1-ECG2, F2-F4, F4-C4, F1-F3, F3-C3, HR, LOC-ROC, P4-O2, P3-O1, Posizione, ROC-LOC, SX1-SX2, SpO2, TERMISTORE, and TORACE (Figure 17.5). The common channels of both bruxism and normal are C4-A1, C4-P4, C3-P3, Dx1-DX2, EMG1-EMG2, ECG1-ECG2, F3-C3, F4-C4, P4-O2, P3-O1, ROC-LOC, and SX1-SX2. We extract the single-channel C4-A1 channel of the EEG signal for the S2 sleep stage. The extracted channel of both subjects such as bruxism and normal are mentioned in Figures 17.6 and 17.7.

The C4-A1 channel of the EEG signal in the S2 sleep stage was filtered using a low pass filter. The cutoff frequencies of the signal are 25 Hz. We applied this filtration on both subjects such as bruxism and normal mentioned in Figures 17.8 and 17.9. After filtration of the channel, we applied the Hanning window techniques in both bruxism and normal data (Figures 17.10 and 17.11). While the estimation of the power spectral densities are mention in Figures 17.12 and 17.13. We used welch techniques for the estimation of the power.

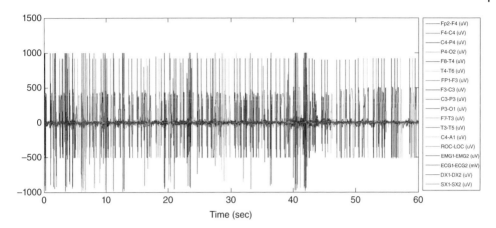

Figure 17.4 The loading of the bruxism data for the EEG signal and the total number of 18 channels of the EEG signal in the S2 snooze stage is present.

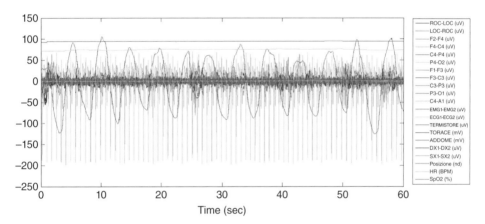

Figure 17.5 Loading of the normal data for the EEG signal in the S2 snooze stage while the total number of 21 channels of the EEG signal is present.

This method changes the signal time series into the segment. We applied this method on bruxism and normal data to find the difference between both subjects for the detection system.

The filtered signal is passing through the hanning window for the both bruxism data and normal data of the S2 snooze stage. The hanning window is less noisy or negligible, so the system is not accurate (Figures 17.10 and 17.11).

The estimating of the PSD by the Welch method. The power signal is estimated by the different frequencies. This method consists of the time series into the segment. It is applied in the windowing signal of both the bruxism data and normal data of the S2 snooze stage (Figures 17.12 and 17.13).

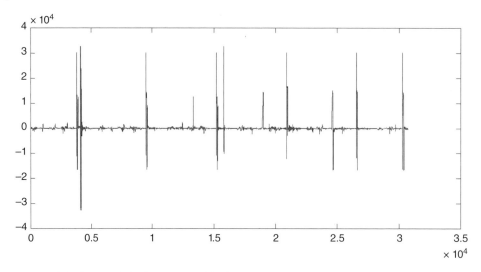

Figure 17.6 Extracted single-channels C4-A1 of the bruxism for the S2 sleep stage.

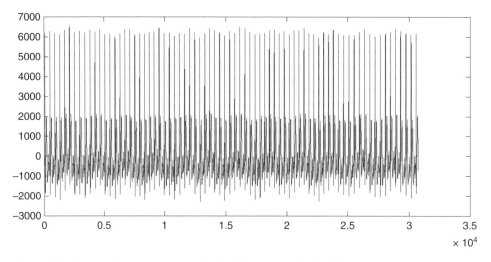

Figure 17.7 Extracted single-channels C4-A1 of the normal for the S2 sleep stage.

17.8 Result

We find the normalized value of the power spectral density. The normalized power specifies the fraction of a specific EEG activity out of the whole power. We achieved the normalized power of the bruxism and normal data of S2 sleep stages. The EEG signals have some waves so we find the normalized value of these waves such as theta, beta, and alpha.

In Table 17.1, the normalized power of the theta wave for the C4-A1 channel is between 0.27037 and 0.35722 for the bruxism data and is 0.2774–0.27919 for the normal data. The difference of the bruxism is 0.8685 and normal is 0.00179. The differences of the normalized

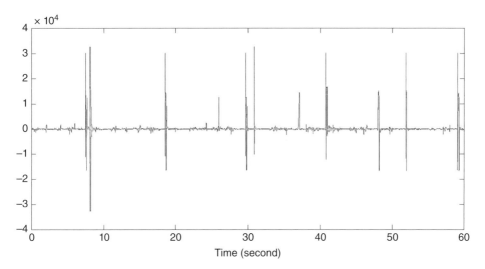

Figure 17.8 Filtered C4-A1 channel of S2 sleep stage for bruxism, we used a low pass filter.

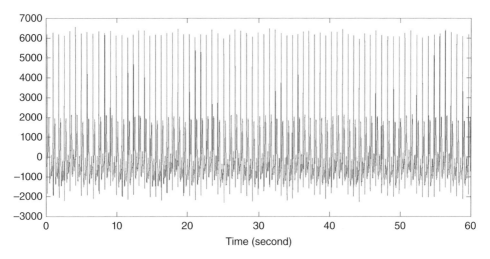

Figure 17.9 Filtered C4-A1 channel of S2 sleep stage for the normal, we used a low pass filter.

power for the C4-A1 channel of the theta wave in the S2 sleep stage of the bruxism are greater than the normal.

In Table 17.2, the normalized power of the beta wave for the C4-A1 channel is between 0.0020038 and 0.035637 for the bruxism data and is 0.00087–0.00087 for the normal data. The difference of the bruxism is 0.0336332 and normal is 0.000244. The differences of the normalized power for the C4-A1 channel of the theta wave in the S2 sleep stage of the bruxism are greater than the normal.

In Table 17.3, the normalized power of the alpha wave for the C4-A1 channel is in between 0.10453 and 0.22804 for the bruxism data and is 0.06829–0.077265 for the normal data. The difference of the bruxism is 0.12351 and normal is 0.008975. The differences of the

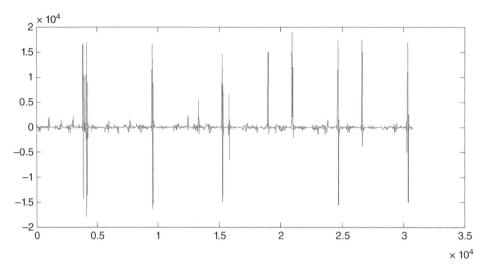

Figure 17.10 Sampled C4-A1 channel of S2 sleep stage for the bruxism using the Hanning window. The Hanning window has negligible noise, so it's helpful to the accuracy of the system.

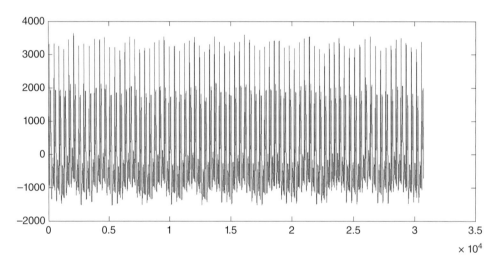

Figure 17.11 Sampled C4-A1 channel of S2 sleep stage for the normal using Hanning window. The Hanning window has negligible noise, so it's helpful to the accuracy of the system.

normalized power for the C4-A1 channel of the theta wave in the S2 sleep stage of the bruxism are greater than the normal.

Figure 17.14 is represented that the normalized values of both subjects for the single-channel C4-A1 of the S2 sleep stage. We obtained that the differences of the normalized power for the bruxism are greater than the normal on all waves of EEG signal. The normalized power of the beta wave for the bruxism and normal is smaller than theta and alpha wave. Additionally, the alpha wave is higher than other waves such as theta and alpha.

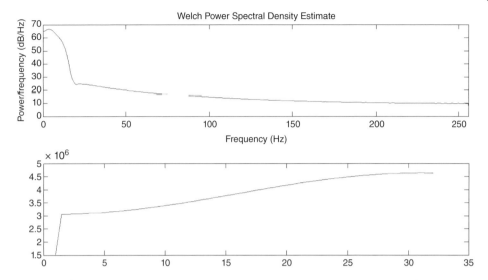

Figure 17.12 It has represented the estimation of the power spectral density using the welch method on bruxism for the S2 sleep stage.

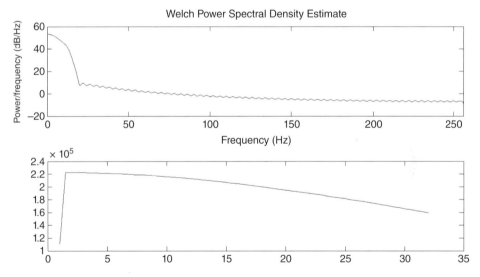

Figure 17.13 It has represented the estimation of the power spectral density using Welch method on normal for the S2 sleep stage.

17.9 Conclusions

Sleep is the physical phenomenon of human beings. The all-living organisms are compulsory to take a good sleep. Because the disturbance of sleep generates many diseases, the bruxism is one of them. We designed a detection system for the bruxism disorder using single-channel C4-A1 of the S2 sleep stage on the human recording. We obtained that the differences of the normalized power for the C4-A1 channel of the EEG signal for the normal

Table 17.1 The comparative analysis between bruxism and a normal human for the C4-A1 channel of the theta wave of the EEG signals in the S2 sleep stage.

S2 sleep stage	Bruxism One	Bruxism Two	Normal Twelve	Normal One
The normalized powers of the theta wave of the EEG signals for the C4-A1 channel.	0.357 22	0.270 37	0.279 19	0.277 4
Differences of the same subjects of the normalized powers	0.086 85		0.001 79	
Remarks	**High Normalized Power of the Bruxism Patient**		**Low Normalized Power of the Normal Human**	

Table 17.2 The comparative analysis between bruxism and normal human for the C4-A1 channel of the beta wave of the EEG signals in the S2 sleep stage.

S2 sleep stage	Bruxism One	Bruxism Two	Normal Four	Normal Sixteen
The normalized powers of the beta wave of the EEG signals for the C4-A1 channel.	0.035 637	0.002 003 8	0.000 87	0.000 244
Differences of the same subjects of the normalized powers	0.033 633 2		0.000 244	
Remarks	**High Normalized Power of the Bruxism Patient**		**Low Normalized Power of the Normal Human**	

Table 17.3 The comparative analysis between bruxism and normal human for the C4-A1 channel of the alpha wave of the EEG signals in the S2 sleep stage.

S2 snooze stage	Bruxism One	Bruxism Two	Normal Three	Normal One
The normalized powers of the alpha wave of the EEG signals for the C4-A1 channel.	0.228 04	0.104 53	0.077 265	0.068 29
Differences of the same subjects of the normalized powers	0.123 51		0.008 975	
Remarks	**High Normalized Power of the Bruxism Patient**		**Low Normalized Power of the Normal Human**	

human is smaller than the bruxism for all waves such as theta, beta, and alpha. We summarized that the alpha wave of bruxism is higher to the beta and theta in both subjects such as bruxism and normal. The future prospects of the research were the detection of bruxism using artificial intelligence techniques.

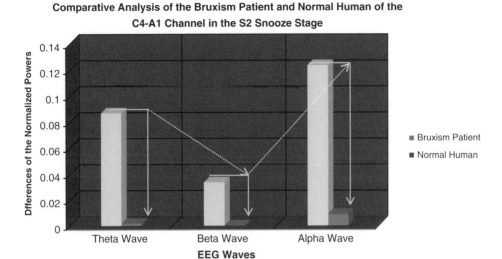

Figure 17.14 Graphical representation for the normalized value of the single-channel C4-A1 in the S2 sleeps stage of both subjects such as bruxism and normal. The differences in the normalized power for bruxism are greater than normal. The normalized power of the beta wave for the bruxism and normal is smaller than theta and alpha wave.

Acknowledgments

We would like to thanks Prof Behra, Prof. Naseem, Prof. Siddiqui, Prof. Shailendra, Prof. Hasin Alam, Prof. Chandel, Prof. Mohd Ahmad, Dr. Dennis, Dr. Ijaz Gul, Dr. Bilal Faheem Chaudhary, Dr. Wazir Ali, and Dr. Deepika Reddy for the useful discussion, proofreading, and motivation of the research chapter. The National Natural Science Foundation of China under grant 61771100 supports this work. The authors do not have any competing interests.

Abbreviations

NREM	Nonrapid eye movement
EEG	Electroencephalogram
EMG	Electromyogram
EOG	Electrooculography
ECG	Electrocardiogram
IMF	Intrinsic mode function
PSD	Power spectral density
PSG	Polysomnography
REM	Rapid eye movement
SGRM	Sparse group representation model
TMJ	Temporomandibular joint
UN	United Nations
US	United States

References

1 P. Blanchet, "Bruxism," in *Encyclopedia of Movement Disorders*, 2010.

2 G. Melo *et al.*, "Bruxism: An umbrella review of systematic reviews," *Journal of Oral Rehabilitation*. 2019.

3 R. V. Murali, P. Rangarajan, and A. Mounissamy, "Bruxism: Conceptual discussion and review," *Journal of Pharmacy and Bioallied Sciences*. 2015.

4 V. Sinisalu and S. Akermann, "Temporomandibular disorders," *Eesti Arst*, 2016.

5 T. Furnish, "Temporomandibular joint disorders," in *Case Studies in Pain Management*, 2014.

6 C. Usha Kumari, A. K. Panigrahy, and N. Arun Vignesh, "Sleep Bruxism Disorder Detection and Feature Extraction Using Discrete Wavelet Transform," 2020.

7 M. B. Bin Heyat, D. Lai, F. I. Khan, and Y. Zhang, "Sleep Bruxism Detection Using Decision Tree Method by the Combination of C4-P4 and C4-A1 Channels of Scalp EEG," *IEEE Access*, vol. 7, pp. 102542–102553, 2019.

8 M. B. Bin Heyat, F. Akhtar, and S. Azad, "Comparative Analysis of Original Wave and Filtered Wave of EEG signal Used in the Prognostic of Bruxism medical Sleep syndrome," *Int. J. Trend Sci. Res. Dev.*, vol. Volume-1, no. Issue-1, pp. 7–9, 2016.

9 G. J. Lavigne, P. H. Rompré, and J. Y. Montplaisir, "Sleep bruxism: Validity of clinical research diagnostic criteria in a controlled polysomnographic study," *J. Dent. Res.*, 1996.

10 A. Kishi and T. Kato, "0686 Sleep Stage Dynamics in Patients with Sleep Bruxism," *Sleep*, 2018.

11 S. Shetty, V. Pitti, C. L. S. Babu, G. P. S. Kumar, and B. C. Deepthi, "Bruxism: A literature review," *Journal of Indian Prosthodontist Society*. 2010.

12 T. Kato, N. M. Thie, J. Y. Montplaisir, and G. J. Lavigne, "Bruxism and orofacial movements during sleep.," *Dental clinics of North America*. 2001.

13 A. Nashed *et al.*, "Sleep Bruxism Is Associated with a Rise in Arterial Blood Pressure," *Sleep*, 2012.

14 N. Huynh, "Sleep-related bruxism," in *The Parasomnias and Other Sleep-Related Movement Disorders*, 2010.

15 M. Partinen, A. Ylikoski, and K. Martikainen, "Sleep difficulties in Parkinson's disease," *Sleep Med.*, 2013.

16 H. Chen and Y. Liu, "Teeth," in *Advanced Ceramics for Dentistry*, 2014.

17 A. R. Sutin, A. Terracciano, L. Ferrucci, and P. T. Costa, "Teeth grinding: Is Emotional Stability related to bruxism?," *J. Res. Pers.*, 2010.

18 J. Kamphuis, P. Meerlo, J. M. Koolhaas, and M. Lancel, "Poor sleep as a potential causal factor in aggression and violence," *Sleep Medicine*. 2012.

19 C. E. Sexton, A. B. Storsve, K. B. Walhovd, H. Johansen-Berg, and A. M. Fjell, "Poor sleep quality is associated with increased cortical atrophy in community-dwelling adults," *Neurology*, 2014.

20 S. M. Alsaadi *et al.*, "Poor sleep quality is strongly associated with subsequent pain intensity in patients with acute low back pain," *Arthritis Rheumatol.*, 2014.

21 A. Y. Izquierdo, F. H. Pascual, and G. C. Monteiro, "Sleep disorders," *Med.*, 2019.

22 S. Chokroverty, "Overview of sleep & sleep disorders," *Indian Journal of Medical Research*. 2010.

23 L. A. Panossian and A. Y. Avidan, "Review of Sleep Disorders," *Medical Clinics of North America.* 2009.

24 *Sleep Disorders and Sleep Deprivation.* 2006.

25 A. Hetzenecker, C. Fisser, S. Stadler, and M. Arzt, "Sleep Apnea," *Dtsch. Medizinische Wochenschrift*, 2018.

26 O. Ludka, "Sleep apnea and cardiovascular disease," *Cas. Lek. Cesk.*, 2019.

27 V. K. Somers *et al.*, "Sleep Apnea and Cardiovascular Disease. An American Heart Association/American College of Cardiology Foundation Scientific Statement From the American Heart Association Council for High Blood Pressure Research Professional Education Committee, Council on ," *Journal of the American College of Cardiology.* 2008.

28 M. R. Mannarino, F. Di Filippo, and M. Pirro, "Obstructive sleep apnea syndrome," *European Journal of Internal Medicine.* 2012.

29 A. C. Stöwhas, M. Lichtblau, and K. E. Bloch, "Obstructive sleep apnea syndrome," *Praxis.* 2019.

30 H. K. Yaggi, J. Concato, W. N. Kernan, J. H. Lichtman, L. M. Brass, and V. Mohsenin, "Obstructive sleep apnea as a risk factor for stroke and death," *N. Engl. J. Med.*, 2005.

31 V. Lavanya, D. Ganapathy, and R. M. Visalakshi, "Obstructive sleep apnea," *Drug Invention Today.* 2019.

32 S. Chowdhuri and M. S. Badr, "Central sleep apnea," in *Geriatric Sleep Medicine*, 2008.

33 D. J. Eckert, A. S. Jordan, P. Merchia, and A. Malhotra, "Central sleep apnea: Pathophysiology and treatment," *Chest*, 2007.

34 S. Javaheri, "Central sleep apnea," *Clinics in Chest Medicine.* 2010.

35 T. I. Morgenthaler, V. Kagramanov, V. Hanak, and P. A. Decker, "Complex sleep apnea syndrome: Is it a unique clinical syndrome?," *Sleep.* 2006.

36 M. T. Khan and R. A. Franco, "Complex Sleep Apnea Syndrome," *Sleep Disord.*, 2014.

37 O. Chowdhury, C. J. Wedderburn, D. Duffy, and A. Greenough, "CPAP review," *European Journal of Pediatrics.* 2012.

38 P. Ryvlin, S. Rheims, and G. Risse, "Nocturnal frontal lobe epilepsy," *Epilepsia.* 2006.

39 L. Nobili *et al.*, "Nocturnal frontal lobe epilepsy," *Curr. Neurol. Neurosci. Rep.*, 2014.

40 O. Farooq, T. Rahman, B. Bin Heyat, M. M. Siddiqui, and F. Akhtar, "An Overview of NFLE," *Int. J. Innov. Res. Electr. Electron. Instrum. Control Eng.*, vol. 4, no. 3, pp. 209–211, 2016.

41 M. B. Bin Heyat, F. Akhtar, and A. Mehdi, "Normalized power are used in the recognition of Insomnia sleep disorder through EMG1-EMG2 Channel," *Int. J. Biol. Med. Res.*, vol. 8, no. 2, pp. 5986–5989, 2017.

42 B. Bin Heyat, F. Akhtar, S. K. Singh, and M. M. Siddiqui, "Hamming Window are used in the Prognostic of Insomnia," in *International Seminar Present Scenario Future Prospectives Res. Eng. Sci. (ISPSFPRES)*, 2017, pp. 65–71.

43 M. B. Bin Heyat, *Insomnia: Medical Sleep Disorder & Diagnosis*, 1st ed. Hamburg, Germany: Anchor Academic Publishing, 2016.

44 M. B. Bin Heyat, F. Akhtar, M. Sikandar, H. Siddiqui, and S. Azad, "An Overview of Dalk Therapy and Treatment of Insomnia in Dalk Therapy Abstract- Treatment of Insomnia in Dalk Therapy," in *National Seminar Research Methodology Ilaj- Bit-Tadbeer*, 2015.

45 M. B. Bin Heyat, F. Akhtar, M. Ammar, B. Hayat, and S. Azad, "Power Spectral Density are used in the Investigation of insomnia neurological disorder," in *XL- Pre Congress Symposium*, 2016, pp. 45–50.

46 H. Bb, F. Akhtar, A. Mehdi, S. Azad, S. Azad, and S. Azad, "Normalized Power are used in the Diagnosis of Insomnia Medical Sleep Syndrome through EMG1-EMG2 Channel," *Austin J. Sleep Disord.*, vol. 4, no. 1, pp. 2–4, 2017.

47 R. Pelayo, "Narcolepsy," in *Encyclopedia of the Neurological Sciences*, 2014.

48 B. R. Kornum *et al.*, "Narcolepsy," *Nature Reviews Disease Primers*. 2017.

49 T. E. Scammell, "Narcolepsy," *New England Journal of Medicine*. 2015.

50 R. Pelayo, "Narcolepsy," in *Encyclopedia of the Neurological Sciences*, 2014.

51 T. Rahman, O. Farook, B. Bin Heyat, and M. M. Siddiqui, "An Overview of Narcolepsy," *Int. Adv. Res. J. Sci. Eng. Technol.*, vol. 3, no. 3, pp. 2393–2395, 2016.

52 D. Lai, M. B. Bin Heyat, F. I. Khan, and Y. Zhang, "Prognosis of Sleep Bruxism Using Power Spectral Density Approach Applied on EEG Signal of Both EMG1-EMG2 and ECG1-ECG2 Channels," *IEEE Access*, vol. 7, pp. 82553–82562, 2019.

53 S. Stevens and C. L. Comella, "Rapid eye movement sleep behavior disorder," in *Parkinson's Disease and Nonmotor Dysfunction: Second Edition*, 2013.

54 J. F. Gagnon, R. B. Postuma, S. Mazza, J. Doyon, and J. Montplaisir, "Rapid-eye- movement sleep behaviour disorder and neurodegenerative diseases," *Lancet Neurology*. 2006.

55 R. K. Malhotra and A. Y. Avidan, "Sleep Stages and Scoring Technique," in *Atlas of Sleep Medicine*, 2014.

56 B. Bin Heyat, Y. M. Hasan, and M. M. Siddiqui, "EEG signals and wireless transfer of EEG Signals," *Int. J. Adv. Res. Comput. Commun. Eng.*, vol. 4, no. 12, pp. 10–12, 2015.

57 M. Bin Heyat and M. M. Siddiqui, "Recording of EEG, ECG, EMG Signal," vol. 5, no. 10, pp. 813–815, 2015.

58 T. Kirschstein and R. Köhling, "What is the source of the EEG?," *Clin. EEG Neurosci.*, 2009.

59 F. L. Da Silva, "EEG: Origin and measurement," in *EEG - fMRI: Physiological Basis, Technique, and Applications*, 2010.

60 M. Toscani, T. Marzi, S. Righi, M. P. Viggiano, and S. Baldassi, "Alpha waves: A neural signature of visual suppression," *Exp. Brain Res.*, 2010.

61 D. L. T. Anderson and A. E. Gill, "Beta dispersion of inertial waves," *J. Geophys. Res. Ocean.*, 1979.

62 R. E. Dustman, R. S. Boswell, and P. B. Porter, "Beta brain waves as an index of alertness," *Science (80-.).*, 1962.

63 J. R. Hughes, "Gamma, fast, and ultrafast waves of the brain: Their relationships with epilepsy and behavior," *Epilepsy and Behavior*. 2008.

64 F. Amzica and M. Steriade, "Electrophysiological correlates of sleep delta waves," *Electroencephalogr. Clin. Neurophysiol.*, 1998.

65 D. L. Schacter, "EEG theta waves and psychological phenomena: A review and analysis," *Biol. Psychol.*, 1977.

66 U. Herwig, P. Satrapi, and C. Schönfeldt-Lecuona, "Using the International 10-20 EEG System for Positioning of Transcranial Magnetic Stimulation," *Brain Topogr.*, 2003.

67 A. Morley, L. Hill, and A. G. Kaditis, "10-20 System EEG Placement," *Eur. Respir. Soc.*, 2016.

68 Trans Cranial Technologies Ltd., "10 / 20 System Positioning Manual," *Technol. Trans Cranial,* 2012.

69 A. Deregibus, T. Castroflorio, A. Bargellini, and C. Debernardi, "Reliability of a portable device for the detection of sleep bruxism," *Clin. Oral Investig.*, 2014.

70 T. Castroflorio, A. Deregibus, A. Bargellini, C. Debernardi, and D. Manfredini, "Detection of sleep bruxism: Comparison between an electromyographic and electrocardiographic portable holter and polysomnography," *J. Oral Rehabil.*, 2014.

71 T. Castroflorio, A. Bargellini, G. Rossini, G. Cugliari, A. Deregibus, and D. Manfredini, "*Agreement between clinical and portable EMG/ECG diagnosis of sleep bruxism*," *J. Oral Rehabil.*, 2015.

72 P. S. Kostka and E. J. Tkacz, "Multi-sources data analysis with sympatho-vagal balance estimation toward early bruxism episodes detection," in *Proceedings of the Annual International Conference of the IEEE Engineering in Medicine and Biology Society, EMBS*, 2015.

73 N. Jirakittayakorn and Y. Wongsawat, "An EMG instrument designed for bruxism detection on masseter muscle," in *BMEiCON 2014 - 7th Biomedical Engineering International Conference*, 2015.

74 A. S. Heyat M.B.B., Lai D., Akhtar F., Hayat M.A.B., "Short Time Frequency Analysis of Theta Activity for the Diagnosis of Bruxism on EEG Sleep," in *Advanced Computational Intelligence Techniques for Virtual Reality in Healthcare. Studies in Computational Intelligence*, K. Gupta D., Hassanien A., Ed. *Springer*, 2020, pp. 63–83.

75 A. R. Hassan and M. I. H. Bhuiyan, "An automated method for sleep staging from EEG signals using normal inverse Gaussian parameters and adaptive boosting," *Neurocomputing*, 2017.

76 A. R. Hassan, "Computer-aided obstructive sleep apnea detection using normal inverse Gaussian parameters and adaptive boosting," *Biomed. Signal Process. Control*, 2016.

77 A. L. Goldberger *et al.*, "Components of a New Research Resource for Complex Physiologic Signals," *Circulation*, 2000.

78 A. L. Goldberger *et al.*, "*PhysioBank, PhysioToolkit, and PhysioNet*," *Circulation*, 2000.

79 A. L. Goldberger *et al.*, "PhysioBank, PhysioToolkit, and PhysioNet: components of a new research resource for complex physiologic signals.," *Circulation*, 2000.

80 J. Barros and R. I. Diego, "On the use of the Hanning window for harmonic analysis in the standard framework," *IEEE Trans. Power Deliv.*, 2006.

81 Y. Li, X. L. Hu, J. Wang, X. H. Liu, and H. T. Li, "FFT algorithm sampled with hanning- window for roll eccentricity analysis and compensation," *J. Iron Steel Res.*, 2007.

82 P. D. Welch, "Welch_1967_Modified_Periodogram_Method," *Trans. Audio Electroacoust.*, 1967.

83 H. R. Gupta and R. Mehra, "Power Spectrum Estimation using Welch Method for various Window Techniques," *Int. J. Sci. Res. Eng. Technol.*, 2013.

18

Handwriting Analysis for Early Detection of Alzheimer's Disease

Rajib Saha[1], Anirban Mukherjee[1], Aniruddha Sadhukhan[1], Anisha Roy[1], and Manashi De[3]

[1]*Department of Computer Science & Engineering, RCC Institute of Information Technology, West Bengal, India*
[2]*Department of Information Technology, RCC Institute of Information Technology, West Bengal, India*
[3]*Department of Computer Science & Engineering, Salt Lake, West Bengal, India*

18.1 Introduction and Background

The umbrella term "neurodegenerative disease (NGD)" is used to refer to conditions that primarily affect the neurons in the human brain. The peripheral nervous system with the muscles, nerve junctions, spinal cord, and motor nerve cells are affected by NGD. The nervous system, including the brain and spinal cord is made of neuron that cannot be reproduced or replaced by the body once they are damaged. Therefore these diseases are incurable. These diseases result in a condition that causes the nerve cells to become progressively damaged and/or die. These damaged or dead neurons are unable to perform efficiently, leading to problems in movement (called ataxias), or mental functioning (called dementias) in patients. Once the disease has progressed beyond a certain level, impairment of cognitive and functional abilities is observed. Behavioral changes are also prevalent in many cases. Based on the type of NGD, currently the most useful and reliable tools for diagnosing NGD are magnetic resonance imagings (MRIs), lumbar puncture or spinal tap, blood tests, etc. Thus clinical assessment is possible, but by the time the disease is recognized it may have progressed and the patient's condition may deteriorate rapidly.

Examples of NGD include Alzheimer's, Parkinson's, and Huntington's disease. Many of these diseases are genetic. Sometimes the cause can be medical conditions like a tumor, a stroke, or alcoholism. It can be caused by exposure to toxins, viruses, and chemicals, whereas sometimes any direct cause cannot be determined. Treatments can only help to treat the symptoms and delay the progress of the disease by relieving pain and helping in increasing mobility [1].

Dementias are responsible for most cases of NGD, with Alzheimer's disease (AD) being the most common. The terms dementia and AD are often used interchangeably, even though AD is a type of dementia. AD mostly affects the older population and is the leading cause of disability among them [2]. AD starts slowly and affects the thought processing, memory, and linguistic abilities of the patient. The first symptoms are generally mild forgetfulness and confusion. Over time the disease progresses as more neurons become

Intelligent Data Analysis: From Data Gathering to Data Comprehension,
First Edition. Edited by Deepak Gupta, Siddhartha Bhattacharyya, Ashish Khanna, and Kalpna Sagar.

damaged or die, resulting in loss of cognitive abilities, thus day-to-day activities can become difficult to perform. Alzheimer's patients develop other symptoms of neural degeneration such as impaired reasoning skills, impaired decision making, aphasia, apraxia, impaired visuospatial abilities, and changes in behavior and overall personality.

There is currently no drug that can reverse the symptoms of this disease. However, there are medications called acetylcholinesterase inhibitors that delay the progression of AD, especially in the early to moderate stages [3]. Therefore, early diagnosis can be very helpful not only for the estimated 27 million patients worldwide but also for their caregivers, by providing cost-effective treatments [1]. But it is a difficult task to recognize AD in its early stages because by the time patients seek medical opinion, they already show signs of memory loss and cognitive impairment. It is important to distinguish between healthy aging and the onset of Alzheimer's. The state of mild cognitive impairment (MCI) is clinically treated as an "in-between" stage. This is due to the fact that the conversion rate for patients with MCI to AD is about 10–30%, whereas only a 1–2% rate with patients not diagnosed with MCI [4].

MCI does not result in cognitive and functional impairments to the levels of severity as displayed by AD but they have a significant chance of progressing into AD.

The exact causes of AD are not known but there are certain risk factors. Age is one risk factor. People older than 65 years are most likely to develop AD. Any immediate family member suffering from Alzheimer's can be a risk. Experts have also identified some genes associated with Alzheimer's. It is important to mention that having one or more of these risk factors does not imply one is suffering from AD. It is believed by the scientist, that the development of abnormal structures called amyloid plaques and neurofibrillary tangles are considered important. Amyloid clumps are protein fragments that can damage the brain's nerve cells. These insoluble, dense protein buildups are found in the hippocampus of the AD patient's brain. This region is responsible for memory management. It helps in converting short-term memory to long-term memory (the fact that hippocampus also helps in spatial navigation can be predicted by the tendency of AD patients). Neurofibrillary tangles are also insoluble substances, which can clog the brain. Microtubules in the brain can be compared to a transport system, which helps the neurons deliver nutrients and information to other cells. The tau proteins, which are responsible for stabilization of these microtubules, are damaged by these tangles. It causes the tau proteins to be chemically altered, resulting in the eventual collapse of the system. Memory loss is believed to be a direct result of this.

Symptoms of AD: Alzheimer's patients display all or a combination of these symptoms frequently and in an ongoing manner, which worsen with each increasing stage of the disease:

1. Memory loss
2. Tasks previously classified as easy and frequently performed get harder and harder to complete
3. Decision making and problem solving are affected
4. Problem in reading, speaking, and/or writing
5. Confusion regarding date, timem and place
6. Behavior and personality changes
7. Uncharacteristic and sudden shyness

Even though it's more prevalent in the senior population, early onset of AD is also possible in people who are in their forties and fifties. The symptoms of early onset AD include trouble remembering words, misplacing personal possessions, lack of concentration, and inability to finish tasks started along with mild memory loss. Mild vision problems might also appear.

Stages of Alzheimer's: Dr. Barry Reisberg of New York University developed a system to break down the trajectory of the progression of this disease into seven stages. This classification is widely used system for AD. Several health care providers and the Alzheimer's Association use this system.

Stage 1: No Impairment

AD cannot be detected at this stage and no signs of dementia are evident. Family history can be used for an early diagnosis.

Stage 2: Very Mild Decline

The patient may notice some forgetfulness or lose track of things. It is hard to distinguish from healthy aging.

Stage 3: Mild Decline

At this stage, someone very close to the person may notice cognitive problems. Performance on memory tests get worse and enable in detecting impaired cognitive function. Difficulty in remembering certain words during conversations, organizing, remembering appointments, planning, remembering names of new acquaintances, and loss of personal items also observed.

Stage 4: Moderate Decline

In stage four of Alzheimer's, the symptoms are evident and the disease is often diagnosed at this stage. It is still considered mild but patients may face difficulty in performing simple calculations, poor short-term and long-term memory, and poor financial management.

Stage 5: Moderately Severe Decline

During the fifth stage of Alzheimer's, patients become dependent on their caregivers for carrying out day-to-day activities. People in stage five of the disease may experience difficulty in recalling certain personal details. Confusion is evident.

There is no significant impairment of functionality. They still mostly recognize close friends and family along with some personal details especially pertaining to their youth or childhood.

Stage 6: Severe Decline

People with the sixth stage of Alzheimer's need help in carrying out most tasks and often have to resort to professional care. They suffer from severe confusion, unawareness of surroundings, are unable to recognize even close friends and family, control bladder or bowel movements, and go through significant change in personality and behavior.

Stages 7: Very Severe Decline

Stage seven is the final stage of Alzheimer's. Patients are unable to communicate or respond, and swallowing also becomes nearly impossible. They need constant assistance and supervision for all daily activity.

But in general, for a broad classification, in this study the three-stage system has been used. The three stages are early or mild, moderate, and severe. The early or mild stage is considered the duration where even though there might be some memory lapses, misplaced personal items, and fumbling for the right word, overall functionality is unaffected or very mildly affected. The moderate stage is the longest stage of the disease. In this stage, the person may display anger, frustration, and other behavioral changes. They will need greater care and often display significant confusion and a sense of being lost. The sleeping pattern might change and trouble may occur regarding control of bladder and bowel movements. Severely affected patients require round the clock care. They lose awareness and cannot respond or communicate properly.

Directly diagnosing AD is difficult because as of now, the only way to confirm is to look at brain tissue after death. Therefore, a series of tests to determine the mental and cognitive abilities and rule out other probable causes of the conditions exhibited by the patients are conducted. The first step is examining family history for genetic links. Several mental, physical, neurological, and imaging tests are conducted. MRIs are used to detect bleeding, inflammation, and structural issues. CT scans are used to detect any abnormality through x-ray imaging. Plaque buildup is detected using positron emission tomography (PET) scans. Blood tests are used to detect genes that might increase the risk factor.

Maintaining a healthy lifestyle by exercising regularly, avoiding substance or alcohol abuse, eating vegan diet, and maintaining an active social life is considered the best way to prevent the onset of AD.

Writing or drawing are complex activities that require cognitive, psychomotor, and biophysical processes. It is a network composed of cognitive, kinesthetic, and motor abilities [5]. Therefore, handwriting can give us a clue about the writer's personality, behavior, and neurophysiologic functioning. It is also important to have good hand–eye coordination, and have good spatial cognitive ability and dexterity. As a result, people suffering from NGD, which affects these functions of the patient's body, have difficulty writing. Significant changes in handwriting is one of the noticeable early features of AD. The cerebral cortex, basal ganglia, and cerebellum are required to carry out the complex task of writing [6].

Handwriting analysis has been associated with AD since the first patient was examined by Alois Alzheimer in 1907. The omission of certain syllables and repetition of others was observed. Agraphia was reported among AD patients in the early stages [7]. Lexicosemantic errors, which transformed to phonological error as the disease progressed, and the dementia, became more severe. Recent studies have been conducted that use dynamic collection of handwriting or analysis, such as [8]. Kinematic measures of the handwriting of persons with MCI compared with those with mild AD and healthy controls to study differences in their handwriting for different written tasks. Goal-oriented tasks showed significant slowdown of motor movements. Handwritten signatures can also be used for an early diagnosis of NGD [9]. It not only represents the first and last names of the person but also provides significant information regarding their writing system and their mental and emotional state. Hence it is used as a biometric and identification for important verifications. The sigma-lognormal model for the signature representation was used for a cost-effective analysis in terms of AD. Movement time and smoothness were the parameters in the study. AD and MCI patients demonstrated slower, shakier, less coordinated, and inconsistent writing, compared to their healthy counterparts.

A study [10] conducted for comparing copying tasks between healthy people, MCI patients and AD patients was able to classify 69–72% of the participants correctly, even though the performance in the MCI group was poor.

Thus the inability to communicate via writing is seen as an early manifestation of AD. Graphology can therefore be used as a diagnostic tool.

Graphology or graphoanalysis is the study of the physical characteristics and patterns in a handwritten document. It can be used to identify the psychological state of the writer. It is also used as a diagnostic tool for aiding in illnesses that affect the motor and cognitive abilities of the patient.

As Alzheimer's is a NGD, handwriting can be used as a tool to diagnose early stages of AD, and monitoring the changes can reflect the stages as the disease progresses. The person may struggle to write their signature and the letters become indecipherable as the disease worsens. Due to impaired cognition, they might also have problems in grammar, spelling, and completing sentences. Dysgraphia has also been found to be a common symptom of AD. Some studies on AD patients have results that demonstrated that handwriting may progressively deteriorate, and that spelling errors may be observed.

Diana Kerwin, neurologist and director of Alzheimer's and Memory Disorders at Texas Health Presbyterian Hospital in Dallas, predicted that it could be due to apraxia, wherein the person forgets how to perform the motor tasks needed to write. Even though the motor system is intact, the instructions from the brain to the hand are impaired and it can affect handwriting. AD can also cause visuospatial impairments in which the brain has difficulty visualizing things and placing them in the correct areas. Handwriting changes can be one of the first observable changes on the onset of AD. Micrographia or excessive space between words and/or letters can also be indicative of a problem.

Handwriting, in the form of signatures and sentences, can be used as a screening mechanism in order to determine the extent of cognitive impairment and the stage of the disease. Patients with advanced AD are often unable to write even a single complete sentence. Handwriting might change to irregular and altered letters with trembles. The writing will also slow down. Copying text also takes a longer time because reading and comprehension is also affected by AD.

AD is associated with breakdown at multiple levels in the writing process. There is yet no single pattern of spelling or impairment associated with AD that can be predicted. Spelling impairment is inevitable, once the disease has progressed to a certain level severity. The most often reported pattern is one of surface dysgraphia in the mild stages of the disease, followed by an increase in nonphonologically plausible errors as the disease progresses. Peripheral dysgraphia is also observed. It is likely to arise from a functional impairment at the level of graphic motor pattern. Production of poorly formed or illegible letters has been reported by several studies. At a moderate stage, the letter writing becomes impaired and illegible [7], and eventually, some patients are unable to write. Copying and cross-case transcription of single letters are used to assess peripheral processes. Subjects were more successful at copying within the case than transcribing across the case (e.g., d to d, F to F, and cross-case transcription, e.g., a to A, T to t).

Area of application of graphology:

1) Personal Relationships: Handwriting analysis can provide deep insights into the personal, social, and professional nature and judge compatibility.
2) Educational: In the field of education for the counseling of students
3) Personnel Selection: In the selection of personnel, handwriting analysis is an invaluable tool for helping to choose the most suitable person for the job.
4) Police Profiling

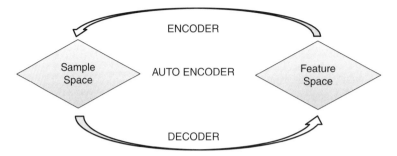

Figure 18.1 This is simplest form of representation of the Encoder architecture.

Variational auto encoder (VAE) is considered a neural network that can encrypt images into a direction in the latent space of z real numbers. For the sample that has been collected from a z-dimensional normal distribution, the direction is assumed to be arbitrary. Then the decoder network decodes the encoded vector representation and obtains the original image. Random samples can be drawn from the distribution and fetches them into the decoder network for which the latent space is z-dimensional normal distribution, from which new images can be obtained that are absent in the data set we trained on.

The architecture of the encoder is represented below (Figure 18.1):

The pivotal facet of the VAE lies in the fact that its latent representation, $y \in \mathbb{P}^K$, is derived from a particular Gaussian distribution (i.e., $p(y) = N(y|\mu, \Sigma)$, where μ denotes mean and Σ denotes covariance). The VAE is trained in an unsupervised manner as the output is basically the reconstructed version of the input. Let us consider the given set of training data is $S = \{s_n\}_{n=1}^N$, and the VAE comprises a probabilistic encoder $q\theta(y|s)$, which finds the latent representation y for the given set of input data s. It also consists of a probabilistic decoder $p\phi(s|y)$, which reconstructs the input data for the specific latent representation, where θ represents the network parameter for encoder and ϕ represents the network parameter for decoder.

Optimization of the variational bound, $\mathcal{L}_{VAE}(\theta, \phi)$, with respect to the parameters θ and ϕ in the encoder and decoder occurs while learning the VAE.

$$\mathcal{L}_{VAE}(\theta, \phi) = \sum_{N=1}^N -E_{y \sim q\theta(y|s_n)}[\log p\phi(s_n \mid y)] + KL(q\theta(y \mid s_n)\|p(y))$$

Here, the first term is the reconstruction error that has been computed by taking the expectation with respect to the distribution of y while the second term denotes regularizer, which is the Kullback-Leibler (KL) divergence between the estimated distribution $q\theta(y|s_n)$ and the true distribution $p(y)$. This divergence is the measurement of the amount of information lost while using q to represent p. The estimation of parameters can be done using an auto-encoding variational Bayes (AEVB) algorithm [11].

The pictorial representation of an encoder and decoder is given in Figure 18.2.

The encoder is said to be a neural network whose input is a data point while output is a hidden representation y, having weights and biases θ. Specifically saying, let us consider y to be a 15 by 15 pixel photo of a handwritten digit. The encoder "encodes" the

Figure 18.2 (a) The encoder compresses data into latent space (y). (b) The decoder reconstructs the data given, which is a hidden representation.

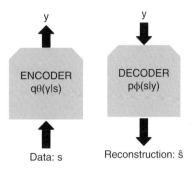

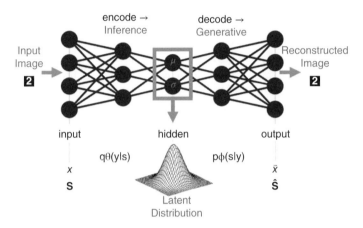

Figure 18.3 Image reconstruction process.

data that is 225-dimensional into a latent (hidden) representation space y, which is much less than 225 dimensions. This is conventionally termed to be a "bottleneck" because the encoder must be capable of efficient compression of the data into this lower-dimensional space.

From Figure 18.2 we can consider or assume the encoder to be denoted as qθ(y|s). We know that the lower-dimensional space is arbitrary: the parameters to qθ(y|s), which is a Gaussian probability density and the output of the encoder. A sampling of this distribution can be done to obtain the noisy values of the representations y (Figure 18.3).

The decoder is also a neural network having representation y as the input while the parameters to the probability distribution of the data is its output, and has weights and biases ϕ. The decoder has been denoted by pϕ(s|y). On the execution with the handwritten digit as example, let it be assumed that the images are black and white and representation of each pixel is in the form of 0 or 1. The probability distribution of a single pixel can be then represented using a Bernoulli distribution. The latent representation of a digit y is given as input to the decoder and the output thus obtained is 225 Bernoulli parameters, one for each of the 225 pixels in the image. Decoding the real-valued numbers in y into 225 real valued numbers between 0 and 1 is done by the decoder. There is the occurrence of loss of information as it goes from a smaller to a larger dimensionality [12].

The requirements of VAE are as follows:

- It is essential to use two tractable distributions:
  - The initial distribution $p(y)$ must be easy to be sampled from
  - The conditional likelihood $(s|y, \theta)$ should be computable
- In practice this implies that the two distributions of the discussion are seldom complex, for example, uniform, Gaussian, or even isotropic Gaussian

18.2 Proposed Work and Methodology

Dementia of Alzheimer's type is a NGD that affects millions of lives around the world. A disease is caused by progressive damage of neurons, which cannot be replaced. Thus this is an incurable disease. The only treatment available for this disease is to control the symptoms. These symptoms include memory loss, cognitive, and motor function impairment. As a result, several physical tasks that are complex in nature and require the proper coordinated involvement of several parts of the brain as well as proper muscle control like writing become difficult to perform. The altering of handwriting is one of the earliest signs of Alzheimer's.

This study aims to collect authentic data samples from patients suffering from various levels of Alzheimer's and to analyze it. The analysis is to extract the common features in the handwriting samples of medically diagnosed Alzheimer's patients and compare it with the features of the healthy controls to isolate the features that can be used as a sign of NGD. This feature extraction process will be done with the help of a deep learning generative model VAE. Early and cost-effective detection of Alzheimer's can help the patient as well as the caregivers to ease the effects.

Step 1: Collection of Data Samples

A sample of individuals with AD and healthy controls were used to collect data. The data set consists of a handwritten paragraph and signed with the date by the authors on white A4 paper. The participants were seated on a chair with attached desks and used ball point pens to write the sample. The same sample paragraph was written by all sample providers. The time required to complete the task, along the stage of the disease, in the case of each patient, was noted. The diagnosis was provided on the consultation of a licensed medical doctor after an interview with family members/caregivers and other medical diagnosis techniques as suggested by the doctor. All the sample providers were informed about the objective of this study and the sample was collected after informed consent.

Step 2: Image Segmentation

Segmentation partitions an image into distinct regions where each pixel in the region has common attributes. For useful image analysis and interpretation, the regions should strongly relate to depicted objects or features of interest. Meaningful segmentation is the first step from low-level image processing transforming a grayscale or color image into one or more other images to high-level image description in terms of features, objects, and scenes. The success of image analysis depends on the reliability of segmentation, but an accurate partitioning of an image is generally a very challenging problem.

Segmentation techniques are either *contextual* or *noncontextual*. The latter takes no account of spatial relationships between features in an image and group pixels together on the basis of some global attribute, e.g., gray level or color. Contextual techniques additionally exploit these relationships, e.g., group together pixels with similar gray levels and close spatial locations.

In handwriting image segmentation, digital handwriting is segmented into three different types of segments, i.e., word segmentation, letter segmentation, and line segmentation, each used for different processing. The line segmentation is done by locating optimal text and gap zones. The words and characters are subsequently located using the same strategy by scanning each line.

Step 3: Image Preprocessing

Image preprocessing, which needs to be done properly so that accurate results are obtained and to minimize the error due to external factors such as noise. Image preprocessing is the technique in which the handwritten sample is translated into a format that can be efficiently processed in further steps. These steps involve binarization, noise removal, line segmentation, word segmentation, and character segmentation. Binarization converts gray scale image into binary image. Noise removal techniques are applied to enhance the quality of the image.

Step 4: Line, Word, and Character Segmentation

For line segmentation, the image is first converted to grayscale and then binarized inversely, such that the background is dark and the text is light. This image is then dilated and the contours are found using the findContours() function in OpenCV library. Each detected contour is stored as a vector of points. The retrieval mode used is RETR_EXTERNAL, which returns only the extreme outer contours. The approximation method used is CHAIN_APPROX_SIMPLE, which compresses horizontal, vertical, and diagonal segments and leaves only their end points. For example, an upright rectangular contour is encoded with four points.

For word segmentation the above method is applied on the segmented lines and the characters are subsequently segmented using the isolated words.

Algorithm for Line Segmentation

STEP 1: Load the sample writings as images

STEP 2: Convert it to grayscale

STEP 3: Convert the grayscale image to binary image using proper threshold and invert the image

STEP 4: Dilate the binary image using kernel of 5×150 matrix of ones

STEP 5: Find contours from the diluted image and consider only the extreme outer flags

STEP 6: Extract the region of interest from the image using those contours and save them as images, which are basically lines in the writings

Algorithm for Word Segmentation

STEP 1: Load the sample lines as images

STEP 2: Convert it to grayscale

STEP 3: Convert the grayscale image to binary image using proper threshold and invert the image

STEP 4: Dilate the binary image using kernel of 5×35 matrix of ones

STEP 5: Find contours from the diluted image and consider only the extreme outer flags

STEP 6: Extract the region of interest from the image using those contours and save them as images, which are basically words in the lines

Algorithm for Character Segmentation

STEP 1: Load the sample words as images

STEP 2: Resize the contour containing the word using bicubic interpolation over 4×4 pixel neighborhood

STEP 3: Convert it to grayscale

STEP 4: Convert the grayscale image to binary image using proper threshold and invert the image

STEP 5: Dilate the binary image using kernel of 5×5 matrix of ones

STEP 6: Find contours from the diluted image and consider only the extreme outer flags

STEP 7: Extract the region of interest from the image using those contours and save them as images, which are basically characters in the words

Step 5: Implementation and Use of VAE

The segmented characters are passed through the VAE system. The characters are then reconstructed and classified to obtain the common features from the handwritten samples. The clustering will be helpful for the purpose of understanding whether the samples have any features that can be used as a parameter in order to help diagnose AD. It can also help in classifying the stage of the AD depending on the handwriting as, with the disease getting progressively worse, the handwriting gets more and more illegible or error prone.

Algorithm for Variational Auto encoder

STEP 1: The image is read

STEP 2: The image is converted to desirable format for required channels and size

STEP 3: The image pixel intensity value is stored in an array

STEP 4: The array is shaped to coincide with the input shape of the VAE model

STEP 5: The data is split into training and test set

STEP 6: The encoder model is constructed using fully connected layers are used in the model to calculate mean and variance

STEP 7: The decoder, a sequential model with fully connected layers, is constructed

STEP 8: The input and output shapes for the encoder and decoder are used to construct the autoencoder

STEP 9: The model is compiled using negative log normal loss function

STEP 10: The middle hidden layer represents the mean and variance layer using KL; the model is trained as a whole and the mean and variance is updated for every batch of data in each epoch using back propagation.

The following flowchart has explained the entire process.

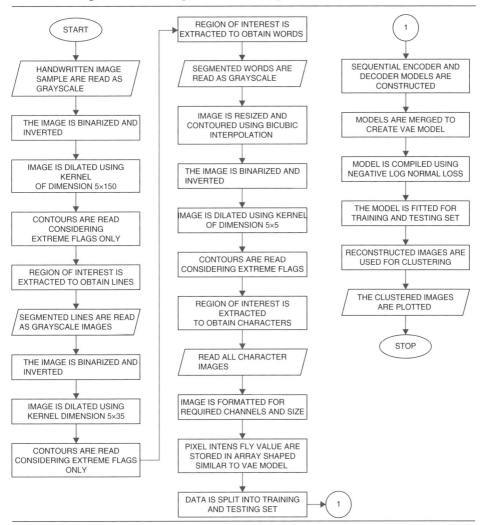

18.3 Results and Discussions

The first step in the preprocessing is line segmentation. The segmentation algorithm used stores the segmented lines for each patient into a separate folder.

Line segmentation samples (Figure 18.4):

After the lines are segmented, it is used as input for the next step, which is word segmentation. The word segmentation algorithm processes the segmented lines and divides it further into smaller segments. It works in a similar fashion and stores the segmented words for each patient (Figure 18.5).

Figure 18.4 Line segment from handwritten sample from patients suffering from AD.

18.3.1 Character Segmentation

Finally, the words are segmented further into characters and stored separately in directories, same as words and lines. These characters are the images that are finally used for the image segmentation process and fed into the VAE system where the images are reconstructed to extract common features. It is done by building training and testing sets, by combining the characters segmented from all the handwritten samples provided by the Alzheimer's patients in early to moderate stages of the disease (Figure 18.6).

The segmented characters were passed through the VAE, to extract the common features. The VAE takes the segmented characters as input and reconstructs the images using only relevant information, by intelligent selective discarding of irrelevant information. This reconstruction as mentioned above is done using an encoder and decoder network. In the last step after 500 epochs, the result over 1050 character training samples, the loss is 0.2573 and validation loss is 0.2648 (Figure 18.7).

The images on reconstruction are clustered in order to loosely group them such that the images that have more in common with each other are closer to each other when depicted as data points in a two-dimensional plane (Figure 18.8).

From the images it can be observed that the character images in the cluster together largely belongs to the same letter. In further observation, it is noticed that some letters are ambiguous and cannot be labeled as any alphabet with 100% certainty. For example, the cluster that largely consists of "e," might also contain "l" or "c." A comparison between some such ambiguous writing in the original sample validates the result obtained from the VAE (Figures 18.9 and 18.10).

During the data collection process it was observed that the healthy controls required 2–3 minutes to write the paragraph while the AD patients in early stages required at least 10 minutes to write the same paragraph. Therefore significant slowing down of the writing process is observed.

From the segmented words we can observe inconsistent pressure application while that is more consistent in case of healthy controls.

Figure 18.5 (a and b) Word segmentation samples produced from the segmented line.

Figure 18.6 Sample of character segmentation obtained using segmented words.

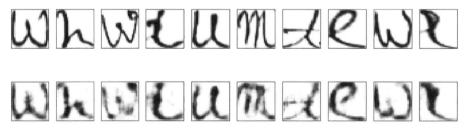

Figure 18.7 Segmented characters reconstructed using VAE.

Another feature of patients affected by AD is that double letters can have unclear/incorrect writing. Problems in spelling phonologically similar words are also evident. As the disease progresses, the patient's motor control gets progressively impaired. The result is shaky, disconnected handwriting (Figure 18.11).

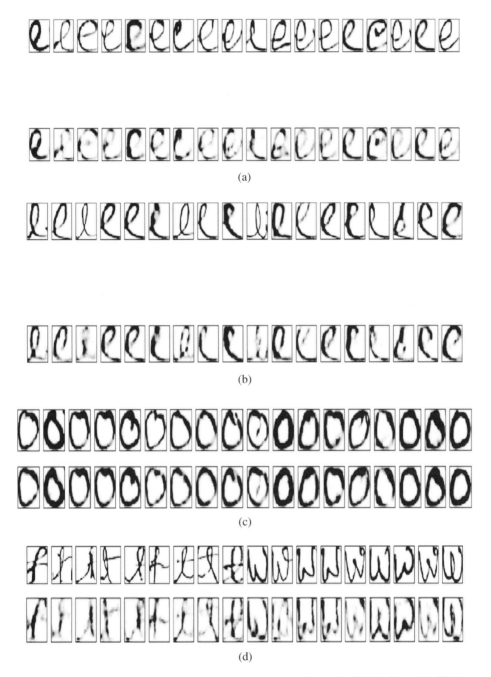

Figure 18.8 Clusters of reconstructed images using VAE. (a) Cluster for "e," (b) Cluster for "l," (c) Cluster for "o," (d) Cluster for "t" and "w."

Figure 18.9   Ambiguous "l" and "e."

Figure 18.10   Ambiguous "c" and "e."

Figure 18.11   Unclear or disconnected writing with spelling errors.

18.4   Conclusion

In this study, a method is proposed to extract the common features for patients suffering from the neurodegenerative AD. This method can help in the early diagnosis of the disease. This is a cost-effective and noninvasive method of diagnostic aid. Any particular pattern has not been identified as an irrefutable sign of AD, but the results from the study can helps test the cognitive abilities of the patient. The study can be further modified to add more functionality like monitoring, to track the gradual progression of the disease and also identify the stage in which the patient is at the time of the sample. This can only continue up to the point where the patient still has some functional ability intact, as in the last stage, where the patient might overcome memory loss, confusion, and impaired functional ability and also be unable to provide sufficient data to analyze using this model.

The study can also be applied to extract features for other NGDs with slight modifications in the implementation to determine whether handwriting analysis can be used as an early diagnostic method or not. This study will also benefit from the availability of a large training set and dynamic data collection. This will help for better image reconstruction. The dynamic collection will provide more precise parameters for more accurate clustering.

References

1 Ranginwala, N.A., Hynan, L.S., Weiner, M.F., and White, C.L.I. (2008). Clinical criteria for the diagnosis of Alzheimer disease: still good after all these years. *The American Journal of Geriatric Psychiatry* 16 (5): 384–388.

2 Alzheimer's Association (2015). Alzheimer's disease facts and figures. *Alzheimer's & Dementia* 11 (3): 332–384.

3 McKhann, G., Drachman, D., Folstein, M., Katzman, R., Price, D., & Stadlan E. M. (1984). Clinical diagnosis of Alzheimer's disease: report of the NINCDS-ADRDA Work Group under the auspices of Department of Health and Human Services Task Force on Alzheimer's Disease.

4 Ward, A., Arrighi, H.M., Michels, S., and Cedarbaum, J.M. (2012). Mild cognitive impairment: disparity of incidenidence and prevalence estimates. *Alzheimer's & Dementia* 8 (1): 14–21.

5 Tseng, M.H. and Cermak, S.A. (1993). The influence of ergonomic factors and perceptual motor abilities on handwriting performance. *American Journal of Occupational Therapy* 47 (10): 919–926.

6 Kandel, E.R., Schwartz, J.H., and Jessell, T.M. (2000). *Principles of Neural Science*, 4e. McGraw-Hill Medical.

7 Platel, H., Lambert, J., Eustache, F. et al. (1993). Characteristics and evolution of writing impairment in alzheimer's disease. *Neuropsychologia* 31 (11): 1147–1158.

8 Werner, P., Rosenblum, S., Bar-On, G. et al. (2006). Handwriting process variables discriminating mild alzheimer's disease and mild cognitive impairment. *Journal of Gerontology: Psychological Sciences* 61 (4): 228–236.

9 Pirlo, G., Cabrera, M.D., Ferrer-Ballester, M.A. et al. (2015). Early diagnosis of neurodegenerative diseases by handwritten signature analysis. In: *New Trends in Image Analysis and Processing, ICIAP 2015 Workshops*, 290–297, Springer.

10 Werner, P., Rosenblum, S., Bar On, G. et al. (2006). Handwriting process variables discriminating mild Alzheimer's disease and mild cognitive impairment. *The Journals of Gerontology. Series B, Psychological Sciences and Social Sciences* 61 (4): P228–P236.

11 D. P. Kingma and M. Welling "Auto-Encoding Variational Bayes", arXiv:1312.6114 (2014)

12 https://jaan.io/what-is-variational-autoencoder-vae-tutorial

Index

Intelligent Data Analysis: From Data Gathering to Data Comprehension,
First Edition. Edited by Deepak Gupta, Siddhartha Bhattacharyya, Ashish Khanna, and Kalpna Sagar.
© 2020 John Wiley & Sons Ltd. Published 2020 by John Wiley & Sons Ltd.